The Dynamic Theory

A New View of Space-Time-Matter

By

Pharis E. Williams

Williams Research

http://www.physicsandbeyond.com/

Library of Congress Cataloging-in-Publication Data

Williams, Pharis E.

The Dynamic Theory-A New View of Space-Time-Matter: physics, particle physics, nuclear physics, electromagnetism, gravity, cosmology, gauge theory, geometry / Pharis E. Williams

Includes bibliographical references and index.

ISBN 978-0-615-44711-7

1. Physics, 2. Nuclear Physics, 3. Particle Physics, 4. Field Theory, 5. Fundamental particle fields, 6. Electromagnetism, 7. Gravity, 8. Cosmology, 9. Dark Matter, 10. Dark Energy, 11. Waves

Cover design by Debbie Balog (See www.dbalog.com)

Send correspondence to the author of this book, see

www.physicsandbeyond.com.

Dedication

Dr. Dan Connor Ross was hired as a visiting professor to teach at the US Naval Academy where he found research reports I had written in the Naval Academy library. He contacted me and I later was able to get him hired by the Los Alamos National Laboratory as a collaborator to assist me in my research. Primarily Dan kept me on track by checking my entire math, but he also played an important role in acting as a sounding board for new thoughts he never knew existed. Dan also provided stimulation to me with his undying confidence in the approach I was taking in the research. He worked toward publishing my research as he formed his publishing company called Los Alamos Books, had a cover sheet developed and began putting the material together.

Cancer stopped Dan's attempt to publish the book he wanted to give the title of this book. Dan's meticulous approach to editing would have been of invaluable assistance in preparing this material for publication. I know this book lacks the grammatical and presentation quality that Dan's editorial talents would have given to it, but I know how much he wished this material to be published and, therefore, this book is dedicated in his memory.

Contents

Forward

A theory is the more impressive the greater the simplicity of its premise, the more different kinds of things it relates, and more extended is its areas of applicability. Therefore classical thermodynamics has made a deep impression upon me. It is the only physical theory of universal content concerning which I am convinced that, within the framework of applicability of its concepts, it will never be overthrown.

Translated from: Einstein's Annalen der Physik Papers: The Complete Collection 1901-1922, WILEY-VCH Verlag GmbH & Co. KGaA, Weinheim, 2005 ISBN 3-527-4064-X

I first came across Pharis Williams' work when I was working on an USAF project to investigate physical theories that might lead to major innovation in our technology. I looked at all the physical theories I could find for the project. Most were just crazy, many where just plain wrong, but a few, just a few, seem to hold promise. The problem with most of the theories I looked at was that they made few testable predictions and gave few ideas of how the theory could be used in useful ways. It is fairly easy to generate theories that, a like Rorschach test, are only useful if you already have ideas in mind. For example, String Theories and Membrane Theories (M theory) are very elegant and mathematically beautiful but real beauty goes beyond appearance. Today some call string theories "not even wrong" since while beautiful in appearance, they produce no predictions that we can test with any technology that we can now contemplate to see if they are right or wrong. With even a wrong theory you can often see what is wrong and what needs to be improved. Yes, many of today's theories and experiments can fit within string theories, but so can countless other non-real worlds and there is no indication as to what is and isn't real for our world. There is no way to tell which solutions they give will be real for our world. It is nice to dream of other universes with other physical laws, but they are just that- dreams with no substance. Theories that generate everything, including an infinity of alternative non-real universes, prove nothing. On the other hand, the Dynamic Theory presented here gives specific and concrete predictions.

When I first found the Dynamic Theory, it was only in a few obscure Los Alamos technical papers. But as I researched more and more physical theories looking for new developments and predictions, this one theory stood out. It seems to have gone without the recognition it deserves even many years later. This book will hopefully correct that and motivate others to pursue this fruitful path. I saw and still see Williams' theory as a fresh theory that generates precise predictions and concepts that, so far, do not conflict with actual experiments- our views and interpretations of them perhaps, but not the actual observable data. His theory is comforting in that it derives general relativity from thermodynamics. I see no other plausible theory that does that. It does take a little while to fully appreciate the approach. But just like the path integrals approach in QED develops by giving up "knowing" exact paths and focusing on the end states, the Dynamic Theory derives its power by focusing on limiting the gauge function by the second law. This gives a new way of looking at the gauge function and ultimately to quantization.

I know of no other theory that can so closely calculate the nuclear energy of low Z isotopes from first principles, generates quantum numbers directly from metric field equations, gives a limit to mass-density conversions, or makes so many concrete predictions with so few assumptions. At first, I was concerned about the views it generated about neutrinos, multiple n photons, gravitational interactions but testing our beliefs should be the role of a new theory. We must look at the observables in our experiments to guide us and not our views and interpretations we hold from our previous theories.

The Dynamic Theory is a 5 dimensional one. Yet, unlike most Kalusa-Klein theories (KK) it does not impose an artificial restraint on the 5th dimension. Here the added dimension is a function of mass and just like the fourth dimensional addition to our understanding imposed a limit to how fast matter could travel, the fifth dimension limits the rate at which that mass can be converted to energy. This alone is an important concept. I fully expect it will ultimately moderate energy released from atomic bombs, limit a star's lifetime and justify the cutoffs to QED. When I look at its EM/gravitational field equations, I see possibilities for explaining why phonons can produce virtual mass particles without interaction with a mass, and why mass-less particle can have no charge. There are still treasures that may be mined from the theory.

It is comforting to know that one theory may span the ideas of mechanical and electromagnetic energy with thermodynamic concepts. When I look at alternative theories such as string theory, I see no way that they can incorporate thermodynamics except as an added assumption. The Dynamic Theory is the benchmark I now use to judge all new theories. As I retire and can spend more time researching mathematical physics and theories, this is the one theory I plan on working on in depth. It, alone, has all the earmarks I have been looking for and expect of a usable the unified theory- thermodynamics, prediction of masses, space curvature, possibilities of rate calculations, experimental guidance to EM and gravity interactions, mechanical entropy, greater understanding of photons, red shifts, peripheral advance, nuclear masses, solution to the self energy problems, non-singular gauge functions, covariant expressions, and even shock waves in explosives. It has made an indelible impression on me.

Dr. Dennis Cravens
Cloudcroft, NM

Part 1 Overview

In the scientific community's search for a fundamental theory the laws of classical thermodynamics have been overlooked. Though these laws have not yet been seen to be violated they do not seem to be a good basis from which to obtain mechanical motion, field theories and quantum mechanics. However, they do j ust that.

It has been shown and will be seen below that the classical laws of thermodynamics require relativity as a subset of the entirety of the universe they describe. See the section titled Relativity from Thermodynamics–Special and General. Another section will show that thermodynamics leads to quantization of paths and electric charge as well as a non-singular gauge potential that produces a n uclear model more accurate than the standard model. See the section titled Non-Singular Potential and Nuclear Physics. A third section will show how thermodynamics, through the time and mass dependence of the gauge function predicts a cosmology that includes red shifts and predictions of effects currently ascribed to dark matter/energy, but excludes a big bang beginning. This is contained in the section titled Red Shifts, Cosmology and Dark Stuff.

These sections together display the unifying properties of classical thermodynamic laws. They show how the second law limits the first law and by this limitation requires both geometry and a set of equations of motion for all dynamics. They show how the combination of stability conditions and the entropy principle for isolated systems produce the limitation to the first law through the appearance of a space-time structure similar to relativity which is required of a system with the speed of light as a limiting velocity. They show how the entropy principle requires two geometrical manifolds for every motion. One manifold is a non-integrable energy manifold with a Weyl geometry. The second manifold is an integrable Riemannian manifold obtained from the first manifold by the use of a gauge function as a geometrical integrating factor. These sections show that, because of the two manifolds, an isolated system whose entropy is constant must satisfy quantum mechanics. It is the five dimensional gauge function, stemming from the five dimensional first law, that produces the field theories and the connection between classical, relativistic and quantum mechanics.

Weyl showed how to use the gauge function to produce field equations and the five dimensional first law demands five dimensional fields that include inductively coupled electric, magnetic and gravitational fields and forces. The imposition of conservation of mass has been shown to require a four dimensional hyper surface, with an Einstein curvature, to be embedded into the five dimensional manifold. A very stable isentropic system has been shown to require the Weyl scale factor to take on a value of unity which means that the length of a vector must return to its original length after completing a closed path. This unity scale factor quantizes the path integral of the gauge potentials. For given gauge potentials the isentropic paths must satisfy quantum mechanics. The gauge function for fundamental particles whose gauge characteristics do not change in space or time must be quantized, non-singular with respect to space, depend upon time and depend upon mass.

Thus, the classical thermodynamic laws, through the stability conditions, entropy principle, and restrictive assumptions such as the isentropic assumption, require the fields, forces and dynamics currently seen in nuclear, atomic, and cosmology. It may, therefore, be seen that thermodynamic laws can and do provide the foundation of

physics as we now know it and is a viable starting point for
a search for a fundamental theory.

Chapter 1 Relativity–Special and General

Today, 105 years after Einstein introduced his special theory of relativity and 92 years after he presented his general theory of relativity, the science community has embraced relativity as the ultimate theory of high velocity motion and gravitational effects. What is not well-known is that classical thermodynamics, which has never been shown to fail yet, may be used to derive both the special and general theories of relativity. This presentation will show how the classical laws of thermodynamics require relativity as a subset of the entirety of the universe they describe. The first law of thermodynamics is a general statement of the conservation of energy in which energy (heat) exchanged between a system and its surroundings is equated to the change of system energy and the sum of the work done by the system. The work done by the system consists of four work terms; thermodynamic work and three mechanical work terms. The second law of thermodynamics places limitations upon the first law and may be stated, as Caratheódory did, in terms of an axiom to replace the engineering method of stating the second law as the Kelvin-Planck and Clausius statements support. The Caratheódory statement is very general and was first used to show that the path dependent first law always has an integrating denominator. This means that there is a unique function such that when the energy exchanged between the system and its surroundings is divided by this function the result is a path independent expression of the change of entropy.

For purely thermodynamic systems, with only a thermodynamic work term, this integrating denominator is a function only of the temperature. Further, this integrating denominator may be used to show that there exists a unique temperature at which no further energy may be exchanged between the system and its surroundings. This unique temperature is called the absolute zero temperature.

On the other hand, if a purely mechanical system is considered, a system with only mechanical work terms, then the integrating denominator is a function of the velocity only. This mechanical integrating denominator may be shown to require a unique velocity at which no further energy may be exchanged between the system and its surroundings. This velocity is unique and the one known velocity with similar characteristics is the speed of light and it may be shown that they are the same. This requirement concerning the unique velocity, or speed of light, establishes Einstein's postulate concerning the constancy of the speed of light upon which he built his special theory of relativity.

Returning to the first law of thermodynamics one notes that the energy exchanged between the system and its surroundings consists of five differentials; the change of system energy, the thermodynamic work and three mechanical work terms. Normally, when teaching theoretical thermodynamics the instructor will invoke the conservation of mass to provide the first of the necessary five equations that will lead to a solution. Of course, the remaining equations are the three mechanical equations of motion and an equation of state. If, however, the conservation of mass is not immediately invoked then it may be seen that the first law is a statement in five dimensions. This brings to mind five dimensional theories that were intended to seek a unification of the electromagnetic and gravitational forces such as the Kulsa-Klein and other five dimensional theories of Einstein and others. However, all of these theories insisted that the fifth dimension must not be a physically real entity. This

2

requirement placed restrictions on the resulting field quantities.

A different approach may be taken. If a f ifth dimension is allowed, considered to be real, and required to be conserved in the same fashion as the conservation of mass, then it may be shown that the conservation of the fifth dimension embeds a four dimensional curved space into the five dimensional manifold whose curvature is given by Einstein's field equations. Therefore, general relativity is obtained as a four dimensional, hyper surface embedded into the five dimensional manifold of the first law.

Special Relativity

In introducing his special theory of relativity Einstein introduces two postulates. The first postulate he elevated to a principle that he called the Principle of Relativity. Within this principle there is the adoption of Newtonian mechanics in addition to the requirement that laws must be the same for all frames of reference. Specifically, Einstein stated that the laws of electromagnetism and optics must be valid for all frames of reference which the equations of mechanics hold good. Within this principle then is contained the assumption that the laws of Newtonian mechanics and Maxwellian electromagnetism must be valid also. Therefore, to show that the special theory of relativity may be derived from another fundamental set of laws it must be shown that these laws produce Newton's equations of motion and Maxwell's electromagnetism and optics in addition to the requirement that these laws be valid in all frames of reference. The second postulate Einstein introduced was his famous postulate stating that the speed of light in empty space is always c independent of the state of motion of the emitting body. Any fundamental set of laws must, therefore, include the constancy of the speed of light in addition to the above requirements. The derivation of Maxwell electromagnetism from thermodynamics will be discussed in a later section and will be assumed to hold in this section.

Arguments that Newtonian and special relativistic mechanics were valid only for frames of references that were in uniform motion with respect to each other prompted Einstein to extend the principle of relativity to a principle that required the laws of physics to apply to all frames of reference regardless of their motion. He then developed a set of field equations that establish the geometry of space curved by the presence of gravitating mass. At the heart of this difference between special relativity and general relativity is geometry. Newton felt that given the correct force laws one was free to choose the geometry in which to seek motion caused by the force. This, of course, requires that one have equations of motion and an additional force law. In his general theory of relativity Einstein replaced the equations of motion and force laws with field equations that produced a geometry in which the force of gravity and the resulting motion was replaced by space curved by gravitating mass and geodesics. This means that any attempt to derive either, or both, relativistic theories from a more fundamental set of laws must include considerations of geometry.

Therefore, it is important that any attempt to find a single set of simple laws that unifies the existing laws and forces must include the following:

1. The laws must apply universally independent of position, time, motion or observer,
2. The laws must provide the equations of motion. This requires a geometry and a variational principle,
3. The laws must provide the force laws, whether electromagnetic, gravitational or another force,

    4. Existing theories must appear as subsets of the totality of these laws.

First Law (Conservation of Energy)

Einstein used his postulate concerning the constancy of the speed of light to show there was a limiting velocity for material objects. This limiting velocity is similar to the limiting aspect of the absolute zero temperature that appears in classical thermodynamics. This gives rise to the question as to whether or not there might be a fundamental connection between the two limiting concepts, one in mechanics and the other in thermodynamics.

It may be remembered that Einstein used the hypothesis of the constancy of the speed of light to modify Newton's equations of motion to establish his special theory of relativity. Therefore, any derivation of the special theory of relativity must not only derive the constancy of the speed of light but must also derive equations of motion. Both of these objectives have been met using the laws of classical thermodynamics as will be discussed below.

The concept of conservation of energy is fundamental to all branches of physics and is the beginning of the basis of thermodynamics and mechanics. In terms of generalized coordinates or independent variables, the notion of work, or mechanical energy, is considered linear forms of the type

$$\mathrm{d}W = F_i(q^1,...,q^n,u^1,...,u^n)\,dq^i \quad (i=1,2,...,n), \quad (1.1)$$

where the forces F_i may be functions of the velocities $(dq^i/dt = u^i)$ as well as the coordinates q^i and the summation convention is used

A system may acquire energy by other means in addition to the work terms; such energy acquisition is denoted $\mathrm{d}E$. The system energy, which represents the energy possessed by the system, is considered to be $U(q^1,...,q^n,u^1,...,u^n)$.

With these concepts, then the First Law, which is the generalized Law of Conservation of Energy, has the form

$$dE = dU - dW = dU - F_i\, dq^i \quad (i=1,...,n). \qquad (1.2)$$

Positive dE is taken as energy added to the system by means other than through the work terms and F_i is taken as the component of the generalized force acting on t he system which caused displacement dq^i.

In an infinitesimal transformation, the First Law is equivalent to the statement that the differential

$$dU = dE + F_i dq^i \qquad (1.3)$$

is exact.

This law is general and does not limit th e dimensionality, motion or observer. Nor is this law sensitive to the geometry chosen in which to express this law. Within the First Law the geometry may be freely chosen.

Second Law

There are processes, or motions, that satisfy the First Law but are not observed in nature. The purpose of the Second Law is to incorporate such experimental facts into the model of dynamics. The statement of the Second Law is made using the axiomatic statement provided by the Greek mathematician Carathéodory, who presented an axiomatic development of the second law of thermodynamics that may be applied to a system of any number of variables. The Second Law may then be stated as follows:

> In the neighborhood (however close) of any equilibrium state of a system of any number of dynamic coordinates, there exist states that cannot

be reached by reversible E-conservative $\left(\eth E = 0 \right)$ processes or motions.

When the variables are thermodynamic variables, the E-conservative processes are known as adiabatic processes.

A reversible process, or motion, is one that is performed in such a way that, at the conclusion of the process, both the system and the local surroundings may be restored to their initial states without producing any change in the rest of the universe.

Consider a system whose independent coordinates are a generalized displacement denoted q, a generalized velocity u (with $u' \equiv du/dt$), and a generalized force F. It can be shown that the E-conservative curve comprising all equilibrium states accessible from the initial state, i, may be expressed by $\sigma(u,q) = \text{constant}$, where σ represents some as yet undetermined function. Curves corresponding to other initial states would be represented by different values of the constant.

Reversible E-conservative curves cannot intersect, for if they did, it would be possible to proceed from an initial equilibrium state i, at the point of intersection, to two different final states f_1 and f_2, having the same q, a long reversible E-conservative paths, which is not allowed by the Second Law. This is an important requirement of the Second Law since this is the foundation of the denial of perpetual motion and we will now show this contention.

When the system can be described with only two independent variables, such as on the E-conservative curve, then if these variables are q and u and F is a generalized force,

$$\eth E = dU - Fdq.$$

Regarding $U = U(q,u)$, then

$$\eth E = \left[\frac{\partial U}{\partial u} \right]_q du + \left[\left[\frac{\partial U}{\partial q} \right]_u - F \right] dq$$

where all quantities on the right hand side are functions of u and q.

An *E*-conservative motion for this system requires that

$$\left[\frac{\partial U}{\partial u}\right]_q du + \left[\left[\frac{\partial U}{\partial q}\right]_u - F\right] dq = 0.$$

Solving for *du/dq* yields

$$\frac{du}{dq} = \frac{-\left[\left[\frac{\partial U}{\partial q}\right]_u - F\right]}{\left[\frac{\partial U}{\partial u}\right]_q}. \tag{1.4}$$

The right hand member is a function of u and q, and therefore, the derivative *du/dq*, representing the slope of an *E*-conservative curve on a (u,q) diagram, is known at all points. Equation (1.4) has therefore a solution consisting of a family of curves, in the phase space, and the curve through any one point may be written $\sigma = \sigma(u,q) = constant$.

A set of curves is obtained when different values are assigned to the constant. The existence of the family of curves $\sigma = \sigma(u,q) = constant$, generated by Equation (1.4) representing reversible *E*-conservative processes follows from the fact that there are only two independent variables and not from any law of physics. Thus, it can be seen that the First Law may be satisfied by any of these $\sigma = constant$ curves. The Second Law requires that these curves do not intersect. Therefore the Second Law, together with the First Law, leads to the conclusion that through any arbitrary initial state point, all reversible *E*-conservative processes lie on a curve, and *E*-conservative curves through other initial states determine a family of non-intersecting curves.

The results of this conclusion are important for mechanics. Consider a system whose coordinates are the

The Dynamic Theory

generalized velocity u, the generalized displacement q and the generalized force F. The First Law is

$$\eth E = dU - Fdq \tag{1.5}$$

where U and F are functions of u and q. Since the (u,q) surface is subdivided into a family of non-intersecting E-conservative curves $\sigma(u,q) = constant$ where the constant can take on various values σ_1, σ_2, ..., and points on the surface may be determined by specifying the value of σ along with q, in all regions where the Jacobian of the transformation does not vanish, so that U, as well as F, may be regarded as functions of σ and q. Then

$$dU = \left[\frac{\partial U}{\partial \sigma}\right]_q d\sigma + \left[\frac{\partial U}{\partial q}\right]_\sigma dq \tag{1.6}$$

and

$$\eth E = \left[\frac{\partial U}{\partial \sigma}\right]_q d\sigma + \left[\left[\frac{\partial U}{\partial q}\right]_\sigma - F\right]dq. \tag{1.7}$$

Since σ and q are independent variables Equation (1.7) must be true for all values of $d\sigma$ and dq.

Suppose $d\sigma = 0$ and $dq \neq 0$. The provision that $d\sigma = 0$ is the provision for an E-conservative process in which $\eth E = 0$. Therefore, the coefficient of dq must vanish. Then, in order for σ and q to be independent and for $\eth E$ to be zero when $d\sigma$ is zero, the equation for $\eth E$ must reduce to

$$\eth E = \left[\frac{\partial U}{\partial \sigma}\right]_q d\sigma,$$

Defining a function λ by $\lambda \equiv \left[\frac{\partial U}{\partial \sigma}\right]_q$ then $\eth E = \lambda d\sigma$.

Integrating Factor

Now, in general, an infinitesimal of the type

$$Pdx + Qdy + Rdz + ... = 0, \tag{1.8}$$

known as a linear differential form, or a Pfaffian expression, when it involves three or more independent variables, does not admit of an integrating factor. This equation is integrable if, and only if, in the neighborhood of any arbitrary point G_o there are points G which are inaccessible from G_o along solution curves Equation (1.7). It is only because of the existence of the Second Law that the differential form for $đE$ referring to a physical system of any number of independent coordinates possesses an integrating factor. The importance for physics is that not only do the two laws require the integrating factor, but they also have been shown to require that the integrating factor be a unique function of velocity only. That is for a system of any number of independent coordinates the transferred energy is given by,

$$đE = \phi(u)\, f(\sigma)\, d\sigma. \qquad (1.9)$$

Since $f(\sigma)d\sigma$ is an exact differential, the quantity $1/\phi(u)$ is an integrating factor for $đE$. Alternatively we can say that the function $\phi(u)$ is an integrating denominator.

It would be nice if there were a simple way of deriving the functional form of $\phi(u)$. In thermodynamics we opted to take the easy way out by assuming that the integrating factor was simply the reciprocal of the temperature as was determined from Carnot's work. However, for mechanical systems we will find the functional form of the integrating factor when we determine the limitation the Second Law imposes upon the First Law.

Constancy of the Speed of Light

The universal character of $\phi(u)$ makes it possible to define an absolute speed. Consider a system of two independent variables q and u, for which two constant speed curves and E-conservative curves may be defined. Suppose there is a constant velocity transfer of energy E between the system and its surroundings at the speed u, from a state b, on an E-conservative curve characterized by the value σ_1, to another state c, on another E-conservative curve specified by σ_2. Then since it is seen that

$$\text{đ}E = \phi(u) \int_{\sigma_1}^{\sigma_2} f(\sigma)\,d\sigma \quad \text{at constant } u. \tag{1.10}$$

For any constant speed process between two other points a to d, at a speed u_3 between the same E-conservative curves the energy transferred is

$$\text{đ}E(u_3) = \text{đ}E_3 = \phi(u_3) \int_{\sigma_1}^{\sigma_2} f(\sigma)\,d\sigma \quad \text{at constant } u_3. \tag{1.11}$$

Taking the ratio of

$$\tag{1.12}$$

$$\frac{\text{đ}E}{\text{đ}E_3} = \frac{\phi(u)\int_{\sigma_1}^{\sigma_2} f(\sigma)\,d\sigma}{\phi(u_3)\int_{\sigma_1}^{\sigma_2} f(\sigma)\,d\sigma} = \frac{\phi(u)}{\phi(u_3)} \frac{a\ fct\ of\ v\ at\ which\ \text{đ}E\ is\ transferred}{a\ fct\ of\ v\ at\ which\ \text{đ}E_3\ is\ transferred}$$

Then the ratio of these two functions is defined by

$$\frac{\phi(u)}{\phi(u_3)} = \frac{\text{đ}E\left(between\ \sigma_1\ and\ \sigma_2\ at\ u\right)}{\text{đ}E_3\left(between\ \sigma_1\ and\ \sigma_2\ at\ u_3\right)} \tag{1.13}$$

or

$$\text{đ}E = \left[\frac{\text{đ}E_3}{\phi(u_3)}\right]\phi(u). \tag{1.14}$$

By choosing some appropriate speed u_3 it follows that the energy transferred at constant speed between two given E-conservative curves decreases as $\phi(u)$ decreases, or the smaller the value of $\text{đ}E$ the lower the corresponding value of $\phi(u)$. When $\text{đ}E$ is zero $\phi(u)$ is also zero. The corresponding velocity u_0 such that $\phi(u_o)$ is zero is the "absolute velocity". Therefore, if a system undergoes a constant velocity motion between two E-conservative curves without an exchange of energy, the velocity at which this takes place is called the absolute velocity.

Just as the absolute temperature in classical thermodynamics is a limiting quantity we may suspect that the absolute speed will also turn out to be a limiting quantity. Because of our experimental evidence that the speed of light behaves as a limiting velocity when electromagnetic forces are involved and the absolute velocity is independent of the force or type of system and is therefore unique, it must be the speed of light. Thus, the first two laws require Einstein's postulate concerning the speed of light.

To be more specific, the absolute velocity is unique for all Galilean frames of reference. There is one such velocity already known and that velocity is the speed of light, c. Therefore, the absolute velocity must be the speed of light and the same for all Galilean observers. This is Einstein's postulate. However, it must be noted that the absolute velocity is not limited to Galilean observers. In the above demonstration of the effect of the integrating denominator as shown in Equation (1.12) constant velocity processes were used as a convenience. Just as the absolute temperature is independent of what paths may be considered so the absolute velocity is independent of what path may be considered.

Mechanical Entropy.

In a system of two independent variables, all states accessible from a given initial state by reversible E-conservative motions lie on a $\sigma(u,q)$ curve. The entire (u,q) space may be thought of as being filled by many non-intersecting curves of this kind, each corresponding to a different value of σ. In a reversible non-E-conservative motion involving a transfer of energy $đE$, a system in a state represented by a point lying on a surface σ will change until its state point lies on another surface $\sigma + d\sigma$. Then $đE = \lambda d\sigma$, where $1/\lambda$, the integrating factor of $đE$, is given by $\lambda = \phi(u) f(\sigma)$, and therefore

$$đE = \phi(u) f(\sigma) d\sigma \text{ or}$$

The Dynamic Theory

$$\frac{\text{đ}E}{\phi(u)} = f(\sigma)d\sigma. \qquad (1.15)$$

Since σ is an actual function of u and q, the right-hand member is an exact differential, which may be denoted by dS; and

$$dS = \frac{\text{đ}E}{\phi(u)} \qquad (1.16)$$

where S is the mechanical entropy of the system and the motion is a reversible one.

The Second Law may be used to prove the equivalent of Clausius' theorem, which is stated here.

Theorem: In any cyclic transformation throughout which the speed is defined, the following inequality holds: $\int \frac{\text{đ}E}{\phi(u)} \leq 0$, where the integral extends over one cycle of the transformation. The equality holds if the cyclic transformation is reversible. Then for an arbitrary transformation $\int_A^B \frac{\text{đ}E}{\phi(u)} \leq S(B) - S(A)$, with the equality holding if the transformation is reversible.

Thus, the Second Law provides an answer to the question that is not contained within the scope of the First Law: "In what direction does a process take place?" The answer is that a process always takes place in such a direction as to cause an increase of the mechanical entropy in the universe. In the case of an isolated system, it is the entropy of the system that tends to increase.

Third Law

There is nothing within the first two laws that allows comparison of the entropy between two systems so a third law is needed. The Third Law may be stated as:

The entropy of a system is a constant when the integrating denominator is zero (the integrating factor is infinite).

Equations of Motion

It was previously shown that the Second Law guarantees the existence of an integrating factor for the First Law where the integrating factor is a function of only the velocity. The Second Law also requires that the entropy must be maximized for an isolated system for which $dE = 0$. This principle becomes a variational principle for isolated systems. A metric may be obtained by considering the stability conditions when q and S are taken as the independent variables.

For example, consider the terms of second order in small displacements beginning with the general condition

$$\delta U - F\delta q - \phi\delta S > 0. \tag{1.17}$$

Choose $U=U(q,S)$, which, because of the identity $\phi dS = dU - Fdq$ is a natural choice for the independent variables, and expand δU in powers of the δq and δS

$$\delta U = \phi\delta S + F\delta q$$

$$+ \frac{1}{2}\left[\frac{\partial^2 U}{\partial q^2}\right]\delta q^2 + 2\left[\frac{\partial^2 U}{\partial q\partial S}\right]\delta q\delta S + \left[\frac{\partial^2 U}{\partial S^2}\right]\delta S^2 \tag{1.18}$$

$$+ \textit{terms of third order...}$$

The inequality Equation (1.17) then shows that in Equation, (1.18) 2nd order terms $+ 3^{rd}$ order terms $+ \ldots > 0$.

Retaining only the second order terms, the criterion of stability is that a quadratic differential form be positive definite;

$$\frac{\partial^2 U}{\partial S^2}(dS)^2 + \frac{\partial^2 U}{\partial S\partial q^\alpha}(dS)(dq^\alpha) + \frac{\partial^2 U}{\partial q^\alpha \partial q^\beta}(dq^\alpha)(dq^\beta) > 0 \, ; \alpha, \beta = 1,2,3. \tag{1.19}$$

Adopting this quadratic form as the metric of a general system whose thermodynamic variables are held fixed, we may then write this metric as

The Dynamic Theory

$$(ds)^2 = h_{ij}dq^i dq^j \; ; \quad i,j = 0,1,2,3, \qquad (1.20)$$

where the summation convention is used and

$$h_{ij} = \frac{\partial^2 U}{\partial q^i \partial q^j}, \qquad (1.21)$$

with $q^0 \equiv S/F_0$, the scaled mechanical entropy, for dimensional correctness.

Thus, the stability conditions provide a metric in the four-dimensional manifold of space- entropy. The arc length s in the space-entropy manifold may be parameterized by choosing $ds \equiv u_0 dt \equiv cdt$, where $u_0 = c$ is the unique velocity appearing in the integrating factor required by the Second Law. There are two reasons for choosing the unique velocity. First, it is the only well-defined velocity we have thus far. Secondly, we may look ahead to the metric of the Special Theory of Relativity. The metric may now be written as

$$c^2(dt)^2 = h_{ij}dq^i dq^j \; ; \quad i,j = 0,1,2,3. \qquad (1.22)$$

Now suppose the systems considered are restricted to only E-conservative systems. Then the principle of increasing mechanical entropy may be imposed in the form of the variational principle

$$\delta \int \sqrt{(dq^0)^2} = 0. \qquad (1.23)$$

In order to use this variation principle, Equation (1.22) may be expanded, solved for (dq^0) and squared to arrive at the quadratic form

$$(dq^0)^2 = \frac{1}{h_{00}}\left[c^2(dt)^2 + 2h_{0\alpha}Adtdq^\alpha - h_{\alpha\beta}dq^\alpha dq^\beta \right], \quad (1.24)$$

where

$$A = \frac{h_{0\alpha}}{h_{00}}u^\alpha \pm \sqrt{\frac{c^2}{h_{00}} + \frac{h_{\alpha\beta}}{h_{00}}u^\alpha u^\beta + \frac{h_{0\alpha}}{h_{00}}(u^\alpha)^2}$$

with $u^\alpha \equiv dq^\alpha/dt$.

By defining $x^0 \equiv ct$, $x^\alpha \equiv q^\alpha$, $\alpha = 1, 2, 3$, then Equation (1.24) may be written as

$$(dq^0)^2 = \frac{1}{f} g'_{ij} dx^i dx^j \; ; \quad i, j = 0, 1, 2, 3 \qquad (1.25)$$

where $f \equiv h_{00}$. This metric obviously reduces, in the Euclidean limit o f constant coefficients, to the metric of Minkowski's space-time manifold of Special Relativity. It is interesting to note that in the metric of Equation (1.25) the difference in the sign of the time and space elements of the metric comes from the fact that the stability conditions are given in terms of space and mechanical entropy while the variational principle was taken to be the Entropy Principle. In this fashion the Second Law guarantees the limiting aspect found in Einstein's Special Theory of Relativity.

In his General Theory of Relativity, Einstein assumed the space-time manifold to be Riemannian. However, this assumption involves the a priori assumption that the scalar product be invariant. This assumption was later questioned by Weyl in his generalization of geometry. From the viewpoint that the adopted postulates should contain the other theories within them then it becomes desirable to determine whether or not these postulates specify the geometry of the $(dq^0)^2$ space-time manifold. More particularly do t he adopted postulates lead to a geometry that includes the geometry of current theories? To arrive at a more general geometry would not be a limitation for it would certainly include the others.

Recalling Equation (1.25), we can define

$$(dq^0)^2 = \frac{1}{f} g'_{ij} dx^i dx^j \equiv \frac{1}{f}(d\sigma)^2 \equiv g_{ij} dx^i dx^j. \qquad (1.26)$$

Now the Second Law guarantees the existence of the function mechanical entropy and that dq^0 be a p erfect differential; therefore $dq^0 = q_i^0 dx^i$, where $q_i^0 \equiv \partial q^0 / \partial x^i$. Then the exactness of dq^0 is stated by $q^0_{i|j} - q^0_{j|i} = 0$.

The exactness of dq^0 required by the Second Law has been used to show that the entropy manifold must be a Riemannian manifold while the energy, or $d\sigma$, manifold must be a Weyl manifold with a non-integrable geometry that Einstein objected to previously. It may be noted here though that the two manifolds are not separable. The gauge function ties them together as a geometric integrating factor. This feature becomes very important in a search for a more fundamental theory and is similar to the path independence of the differential element of the entropy and the path dependence of the energy exchanged.

Equations of motion are then seen to be the result of the First and Second Laws in that they combine to produce a metric through the stability conditions and a variational principle in either the Principle of Increasing Entropy for an isolated system or Decreasing Free Energy for a non-isolated system. In the above an isolated system would require that the entropy increase. This would require equations that seek out the path that maximizes the entropy. Since geodesic equations are the same whether maximizing or minimizing the arc length we find that the relativistic equations of motion are obtained in the entropy manifold. Thus, we find that the laws of classical thermodynamics require both Einstein's postulate concerning the constancy of the speed of light and his equations of motion which include both specifying the geometry and a variational principle. This, then, displays that Einstein's special theory of relativity is a subset of the laws of classical thermodynamics.

However, it should be noted that both the First and Second Laws must both be satisfied simultaneously. This means that we might become confused as to which set of equations determines the motion. For example, if we use only one mechanical force and write the First Law as in Equation (1.15) we find that an isolated system must have

$$đE = 0 = dU - Fdq$$
$$\Rightarrow dU = Fdq$$

(1.27)

Then should U be the classical kinetic energy and a function only of the velocity we would have

$$\frac{dU}{dv}\frac{dv}{dt} = F\frac{dq}{dt}$$

(1.28)

$$\Rightarrow mva = Fv \Rightarrow F = ma$$

In Equation (1.28) we see the classical equations of motion given by Newton. Both the First Law and the Second Law must be satisfied. Therefore, we have what we might call First Law equations of motion as seen in Equation (1.28) and Second Law equations of motion are given by the geodesics obtained when maximizing the entropy in Equation (1.26). Both sets of equations of motion must be simultaneously satisfied.

General Relativity

Many researchers tried to expand upon Einstein's general theory of relativity as a m eans of seeking a unification of the gravitational and electromagnetic fields. Some researchers, including Einstein himself, sought to find success by investigating the use of five dimensions. One researcher, Hermann Weyl, sought to unify these fields by relaxing the invariance of the self, inner product of a vector. While Weyl did not succeed in unifying the electromagnetic and gravitational fields, he did show that Maxwellian electromagnetism could be derived using a gauge function and its first order derivatives the gauge potentials. Weyl's work has provided the basis of gauge field research since 1929. These two methods of seeking a unification of fields can be combined to derive Einstein's General Theory of Relativity.

Weyl's Gauge Principle

In order to acquaint or remind the reader of Weyl's Gauge Principle we must start with Weyl's scale factor where the length of a vector is allowed to change from point to point within the space. However, the allowed change to the vector length is governed by differential geometry and parallel displacement of the vector. These considerations require that the vector length l may change as it moves around in the space in accordance with:

$$l = l_0 \exp\left[\frac{1}{\gamma} \int \phi_j \, dx^j \right]. \tag{1.29}$$

The integrand of Equation (1.29) has been called the gauge potential since Weyl published his work and is defined by:

$$\phi_i \equiv \frac{1}{2} \frac{\partial \ln f}{\partial x^i}. \tag{1.30}$$

The electromagnetic field tensor is formed using the derivatives of the gauge potentials and is given by:

$$F_{ij} \equiv \phi_{i,j} - \phi_{j,i}. \tag{1.31}$$

The gauge field tensor given by Equation (1.31) has sixteen components when the indices range over the four dimensions of space and time. Weyl showed that these equations were the Maxwell equations of electromagnetism.

Five-Dimensional Gauge Fields with Conservation of the Fifth Dimension

Let a five-dimensional manifold have some physical reality in all five dimensions. The first four are the four dimensions of space-time as used in Einstein's general theory of relativity. The physical manifestation of the fifth dimension is to be learned during this investigation. Whatever the fifth dimension turns out to be, requiring it to be conserved will embed a four-dimensional hyper surface into the five-dimensional manifold.

Relativity-Special and General

Just as Weyl showed the gauge fields may be derived from gauge potentials the components of the 5-dimensional field tensor may be written in matrix form as:

$$F_{ij} = \begin{bmatrix} 0 & E_1 & E_2 & E_3 & V_0 \\ -E_1 & 0 & B_3 & -B_2 & V_1 \\ -E_2 & -B_3 & 0 & B_1 & V_2 \\ -E_3 & B_2 & -B_1 & 0 & V_3 \\ -V_0 & -V_1 & -V_2 & -V_3 & 0 \end{bmatrix}. \qquad (1.32)$$

The hyper surface field tensor will be given by $F_{\alpha\beta} = F_{ij} y_\alpha^i y_\beta^j$ where $y_\alpha^i = \partial y^i / \partial x^\alpha \, \delta_\alpha^i$ for the surface indices $i = 0,1,2,3$ and $y_\alpha^4 = \partial y^4 / \partial x^\alpha$. The space indices, i,j,k range over $0,1,2,3,4$ while the surface indices α,β,η,ν only take on values of $0,1,2,3$. The coefficients of the surface metric are given by the metric sum:

$$g_{\alpha\beta} = a_{ij} y_\alpha^i y_\beta^j = a_{\alpha\beta} + h_{\alpha\beta} \quad \text{where} \quad h_{\alpha\beta} = 2a_{\alpha 4} y_\beta^4 + a_{44} y_\alpha^4 y_\beta^4. \qquad (1.33)$$

The space energy-momentum tensor for matter under the influence of gauge fields is given by:

$$T_{sp}^{ij} = \gamma u^i u^j + \frac{1}{c^2}\left[F_k^i F^{kj} + \frac{1}{4} a^{ij} F^{kl} F_{kl} \right], \qquad (1.34)$$

and may be written in terms of the surface metric when the components due to the fifth dimension are added and is:

$$T_{sp}^{\alpha\beta} = \gamma u^\alpha u^\beta + \frac{1}{c^2}\left[F_k^\alpha F^{k\beta} + F_4^\alpha F^{4\beta} + \frac{1}{4}(g^{\alpha\beta} - h^{\alpha\beta})(F^{\mu\nu} F_{\mu\nu} + F^{4\nu} F_{4\nu}) \right], \qquad (1.35)$$

since:

$$u^4 \equiv \frac{dy^4}{dt} = \frac{\partial y^4}{\partial t} + \overline{\nabla}\cdot(y^4\overline{u}) = 0, \qquad (1.36)$$

is the statement required by the conservation of the fifth dimension. The surface energy-momentum tensor may now be found within the space tensor and written:

$$T^{\alpha\beta} = T_{sp}^{\alpha\beta} - \frac{1}{c^2}\left[F_4^\alpha F^{4\beta} - \frac{1}{2} h^{\alpha\beta} F^{4\nu} F_{4\nu} \right]. \qquad (1.37)$$

The Dynamic Theory

The form of this expression for the surface energy-momentum tensor suggests writing

$$CT^{\alpha\beta} \equiv G^{\alpha\beta} \equiv R^{\alpha\beta} - \frac{1}{2}g^{\alpha\beta}R. \qquad (1.38)$$

Equations (1.38), of course, look like the field equations of Einstein's general theory of relativity and are the reason the conservation of the fifth dimension was used to express the surface energy-momentum tensor. When Equations (1.38) are taken to be Einstein's field equations,

$$G^{\mu\nu} = -\frac{8\pi K}{c^2}T^{\mu\nu}, \qquad (1.39)$$

where K is the gravitational constant. Then the field components V_0, V_1, V_2, and V_3 must be related to the gravitational field and the fifth gauge potential must be related to the gravitational potential. Therefore, the physical reality of the fifth dimension is mass density.

From the above it may be seen that a five dimensional manifold of space-time-mass may have a four dimensional hyper surface embedded into it by requiring mass to be conserved. The field equations describing the curved hyper surface due to conservation of mass are Einstein's field equations of his general theory of relativity. This shows that Einstein's general theory of relativity may be derived from a five dimensional manifold of space-time-mass using gauge fields and the conservation of mass.

Phenomena predicted by Einstein's general theory of relativity can be seen in two different ways. One view is as Einstein intended; that is, as motion on a curved four dimensional space-time hyper surface in the absence of forces. The second view is from a five dimensional manifold of space-time-mass together with the restriction of mass conservation. In this view motion is seen as the result of forces due to gauge fields without a curvature in the manifold. Conservation of mass is the restriction that embeds Einstein's curved space-time hyper surface into the five dimensional manifold.

Summary

The laws of classical thermodynamics require that there exist an integrating factor for the 1st Law, the integrating factor be a function of velocity only, there exists a unique limiting velocity, motion for an isolated system must cause an increase in entropy in a relativistic space-time manifold, motion for a non-isolated system must minimize free energy, there must be two geometrical manifolds associated with motion of a system, one manifold has a non-integrable Weyl geometry, one manifold has an integrable Riemannian geometry, there exists a g auge function that acts as a geometrical integrating factor for the Weyl manifold, and the gauge function produces Maxwellian electromagnetism. Therefore, the laws of classical thermodynamics require Einstein's special theory of relativity. The laws also require that the curvature of a four dimensional hyper surface embedded into a five dimensional space-time-mass manifold by the requirement that mass be conserved must be given by Einstein's general theory of relativity. Then the laws of classical thermodynamics produce Newtonian and relativistic mechanics as well as Maxwell's electromagnetism.

Chapter 2 Non-Singular Potential and Nuclear Physics

Historically the Maxwell electromagnetic field equations arose as a distillation of empirical laws such as Coulomb's law. However, to recover Coulomb's law from Maxwell's equations requires the integration of a non-physical Dirac function to get the electrostatic potential. By reversing the approach this problem may be avoided, quantum mechanics derived, and new insights to nuclear physics obtained at the same time. In 1918 Weyl introduced his geometrical description of electromagnetism wherein his scale factor led to the gauge function and Einstein argued against it. In 1926 London showed that if Weyl's scale factor was required to be fixed at unity and one was given an electrostatic potential the only paths allowed were those satisfying Schrödinger's wave equation. This showed that quantum mechanics was required for a unity scale factor.

The requirement of a unity scale factor and the five dimensions given by the first law of thermodynamics may be used to derive an electrostatic potential for a particle whose gauge characteristics are independent of where and when they exist. T wo interesting features of this electrostatic potential are that these particles must have a quantized electric charge and that they must be non-singular. A quantized charge is not new as all known charged particles come with their charge quantized. The non-singular feature is new from the point of view that the electrostatic potential determined from Maxwell's equations is a singular potential that varies as the inverse of the distance from the particle. The non-singular potential has a long range inverse relation to the separation, but there is a short range maximum absolute value and a return to zero as the separation vanishes. This non-singular potential produces a force between particles that changes its long range character when separations go below the separation of the maximum value separation. This change in character leads to nuclear physics.

The non-singular character of the potential leads to the development of two- and three-particle Schrödinger and Dirac quantum mechanics that produces a new model of nuclear physics. The two-particle Dirac equation leads to the fields contained in the Yang-Mills equations while the three-particle fields satisfy the SU-3 group relations. The resulting nuclear model predicts the low-Z nuclear masses much more accurately than does the semi-empirical nuclear mass formula. The non-singular potential in the two-particle system also predicts the half life of the neutron.

In addition to the non-singular character of the potential there are two other features that, while they do not influence the nuclear model much, they do play a larger role in gravitational physics. One feature is that the gravitational field is time dependent and the other feature is that the electric field depends upon the mass. The time dependence shows up in red shifts and dark matter/energy. However, the only thing that the mass dependence is known to appear in to date is the ratio of the electromagnetic to the gravitational force where the ratio is established by the time and mass dependences of the gauge function. The value of the ratio of the electrostatic to gravitational force is another aspect of the inductive coupling between the electromagnetic and the gravitational fields.

Quantization

The two thermodynamic laws require that there be two geometric manifolds tied by a geometric integrating factor. An interesting result of these two manifolds occurs when one wishes to consider the most stable of all systems, a constant entropy system. When one looks at the metric for an isentropic state of an isolated system one finds that the condition which the German physicist London imposed upon Weyl's theory in 1927 is required. Namely, one finds that in order to satisfy the isentropic condition the line integral formed by the gauge potentials and the differentials of the metric variables must be quantized, or since $\dj E = 0$, then in the energy space the final arc element must equal the initial arc element so that $(d\sigma)=(d\sigma)_0$ and

$$e^{\int \phi_j dx^j} = 1. \tag{2.1}$$

This is the unity scale factor condition and is satisfied only if

$$\int \phi_j dx^j = 2\pi i N \tag{2.2}$$

where N is an integer and i is the square root of minus one. Therefore, requiring an isentropic state requires Weyl's Quantum Principle of 1929, or, since London showed this condition may be used to derive Schrödinger's equations, an isentropic state means quantum mechanics must be used.

Weyl first proposed his scale factor in his 1918 attempt to embed electromagnetism into geometry (Weyl, 1918). Schrödinger noticed that for a large number of systems satisfying the Bohr-Somerfield quantization conditions, the exponent of the non-integrable Weyl scale factor:

$$\ell = \ell_0 \exp\left[\frac{1}{\gamma}\int \phi_j dx^j\right], \tag{2.3}$$

became quantized. He also showed that, if the unit of quantization was taken to be imaginary with a magnitude of Planck's constant, then the Weyl factor would be unity. London showed that the Weyl scale factor was proportional to the Schrödinger wave function and that, if one knew the gauge potentials appearing in the exponent of the Weyl factor and asked what paths are allowed if the exponent were to remain quantized, the paths allowed were those given by Schrödinger wave mechanics. Though London's reinterpretation was tentative, Weyl seized upon it a nd presented a formulation that was complete and went further to propose his scale factor as a principle since electromagnetism could be *derived* from the gauge potentials. The analysis would include the fact that Weyl introduced a special case of Nother's theorem, which displayed the analogy between energy-momentum and electromagnetic conservation laws and, thereby, made the result familiar to physicists in the context of field theory.

Many scientists have engaged in gauge theory research as a result of Weyl's initial work. These researchers have given numerous invaluable contributions to the field. However, the synopsis above is sufficient to display that the basis of quantum gauge theory began with the concept of quantized motion. London showed that the solutions to Schrödinger's wave equation determined the paths that were allowed, while Weyl's scale factor remained unity and its exponent quantized. This provided a description of quantized paths or motion; namely, quantum mechanics.

The exponent of the scale factor may be thought of as having three parts: an integrand (one or more of the respective gauge potentials), a path (which is given by the differentials) and an integral value (the result of integrating the integrand over the path.) Thus, while London showed that the equations of quantum mechanics gave the paths allowed by a quantized exponent, provided the gauge potentials were known, another question might be asked of the exponent with equal expectations with regard to the descriptions of physical phenomena.

Before presenting an additional question that Weyl's Principle may address, it must first be pointed out that all known particles with electric charge exhibit the property of quantization of electrostatic charge. It is these quantized potentials and the interactions of a particle with electromagnetic fields that establishes the particle's identity. If this identity is to persist in time and through movement in space, the identifying gauge potentials must be independent of the motion. The atomic states reflect the interaction of two, or more, particles. Here the question concerns the identity of a particle. In the atomic states the interacting particles do not change their gauge properties during the motion, or time, involved in this interaction. London found that the Weyl Principle pointed to an infinite number of states that were independent of the path. The question to be asked now is, "Can the Weyl Quantum Principle also point to the gauge potentials that may be possessed by particles such that these properties are independent of the path?"

Weyl's Principle demands that the scale, or gauge, Equation (2.3), of the manifold be unity with $\ell = \ell_o$ for any path. Obviously, then:

$$e^{\int \phi_j dx^j} = 1 = e^{a+ib} = e^a(\cos b + i \sin b), \qquad (2.4)$$

for which $a = 0$ and $b = \cos^{-1}(1) = \sin^{-1}(0) = 2\pi N$ with $N = 0, \pm1, \pm2, \ldots$. Thus, the gauge potentials sought must satisfy:

$$\int \phi_j dx^j = i2\pi N, \quad \text{where } \varphi_j \equiv \frac{1}{2}\ln\left(\frac{\partial f}{\partial x^j}\right), \qquad (2.5)$$

and the i on the right hand side indicates an imaginary value, f is the gauge function and the summation convention is to be applied to the j's. Here N has been used to distinguish the gauge potential quantum number from the previous orbital, or path, quantum number, n. If the gauge potentials are to be independent of the dx^j then all of the path elements may be chosen to be zero except a single arbitrary dx^j. Therefore, the gauge potentials must satisfy the condition that:

$$\int \phi_j dx^j = i2\pi N_j, \qquad (2.6)$$

where now the summation convention is not used and each gauge potential is quantized by the value N_j. When ϕ_0 is identified with the electrostatic potential, as is done in quantum mechanics, the quantum number, N_0, quantizes the charge of the electrostatic potential. The components of the five dimensional gauge vector potential, (ϕ_1, ϕ_2, ϕ_3, ϕ_4), must also be quantized. The five components of the gauge potential must then be quantized as ($N_0\phi_0$, $N_1\phi_1$, $N_2\phi_2$, $N_3\phi_3$, $N_4\phi_4$). Therefore, the same quantum condition which Schrödinger, London and Weyl used to quantize the paths allowed by a unity scale factor also quantizes the gauge potentials allowed for particles!

Non-singular Gauge Potential

Any potential $\phi_{j=}$ $N_j\phi_j$ allowed by the quantum condition and the gauge fields which come from the gauge potentials, that is:

$$F_{jk} \equiv \frac{\partial \phi_j}{\partial x^k} - \frac{\partial \phi_k}{\partial x^j}, \tag{2.7}$$

must also satisfy the five dimensional extended Maxwell field equations.

If, however, one is operating in a five dimensional manifold of space, time and mass where Weyl's Gauge Principle in five dimensions produces the gauge field tensor that looks like

$$F_{ij} = \begin{vmatrix} 0 & iE_1 & iE_2 & iE_3 & iV_0 \\ -iE_1 & 0 & B_1 & -B_2 & V_1 \\ -iE_2 & -B_1 & 0 & B_3 & V_2 \\ -iE_3 & B_2 & -B_3 & 0 & V_3 \\ -iV_0 & -V_1 & -V_2 & -V_3 & 0 \end{vmatrix}. \tag{2.8}$$

and the field equations given by Equations (2.7) yields the extended Maxwell field equations

$$\overline{\nabla} \bullet \overline{B} = 0 \qquad [a]$$

$$\frac{1}{c}\frac{\partial \overline{B}}{\partial t} + \overline{\nabla} \times \overline{E} = \overline{0} \qquad [b]$$

$$\overline{\nabla} \times \overline{B} - \frac{1}{c}\frac{\partial \overline{E}}{\partial t} + a_0 \frac{\partial \overline{V}}{\partial \gamma} = \frac{4\pi \overline{J}}{c} \qquad [c]$$

$$\overline{\nabla} \bullet \overline{E} + a_0 \frac{\partial V_0}{\partial \gamma} = 4\pi\rho \qquad [d]$$

$$\frac{\partial \rho}{\partial t} + \overline{\nabla} \bullet \overline{J} + a_0 \frac{\partial J_4}{\partial \gamma} = 0 \qquad [e]$$

$$\overline{\nabla} \times \overline{V} + a_0 \frac{\partial \overline{B}}{\partial \gamma} = \overline{0} \qquad [f]$$

$$\overline{\nabla} V_0 + \frac{1}{c}\frac{\partial \overline{V}}{\partial t} = a_0 \frac{\partial \overline{E}}{\partial \gamma} \qquad [g]$$

$$\overline{\nabla} \bullet \overline{V} + \frac{1}{c}\frac{\partial V_0}{\partial t} = -\frac{4\pi J_4}{c} \qquad [h] \qquad (2.9)$$

In order to find the allowed gauge function and potentials the quantized gauge potential must satisfy the extended Maxwell field equations. After considerable operations the allowed gauge function and fields may be found. The gauge function is

$$\ln f^{\frac{1}{2}} = \left(\frac{r_o}{r}\right) e^{-\left(\frac{\lambda_N}{r}\right)} \sin^{\lambda_o}\theta\, e^{-H_o t} e^{-K_\gamma \gamma}$$

$$\lambda_o \ll 1 \Rightarrow \sin^{\lambda_o}\theta \cong 1$$

$$N_0 = N_4$$

$$K_\gamma = \frac{H_o}{a_o c} \tag{2.10}$$

$$Z = \left(N_0 - N_1\right)$$

$$J_4 = -\frac{a_o c^2 K_\gamma}{H_o}\rho$$

$$\text{and } N_1 = N_2 = N_3$$

The field tensor for the fundamental particle is given by

$$\frac{F_{ij}}{\left(\dfrac{e}{4\pi\varepsilon_o}\right)} \approx
\begin{vmatrix}
0 & Z\left(\frac{r_o H_o}{cr^2}\right)\left(1-\frac{\lambda_N}{r}\right)e^{-\left(\frac{\lambda_N}{r}\right)} & Zr_o\lambda_0\left(\frac{1}{cr^2}\right)e^{-\left(\frac{\lambda_N}{r}\right)}\cot\theta & 0 & 0 \\
-Z\left(\frac{r_o H_o}{cr^2}\right)\left(1-\frac{\lambda_N}{r}\right)e^{\left(\frac{\lambda_N}{r}\right)} & 0 & 0 & 0 & -Ze\frac{a_o c}{H_o}r_o K_\gamma\left(\frac{1}{r^2}\right)\left(1-\frac{\lambda_N}{r}\right)e^{-\frac{\lambda_N}{r}} \\
-Zr_o\lambda_0\left(\frac{1}{cr^2}\right)e^{\left(\frac{\lambda_N}{r}\right)}\cot\theta & 0 & 0 & 0 & -Ze\frac{a_o c}{H_o}\lambda_0 r_o K_\gamma\left(\frac{1}{r^2}\right)e^{-\left(\frac{\lambda_N}{r}\right)}\cot\theta \\
0 & 0 & 0 & 0 & 0 \\
0 & Ze\frac{a_o c}{H_o}r_o K_\gamma\left(\frac{1}{r^2}\right)\left(1-\frac{\lambda_N}{r}\right)e^{-\frac{\lambda_N}{r}} & Ze\frac{a_o c}{H_o}\lambda_0 r_o K_\gamma\left(\frac{1}{r^2}\right)e^{\left(\frac{\lambda_N}{r}\right)}\cot\theta & 0 & 0
\end{vmatrix} \tag{2.11}$$

This forms a complete solution set for the fundamental particle.

An interesting result of this solution of the gauge function and fields for a fundamental particle is the prediction of the ratio of the electromagnetic and gravitational forces. The electric field is

$$E_r = Z\left(\frac{e r_o H_o}{4\pi\varepsilon_o cr^2}\right)\left(1-\frac{\lambda_N}{r}\right)e^{-\left(\frac{\lambda_N}{r}\right)}. \tag{2.12}$$

Equation (2.12) would be the same as the experimental value of the electric field if

$$r_o = \frac{c}{H_0}. \tag{2.13}$$

The electrical force between identical particles would then be given by

The Dynamic Theory

$$F_e = eE_r \approx Z\left(\frac{e^2 r_o H_o}{4\pi\varepsilon_o cr^2}\right)\left(1-\frac{\lambda_N}{r}\right)e^{-\left(\frac{\lambda_N}{r}\right)}$$

$$= Z\left(\frac{e^2}{4\pi\varepsilon_o r^2}\right)\left(1-\frac{\lambda_N}{r}\right)e^{-\left(\frac{\lambda_N}{r}\right)} \qquad (2.14)$$

The gravitational field is

$$V_r = \frac{a_o cK_\gamma}{H_o}E_r$$

$$\approx Z\frac{a_o cK_\gamma}{H_o}\left(\frac{e}{4\pi\varepsilon_o r^2}\right)\left(1-\frac{\lambda_N}{r}\right)e^{-\left(\frac{\lambda_N}{r}\right)}e^{-H_o t} \qquad (2.15)$$

Consider the gravitational charge density

$$\frac{J_4}{c} = -\frac{a_o cK_\gamma}{H_o}e \qquad (2.16)$$

Integrating Equation (2.16) produces the gravitational charge, or

$$\frac{M}{c} \equiv \iiint_{vol}\frac{J_4}{c}dvol = -\frac{a_o cK_\gamma}{H_o}\iiint_{vol}\rho dvol$$

$$\approx -\frac{a_o cK_\gamma}{H_o}e \qquad (2.17)$$

Now we can form the gravitational force by multiplying Equation (2.15) by (2.17) and obtain

$$F_g = mV_g = -\frac{a_o cK_\gamma}{H_o}eZ\frac{a_o cK_\gamma}{H_o}\left(\frac{e}{4\pi\varepsilon_o r^2}\right)\left(1-\frac{\lambda_N}{r}\right)e^{-\left(\frac{\lambda_N}{r}\right)}e^{-H_o t}$$

$$= -Z\frac{a_o^2 c^2 K_\gamma^2}{H_o^2}\left(\frac{e^2}{4\pi\varepsilon_o r^2}\right)\left(1-\frac{\lambda_N}{r}\right)e^{-\left(\frac{\lambda_N}{r}\right)}e^{-H_o t} \qquad (2.18)$$

Forming the ratio of the electrical force and the gravitational force produces

$$F_{ratio} \approx \frac{Z\left(\dfrac{e^2}{4\pi\varepsilon_o r^2}\right)\left(1-\dfrac{\lambda_N}{r}\right)e^{-\left(\frac{\lambda_N}{r}\right)}}{Z\dfrac{a_o^2 c^2 K_\gamma^2}{H_o^2}\left(\dfrac{e^2}{4\pi\varepsilon_o r^2}\right)\left(1-\dfrac{\lambda_N}{r}\right)e^{-\left(\frac{\lambda_N}{r}\right)}e^{-H_o t}} \qquad .(2.19)$$

$$\approx \frac{H_o^2}{a_o^2 c^2 K_\gamma^2}$$

Comparing this to the classical ratio we find

$$F_{ratio} \approx \frac{H_o^2}{a_o^2 c^2 K_\gamma^2} = \frac{\dfrac{e^2}{4\pi\varepsilon_o}}{Gm^2}. \qquad (2.20)$$

Our solution for the gauge function and fields is only good for fundamental particles and not for a composite particle. Therefore, we should only use a particle such as the electron in Equation (2.20). When the electron is used the force ratio is

$$F_{ratio} \approx \frac{H_o^2}{a_o^2 c^2 K_\gamma^2} = \frac{e^2}{4\pi\varepsilon_o Gm_e^2} = 4.17\times10^{42}$$

$$\Rightarrow a_o^2 c^2 K_\gamma^2 = \frac{H_o^2}{4.17\times10^{42}} \qquad .(2.21)$$

$$\Rightarrow a_o c K_\gamma = \frac{H_o}{2.04163\times10^{21}} \approx 1.12\times10^{-39} \ 1/\text{sec}.$$

This shows that it is the ratio of the dependence of the gravitational field on time to the dependence of the electric field on mass density that sets the ratio of the electric to gravitational force. Further, this shows that the dependence of the electric field upon mass density is very weak, but non-zero.

Nuclear Physics

The non-singular fields of Equation (2.11) indicate that interactions between two particles for which $\lambda_1 \neq \lambda_2$ will not satisfy Newton's Third Law. See Figure 2.1. Therefore any investigation into interactions between particles with these non-singular potentials must do so without imposing Newton's Third Law.

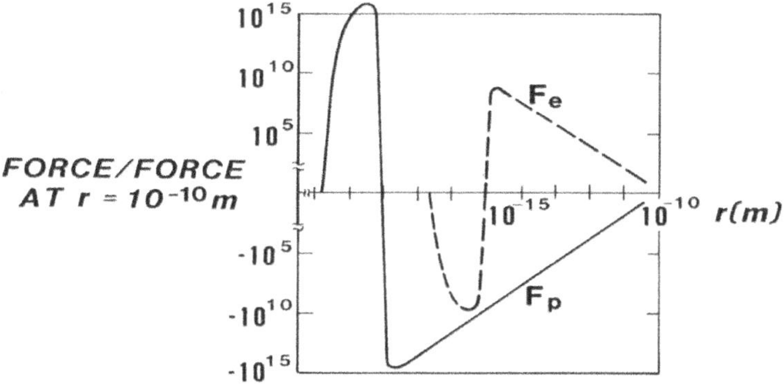

Figure 2.1 Forces on un-like particles.

Motion OF and ABOUT the Center of Mass

It has been shown that, while it is easier to study multiple body problems when Newton's Third Law holds, it is not mandatory. The two-body problem may be written in terms of the motion of each body or as the motion OF the center of mass and motion ABOUT the center of mass. While it may be tempting to use the motion of each body it is easier to compare with classical solutions of motions when using the motions OF and ABOUT the center of mass. It is then easy to see the conditions for which Newton's Third Law holds for when it holds there will be no motion OF the center of mass.

Newtonian Motion OF and ABOUT the Center of Mass

An investigation of the equations of motion may start with the well-known change of coordinates that allows one to analyze the motion in terms of the center of mass,

$$\bar{R} = \frac{m_1 \bar{r}_1 + m_2 \bar{r}_2}{m_1 + m_2},$$

(2.22)

$$\bar{r} = \bar{r}_1 - \bar{r}_2,$$

where the small r's represent vectors to each body or between the bodies, and the capital R represents the vector to the center of mass of the two bodies. Of course the inverse transformations associated with the transformation contained in Equations (2.22) exist and are the usual ones associated with the center of mass (COM) approach.

By using the COM approach and writing the low velocity limit of the equations of motion (EOM), we find:

$$m_1 \ddot{r}_1 \hat{r}_1 = \bar{F}_1^i + \bar{F}_1^e,$$

$$m_2 \ddot{r}_2 \hat{r}_2 = \bar{F}_2^i + \bar{F}_2^e,$$

(2.23)

where the superscripts i and e represent internal and external forces respectively and the $\wedge$ denotes a unit vector. By using the standard definition for the total mass, $M = m_1 + m_2$, and the reduced mass, $\mu = (m_1 m_2)/M$, and setting all external forces to zero, Equations (2.23) may be put into the form:

$$M \ddot{R} \hat{R} = \bar{F}_1^i + \bar{F}_2^i,$$

$$\mu \ddot{r} \hat{r} = \frac{\mu}{m_1} \bar{F}_1^i - \frac{\mu}{m_2} \bar{F}_2^i.$$

(2.24)

The EOM, Equations (2.24), display the effect of Newton's Third Law on the two-body problem. Should Newton's Third Law hold in both magnitude and direction then the first equation shows that the force on the COM vanishes while the two forces, which must remain separate when Newton's Third Law does not hold, becomes a single-force statement without any reference to the mass of the bodies. Further, the first equation gives the motion *OF* the COM while the second equation gives the motion *ABOUT* the COM.

The Dynamic Theory

The force on body one due to the presence of body two is:

$$\overline{F}_1^i = -Z_1 e \overline{\nabla} V_2^i = -\left(\frac{Z_1 Z_2 e^2}{4\pi\varepsilon_0}\right)\frac{1}{r^2}\left(1 - \frac{\lambda_2}{r}\right)e^{\left(-\frac{\lambda_2}{r}\right)}\hat{r}, \quad (2.25)$$

while the force on body two due to the presence of body one is:

$$\overline{F}_2^i = -Z_2 e \overline{\nabla} V_1^i = \left(\frac{Z_1 Z_2 e^2}{4\pi\varepsilon_0}\right)\frac{1}{r^2}\left(1 - \frac{\lambda_1}{r}\right)e^{\left(-\frac{\lambda_1}{r}\right)}\hat{r}. \quad (2.26)$$

Equations (2.25) and (2.26) exhibit the property that at large separations they approximately obey Newton's Third Law, but as the separation approaches the larger of the λ's they begin to severely depart from Newton's Third Law. Therefore, these two forces cannot be combined into a single central force as is done in classical mechanics, nor can their potentials be combined into a single potential.

By using the equations of motion, Equations (2.24), with the force laws, Equations (2.25) and (2.26), the Conservation of Energy, and transferring to the cylindrical coordinates typical of motion for central forces, it may be shown that the energy, which is a constant of the low velocity motion, becomes:

$$E = K + V + K_c + V_c, \quad (2.27)$$

where $K+V$ is the energy *ABOUT* the COM and K_c+V_c is the energy *OF* the COM. In Equation (2.27), the parts are given by:

$$K = \frac{\mu k}{2r}\left[\left(1 - \frac{\lambda_1}{r}\right)\frac{e^{-\frac{\lambda_1}{r}}}{m_2} + \left(1 - \frac{\lambda_2}{r}\right)\frac{e^{-\frac{\lambda_2}{r}}}{m_1}\right]$$

$$V = -\frac{\mu k}{r}\left[\frac{e^{-\frac{\lambda_1}{r}}}{m_2} + \frac{e^{-\frac{\lambda_2}{r}}}{m_1}\right] \tag{2.28}$$

$$K_c = \frac{kR}{2r^2}\left[\left(1 - \frac{\lambda_1}{r}\right)e^{-\frac{\lambda_1}{r}} - \left(1 - \frac{\lambda_2}{r}\right)e^{-\frac{\lambda_2}{r}}\right]$$

$$V_c = -\frac{k}{r}\left[e^{-\frac{\lambda_1}{r}} - e^{-\frac{\lambda_2}{r}}\right],$$

where $k = (Z_1 Z_2 e^2)/(4\pi\varepsilon_0)$.

Quantum Mechanical Motion OF and ABOUT the Center of Mass

Both non-relativistic and relativistic quantum mechanics may also be developed without the assumption of Newton's Third Law. Some surprising results occur as the result of this development.

Non-Relativistic Quantum Mechanics

An expanded Schrödinger-like wave equation may be developed using similar assumptions that Schrödinger used except that Newton's Third Law is not assumed nor applied. When this is done, and the required linear behavior and free system limit is met, the resulting wave equation ends up being:

$$-\frac{\hbar^2}{\mu}\frac{\partial^2\psi(x,X,t)}{\partial x^2} + V(x,t)\psi(x,X,t)$$

$$-\frac{\hbar^2}{2M}\frac{\partial^2\psi(x,X,t)}{\partial X^2} + V_c(x,t)\psi(x,X,t) = i\hbar\frac{\partial\psi(x,X,t)}{\partial t}. \tag{2.29}$$

The time-independent, expanded wave equation in cylindrical coordinates is then:

$$-\frac{\hbar^2}{2\mu}\frac{\partial^2\psi(\overline{r},\overline{R})}{\partial\overline{r}^2}+V(\overline{r},\overline{R})\psi(\overline{r},\overline{R})$$

$$-\frac{\hbar^2}{2M}\frac{\partial^2\psi(\overline{r},\overline{R})}{\partial\overline{R}^2}+V_c(\overline{r},\overline{R})\psi(\overline{r},\overline{R})=E\psi(\overline{r},\overline{R}),$$

$$(2.30)$$

where the potentials are those in Equations (2.28).

Relativistic Quantum Mechanics (Weak Force)

Relativistic equations may also be developed without the imposition of Newton's Third Law through a long, laborious process that is virtually the same as the process used in standard relativistic quantum mechanics. Defining $B^{\beta}{}_{\theta\mu\nu}=\partial_{\theta\mu}A^{T\beta}{}_{\nu}-\partial_{\theta\nu}A^{T\beta}{}_{\mu}$ the Klein-Gordon equation with fields for the motion OF and ABOUT the center of mass for two particles may be written as:

$$[(i\partial_{\theta\mu}I^{\theta}-e_{\theta}A_{\mu}^{T\theta})(i\partial_{\theta}^{\mu}I^{\theta}-e_{\theta}A^{T\theta\mu})$$

$$-(m_{\theta}I^{\theta})^2-i\sigma^{\mu\nu}e_{\beta}B^{\beta}_{\theta\mu\nu}I^{\theta}]\psi(x)=0,$$

$$(2.31)$$

with the usual definition for $A^{T\beta}{}_{\mu}$ and I^{θ} is the 1 by 2 identity matrix. The $B^{\beta}{}_{\theta\mu\nu}$ as defined are the Yang-Mills fields normally ascribed to the weak nuclear forces.

Relativistic Quantum Mechanics (Strong Force)

Three-body problems may be determined by defining $D^{\eta}{}_{\theta\mu\nu}=\partial_{\theta\mu}C^{\eta}{}_{\nu}-\partial_{\theta\nu}C^{\eta}{}_{\mu}$ with the definition $C^{\theta}{}_{\mu}=A^{\beta}{}_{\mu}+A^{\gamma}{}_{\mu}$, where θ, β, and γ are cyclic. Equation (2.31) becomes:

$$\left[\begin{array}{c}\left(i\partial_{\theta\mu}I^{\theta}-e_{\theta}C_{\mu}^{\theta}\right)\left(i\partial_{\theta}^{\mu}I^{\theta}-e_{\theta}C^{\theta\mu}\right)\\ -\left(m_{\theta}I^{\theta}\right)-i\sigma^{\mu\nu}e_{\eta}D^{\eta}_{\theta\mu\nu}I^{\theta}\end{array}\right]\psi(x)=0 \quad (2.32)$$

where θ, β, and η range from 1 to 3 and θ, β, and η are cyclic and the I^{θ} is the 1 by 3 identity matrix.

Equation (2.32) represents the Klein-Gordon equation for three particles when one particle may be different from the other two. However, for three like particles it remains valid, but some terms vanish due to the equal and opposite forces.

Also, Equation (2.32) is the equation describing the strong nuclear forces that bind two protons to a single electron to form the deuterium nucleus. Of course this occurs when the separation of all three particles is approximately of nuclear separations. This may be seen in Figure 2.1 when using a separation of approximately 1 Fermi where the force on the electron is virtually zero while the force on the proton is still strong. It may also be seen that the three-body system of two protons in orbit around a single electron is a symmetrical system wherein the COM has no motion.

Nuclear Model

The suggestion that a neutron consists of a proton in orbit around an electron described by Equation (2.31) is very different from the standard nuclear model and raises many questions such as, "Isn't this a violation of the Heisenberg uncertainty principle?" and "Doesn't such motion violate spin conservation?" These questions have already been answered. The answer to both questions comes from the fact that the unit of action determines both the uncertainty and the spin and the unit of action depends upon the geometry. While a vector curvature determined by the mass of the Earth would be too small to notice the gauge function has a noticeable effect at nuclear separations. For instance, the proton orbit in the neutron has a spin of approximately 2/3 of Planck's constant while the electron spin is only 1/1,832 times the proton spin. There is no violation of Heisenberg's uncertainty principle or spin conservation.

Weyl's Gauge Principle and the restriction to a unity scale factor not only require the electromagnetic gauge fields, but also require that the electrostatic potential be quantized and non-singular. The non-singular character of the potentials violates Newton's Third Law of equal and opposite forces. However, when quantum mechanical procedures are applied in the absence of Newton's Third

Law the relativistic quantum mechanical equations of the motion OF and ABOUT the center of mass contain the Yang-Mills fields and the SU3 group fields of the standard model. Thus, while the new view and approach are completely different from the standard model, the standard model is not violated for its equations are contained within the new view. Plus, the new view adds predictions not possible in the standard model.

Another way of visualizing the neo-coulombic force is to make a plot of it and compare it with a plot of the columbic force. Figure 2.2 compares these two forces plotted with the separation variable in fermions and normalized so that the columbic force at one Fermi separation is unity. Note that this plot compares the forces for like particles to ensure that λ is the same for both particles. Figure 2.2 shows that the neo-coulombic force is virtually indistinguishable from the Coulomb force for separations greater than approximately 10λ, which is an approximately atomic separation. However, at a separation of exactly λ, the force is identically zero. In terms of the classical notion of nuclear forces, we would say that at separations greater than 10λ, the nuclear force is negligible, whereas at a separation of λ the magnitude of the nuclear force was equal to the magnitude of the coulombic force. The neo-coulombic force becomes an attractive force for separations less than λ. This is exactly the behavior to be expected of a non-singular potential. For a potential to be non-singular it must tend to zero as r goes to zero. Such a potential which tends to zero for r tending to zero and for r tending to ∞ must have a maximum absolute value in between. At that maximum the force, being determined by the slope of the potential, will go to zero and will be of the opposite signs on each side of the zero.

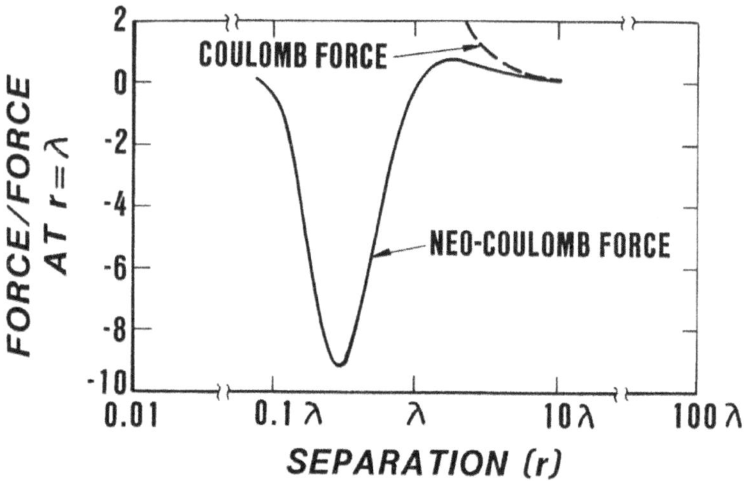

Figure 2.2. Comparison of coulomb and non-coulombic forces at short range.

Now let us look at the force between unlike particles, say a proton and an electron. Consider the electron and proton to be placed on a horizontal surface separated by a distance, r, with the proton to the right of the electron. Thus, the long range attractive forces between these two particles will cause the proton to experience a force to the left while the electron will experience a force to the right. We may than write the force on the proton that is due to the positive charge of the proton being in the electron field as

$$F_p = q_p E_e = \frac{-e^2}{4\pi\varepsilon_o r^2}\left(1 - \frac{\lambda_e}{r}\right)e^{\frac{-\lambda_e}{r}} \tag{2.33}$$

where the electron field involving the electron lambda has been accounted for. The electron force owing to the electron charge being in the proton field is given by

$$F_e = q_e E_p = \frac{e^2}{4\pi\varepsilon_o r^2}\left(1 - \frac{\lambda_p}{r}\right)e^{\frac{-\lambda_p}{r}} \tag{2.34}$$

Figure 2.3 plots both these forces as a function of the separation, r, where, $\lambda_p = 10^{-15}$ m, or $\lambda_p = 1$ Fermi has been assumed. The electron-electron scattering data show that the electron-electron interaction behaves in a coulombic manner even when separations are approximately 0.01-0.1 Fermi. To be consistent with this data, we have assumed $\lambda_e = 10^{-3}$ Fermi.

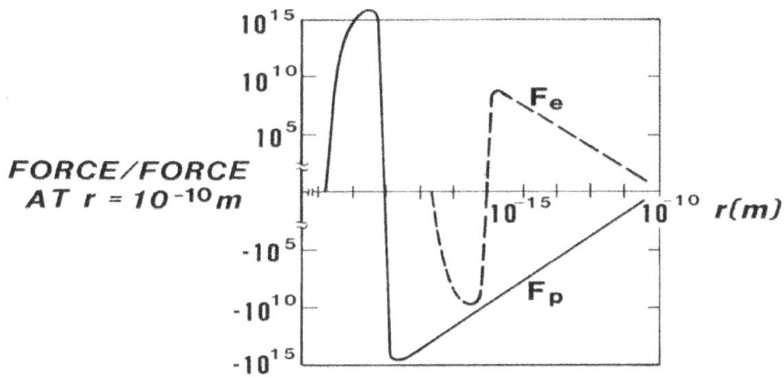

Figure 2.3. Unlike particle forces

From the plot of the force on the proton and the force on the electron in Figure 2.3, we see that for separations less than about 10 Fermi's the forces become extremely unsymmetrical. This immediately and visually demonstrates that the neo-coulombic exponential force violates Newton's third law requiring that the force on the proton be equal in magnitude and opposite indirection to the force on the electron. The question arises whether or not a violation of Newton's third law has ever been seen as the result of an interaction between an electron and a proton? The answer, based on a neutron disintegration from which a proton and electron emerge, is definitely yes; Newton's third law was seen to be violated. To reinstate Newton's third law in neutron disintegration and all other beta decay, Pauli postulated the existence of the neutrino. Fermi later developed his theory of weak interactions, from which appeared the necessity to talk of a fourth force in nature.

It was previously shown that the neo-coulombic force, which requires distinct λ for distinct fundamental particles, leads to the Yang-Mills equations for describing the weak forces. Does this mean the neutrino does not exist? What about the experimental evidence submitted in support of the capture of a free neutrino?

If we again consider the plots of the proton and electron forces in Figure 2.3 we see that, at atomic separations and greater distances, the forces obey Newton's third law and the difference between the neo-coulombic and columbic forces is so small that it could not be detected in atomic or macroscopic phenomena. But as the separation becomes smaller, the picture begins to change. When the r approaches λ_p, the electron is no l onger attracted to the proton as strongly as the proton is attracted to the electron. If the separation is exactly λ_p, then the electron is indifferent to the proton's presence. The proton, on t he other hand, is still very much attracted to the electron. If for the moment, we ignore the interpretation of Heisenberg's uncertainty principle that would say it cannot be, then we could easily imagine a circular proton orbit around a stationary electron, during which the proton stays at a radius of λ_p from the electron. The electron should be stationary during such motion because it would experience no force.

We now consider a separation between the electron and proton, which is some simple fraction of λ_p. Here, we find the electron repulsed by the proton, but the proton is still attracted to the electron. Notice that the force on both particles, from our initial positioning of the proton on t he right, is to the left. If both particles were given an angular momentum such that they were placed into synchronized circular orbits, then because their synchronous motion always results in the force on both particles being directed along the line separating them and from the proton toward the electron or from the electron away from the proton then, again ignoring arguments from the uncertainty principle, circular orbits in which the electron is in a small orbit about a space point could be imagined, where the proton is in a much larger orbit about the same space point.

Let us follow this picture a little farther and write simple Newtonian like force laws for this situation. The situation envisioned is presented in Figure 2.4. The electron position is given by r_e from the origin, and the position of the proton is given by r_p. The separation between them is

$$r = r_p - r_e . \tag{2.35}$$

Because the force is always directed along the line separating the two particles, we may write the radial equation of motion for the proton as

$$m_p \left(\frac{v_p^2}{r_p} \right) = \frac{-e^2}{4\pi\varepsilon_o r^2} \left(1 - \frac{\lambda_e}{r} \right) e^{\frac{-\lambda_e}{r}} \tag{2.36}$$

where the assumed circular motion has been taken into account and v_p is the tangential proton velocity. The electron equation of motion is given by

$$m_e \left(\frac{v_e^2}{r_e} \right) = \frac{e^2}{4\pi\varepsilon_o r^2} \left(1 - \frac{\lambda_p}{r} \right) e^{\frac{-\lambda_p}{r}} \tag{2.37}$$

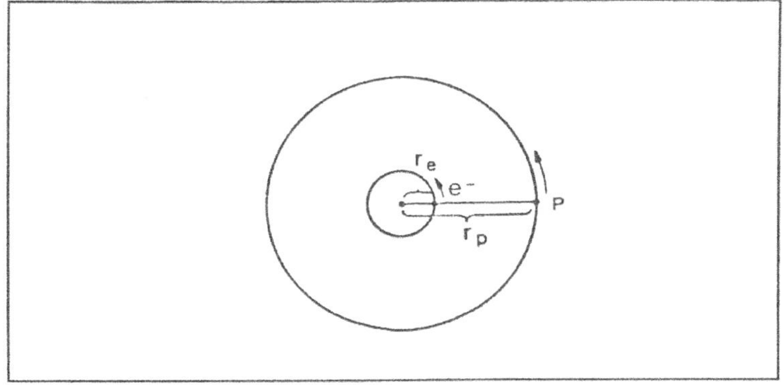

Figure 2.4. Circular orbits for the electron and proton in a neutron.

The right-hand side of Equations (2.36) and (2.37) are both functions of the separation, r, whereas the two left-hand sides are individually functions of r_e and r_p. A solution is possible only when the three equations, Equations (2.35), (2.36), and (2.37), are solved simultaneously.

An alternative approach is to add Equation (2.36) and (2.37) then obtain the equation of motion for the center of mass,

$$m_p\left(\frac{v_p^2}{r_p}\right) + m_e\left(\frac{v_e^2}{r_e}\right) = \frac{e^2}{4\pi\varepsilon_o r^2}\left[\left(1-\frac{\lambda_p}{r}\right)e^{\frac{-\lambda_p}{r}} - \left(1-\frac{\lambda_e}{r}\right)e^{\frac{-\lambda_e}{r}}\right] \quad (2.38)$$

or

$$M\left(\frac{V^2}{R}\right) = \frac{e^2}{4\pi\varepsilon_o r^2}\left[\left(1-\frac{\lambda_p}{r}\right)e^{\frac{-\lambda_p}{r}} - \left(1-\frac{\lambda_e}{r}\right)e^{\frac{-\lambda_e}{r}}\right] \quad (2.39)$$

where $R = \left(m_p r_p + m_e r_e\right)/\left(m_p + m_e\right)$ and $M = m_p + m_e$.

From Equation (2.39) we see that bound states, where the center of mass is in motion as the result of the asymmetrical force, may only occur when the separation is less than λ_p.

All of these equations of motion exhibit a feature not usually found in equations of motion. That is, because the force depends on the separation, r, between the particles and not strictly on the position, r_e, for the election, then the usual integration of the force over a change of position, which produces the potential energy, cannot be readily done because,

$$V\left(r_e\right) = -\int F_e dr_e$$

$$= -\int\left[\frac{e^2}{4\pi\varepsilon_o r^2}\left(1-\frac{\lambda_p}{r}\right)e^{\frac{-\lambda_p}{r}}\right]dr_e \quad (2.40)$$

However, we may now use the process of determining the motion of the center of mass and the motion about the center of mass rather than the motion of the individual particles. We have already obtained this solution above and found that the energies were given by

$$E = K + V + K_c + V_c,$$
(2.41)

with

$$K = \frac{\mu k}{2r}\left[\left(1 - \frac{\lambda_p}{r}\right)\frac{e^{-\frac{\lambda_p}{r}}}{m_e} + \left(1 - \frac{\lambda_e}{r}\right)\frac{e^{-\frac{\lambda_e}{r}}}{m_p}\right]$$

$$V = -\frac{\mu k}{r}\left[\frac{e^{-\frac{\lambda_p}{r}}}{m_e} + \frac{e^{-\frac{\lambda_e}{r}}}{m_p}\right]$$
(2.42)

$$K_c = \frac{kR}{2r^2}\left[\left(1 - \frac{\lambda_p}{r}\right)e^{-\frac{\lambda_p}{r}} - \left(1 - \frac{\lambda_e}{r}\right)e^{-\frac{\lambda_e}{r}}\right]$$

$$V_c = -\frac{k}{r}\left[e^{-\frac{\lambda_p}{r}} - e^{-\frac{\lambda_e}{r}}\right],$$

In Equation (2.42) the subscript c refers to the center of mass value while the absence of a subscript refers to motion about the center of mass.

Heisenberg's Uncertainty Principle and the Gauge Function.

The suggestion that bound states of electrons and protons might exist where the orbits are of the approximate order of magnitude of nuclear dimensions, is essentially a return to the notion that a neutron might be such a state. This idea gave way under arguments of conservation of momentum and Heisenberg's Uncertainty Principle to the view that electrons are forbidden to be found within the nucleus. Therefore, let us take another look at those fundamental tenets of quantum mechanics, the Poisson brackets.

The classical Poisson bracket is defined by

$$\{F,G\} = \sum_j \left(\frac{\partial F}{\partial q_j} \frac{\partial G}{\partial p_j} - \frac{\partial F}{\partial p_j} \frac{\partial G}{\partial q_j} \right) \tag{2.43}$$

where F and G are any two functions of the canonically conjugate variables q_j and p_j. The special relations that occur when F and G are q_j and p_j, respectively, are especially important in quantum mechanics; these are, classically:

$$\{q_j, q_k\} = 0$$

$$\{p_j, p_k\} = 0 \tag{2.44}$$

$$\{p_j, p_k\} = \delta_{jk},$$

where δ_{jk} is the Kronecker delta. The classical Poisson brackets of Equation (2.44) are obtained when Euclidean spaces are assumed. However, the definition of Poisson brackets remains valid for general metric spaces, when the notion of covariant differentiation is used. If we now consider the momenta, expressed in a general coordinate system, the covariant components, $p_j = mg_{ij}\dot{x}^i$ and

$p_j = g^{ji} p_i = m\dot{x}_j$ are the contravariant components. Covariant differentiation must be carried out with respect to contravariant vector components. There, in a general space the canonically conjugate variables to be considered are x^j and p^k, and the Poisson bracket of the position and momenta becomes

$$\{x^j, p^k\} = \left[\frac{\partial x^j}{\partial x^l} + \left\{ {}_{s\ l}^{\ j} \right\} x^s \right] \frac{\partial p^k}{\partial p^l} - \frac{\partial x^j}{\partial p^l} \left[\frac{\partial p^k}{\partial x^l} + \left\{ {}_{n\ l}^{\ k} \right\} p^n \right] \tag{2.45}$$

$$= \left[\delta_{il} + \left\{ {}_{s\ l}^{\ j} \right\} x^s \right] \delta_{lk}$$

or

$$\left\{ x^j, p^k \right\} = \delta_{ik} + \left\{ {}^{\ j}_{s\ k} \right\} x^s \tag{2.46}$$

Quantum mechanics adopts the operator,

$$\left(\frac{\hbar}{i} \right) \frac{\partial}{\partial x^j} \to p_j \tag{2.47}$$

for the momentum. This, in general case, becomes the covariant operator

$$\left(\frac{\hbar}{i} \right) (\ )_{,j} \to p_j . \tag{2.48}$$

The operator for the contravariant momentum components is then

$$\left(\frac{\hbar}{i} \right) g^{jl} (\ )_{,l} \to p^j . \tag{2.49}$$

Now if we look at the quantum Poisson bracket, where the operators are operating on a scalar ψ, then

$$\left[x^j, p^k \right] \Psi = \left[x^j \left(\frac{\hbar}{i} \right) g^{kl} \frac{\partial \Psi}{\partial x^l} - \left(\frac{\hbar}{i} \right) g^{kl} \left(x^j \Psi \right)_{,l} \right]$$

$$= x^j \left(\frac{\hbar}{i} \right) g^{kl} \frac{\partial \Psi}{\partial x^l} - g^{kl} \left(\frac{\hbar}{i} \right) \left[\frac{\partial x^j}{\partial x^l} + \left\{ {}^{\ j}_{s\ l} \right\} x^s \right] \Psi \tag{2.50}$$

$$- \left(\frac{\hbar}{i} \right) g^{jl} x^j \frac{\partial \Psi}{\partial x^l}$$

$$= i\hbar \left[\delta_{jl} + \left\{ {}^{\ j}_{s\ l} \right\} x^s \right] \Psi$$

This may be written in terms of the classical Poisson bracket, Equation (2.45), as

$$\left[x^j, p^k \right] \Psi = i\hbar g^{kl} \left\{ x^j, p^l \right\} \Psi . \tag{2.51}$$

If the space is Euclidean, then the g^{kl} become the Kronecker delta and the Christoffel symbols vanish and the quantum Poisson bracket of Equation (2.50) becomes

$$\left[x^j, p_k \right] \Psi = i\hbar \delta_{jk} \Psi \tag{2.52}$$

Non-Singular Potential and Nuclear Physics

because $p^k = g^{kl} p_l = \delta^{kl} p_l = p_k$. However, from Equations (2.48) and (2.49), we see that the metric does play a role in the quantum operators. This should also be seen in the use of the operators in the Schrodinger Hamiltonian operator, because

$$p_j p^j = \frac{\hbar}{i} \frac{1}{\sqrt{g}} \frac{\partial}{\partial x^j} \left(\sqrt{g} \, p^j \right)$$

$$= \frac{\hbar}{i} \frac{1}{\sqrt{g}} \frac{\partial}{\partial x^j} \left(\sqrt{g} \left(\frac{\hbar}{i} \right) g^{jl} \frac{\partial}{\partial x^l} \right) \qquad (2.53)$$

$$= \frac{\hbar^2}{\sqrt{g}} \frac{\partial}{\partial x^j} \left(\sqrt{g} \, g^{jl} \frac{\partial}{\partial x^l} \right)$$

becomes the operator to be used in a general space and, of course, is the operator currently used in applying Schrodinger's equation to the hydrogen atom. The geometrical effect may be seen also in Dirac's equation by using the same procedure.

Now, of what benefit is this discussion of geometrical effect upon quantum mechanics in considering the neo-coulombic force? Recall that the neo-coulombic force came from a gauge function in a Weyl space. A gauge function has a geometrical effect that could be thought of as effectively changing the unit of action in quantum mechanics. To see the basis for this statement, let us recall the quantum Poisson bracket operations on a scalar, Equation (2.50),

$$\left[x^j, p^k \right] \Psi = i\hbar \left[\delta_{jl} + \left\{ \begin{smallmatrix} j \\ s \ l \end{smallmatrix} \right\} x^s \right] \Psi \qquad (2.54)$$

and let us define

$$\hbar' \delta_{jk} = \hbar g^{kl} \left[\delta_{jl} + \left\{ \begin{smallmatrix} j \\ s \ l \end{smallmatrix} \right\} x^s \right] \qquad (2.55)$$

then we can write

$$\left[x^j, p^k \right] \Psi = i\hbar' \delta_{jk} \Psi \qquad (2.56)$$

which has the same form now used in quantum mechanics but the effective unit of action $\hbar'$ depends on the geometry as seen by Equation (2.55).

Neutron units of action

While we have not yet displayed an analytical expression for $\hbar'$, this absence of an analytical expression for the effective unit of action does not completely stop us from considering the possibility that a neutron may be a proton in a large orbit about an electron in a small orbit. We may, for the moment, acknowledge the difficulty of obtaining an analytical expression for $\hbar'$ by allowing the $\hbar'$, or the unit of action, for the proton and electron to be a function of their orbit, and we may designate $\hbar'_e$ to be the effective unit of action for the electron orbit in a neutron and $\hbar'_p$ to be the unit of action for the neutron's proton orbit. If the effective unit of action depends upon the orbit, as it appears here that it must, then the interpretation that Heisenberg's Uncertainty Principle rules out the possibility of an electron being contained within nuclear discussions is inapplicable.

Another argument against the neutron being an electron and proton in nuclear sized orbits is based on an argument that the principle of angular momentum cannot be conserved. The neo-coulombic forces, which require that the force between the electron and proton be directed on a line between them, require that angular momentum be conserved as was shown earlier. However, the effective unit of action for the electron orbit requires that, in the neutron the orbital angular momentum would be given by $\hbar'_e$ and its intrinsic spin angular momentum would be $\hbar'_e/2$. Similarly, for the proton the orbital angular momentum would be $\hbar'_p$ and the spin $\hbar'_p/2$.

After the neutron decays, the angular momentum is the sum of the two particles' intrinsic spin angular momenta, which is given by $\hbar$ because both particles are free and, therefore, each has an intrinsic spin angular momentum of $\hbar/2$. Therefore, the conservation of angular momentum is expressed as

$$\frac{1}{2}\left(\pm\hbar_e \pm\hbar_p\right)+\hbar_e +\hbar_p =\hbar . \tag{2.57}$$

Experimental evidence of orbital and/or spin angular momentum is contained in the experimental magnetic moments. If we equate the intrinsic and orbital magnetic moments of the electron and proton while they are in the orbital configuration to the experimental value of the neutron's magnetic moment we have

$$\pm\frac{2}{2}\frac{\hbar_e}{\hbar}\mu_B \pm\frac{2}{2}\frac{\hbar_p}{\hbar}\mu_n +\frac{\hbar_e}{\hbar}\mu_B +\frac{\hbar_p}{\hbar}\mu_n =-1.91315\mu_n \tag{2.58}$$

where μ_B is a Bohr magneton and μ_n is a nuclear magneton.

Equations (2.57) and (2.58) represent two equations in the two unknowns, $\hbar'_e$ and $\hbar'_p$, which may be solved to obtain the effective units of action for the electron and proton orbits making up a neutron such that angular momentum is conserved during neutrons' decay and that the correct magnetic moment of the neutron is ensured. Substituting the experimentally measured values of intrinsic magnetic moments for the electron and proton into Equation (2.58) would produce a more accurate solution because it would then contain the anomalous magnetic moments. Then we would have

$$\pm2.002319\frac{\hbar_e}{\hbar}\mu_B \pm2.79275\frac{\hbar_p}{\hbar}\mu_n +\frac{\hbar_e}{\hbar}\mu_B +\frac{\hbar_p}{\hbar}\mu_n =-1.91315\mu_n \tag{2.59}$$

The only simultaneous solution of Equations (2.57) and (2.59), for which $\hbar'_e$ and $\hbar'_p$ are both positive, are

$$\hbar'_e =8.0517x10^{-4}\hbar$$
$$\hbar'_p =0.66586\hbar \tag{2.60}$$

The values of the effective units of action for the proton and electron given in Equation (2.60) show that angular momentum is conserved during the decay of a neutron when the neutron is considered to be a proton in orbit around an electron under the neo-coulombic force.

The third major argument against a neutron being states of the electron and proton orbits stems from the experimental evidence on the violation of Newton's Third Law during decay. That is, the energy of the electron emerging after decay is inconsistent with the equal and opposite columbic forces between an electron and a proton. Here, we find that the neo-coulombic forces are unequal in magnitude and opposite in direction; thus the energy of an electron emerging as the result of decaying from such an orbit cannot be consistent with Newton's third law. There is a fourth argument against this picture of a neutron: the possible existence of the neutrino. The above picture of the neutron produces no ne ed to postulate the existence of neutrinos. What then can be said about the experimental evidence that has been put forward in support of the capture of free neutrinos? The answer to this question will be presented in the later section on f ive dimensional wave solutions.

The equations of motion for the center of mass for the two-particle system of a proton and an electron show that a bound orbit of the proton around the electron exists at nuclear separations. The center of mass of the neutron is bound inside a positive energy well from which it ma y escape by quantum tunneling (See Figure 2.5). The rate of neutron disintegration may be calculated and shown to compare with the experimental half-life.

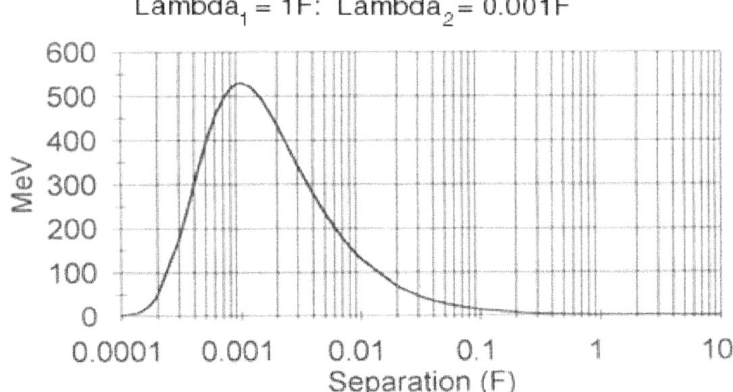

Figure 2.5. The positive energy well in which the COM is trapped in a neutron.

Nuclear Masses

The difficulty produced by the asymmetry of forces that arises in the interaction of an electron with a proton may be avoided if two protons are considered to be in orbit about the single electron. If we think of a snapshot of such a case we would find that the situation depicted in Figure 2.6 allows us to visualize the forces.

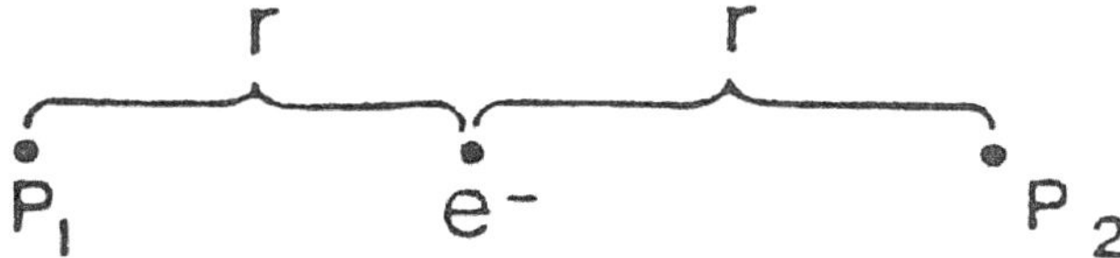

Figure 2.6. Two protons in orbit around an electron.

The force on the electron would be zero because it has a proton on each side diametrically opposed to one another. The force on each proton will be made up of two parts; one, the force that is due to the presence of the

The Dynamic Theory

electron, and the other, owing to the other proton. The symmetry guarantees that each proton will experience an identical force, if circular orbits are assumed, toward the center of rotation. The force on the proton on the left would be

$$F_{p1} = \frac{-e^2}{4\pi\varepsilon_o r^2}\left[\left(1-\frac{\lambda_e}{r}\right)e^{\frac{-\lambda_e}{r}} - \frac{1}{4}\left(1-\frac{\lambda_p}{(2r)}\right)e^{\frac{-\lambda_p}{2r}}\right]. \qquad (2.61)$$

To be sure, quantum mechanical procedure should be used; however, it may be beneficial to begin by assuming circular orbits similar to Bohr's initial approach to atomic structure. This may indicate the potential utility of the force in Equation (2.61), as well as perhaps identifying procedures to be used later.

Any nuclear orbits should probably be relativistic; therefore, in cylindrical coordinates, where the velocity for motion in a plane is given by

$$\bar{v} = \dot{r}\hat{r} + r\dot{\theta}\hat{\theta},$$

then we have

$$\gamma = \sqrt{1-\frac{v^2}{c^2}} = \sqrt{1-\frac{\left(\dot{r}^2 + r^2\dot{\theta}^2\right)}{c^2}}$$

For circular orbits, this becomes

$$\gamma = \sqrt{1-\frac{r^2\dot{\theta}^2}{c^2}}. \qquad (2.62)$$

Thus, the relativistic equations of motion for the proton become

$$\frac{d}{dt}\left[\frac{m_p\left(\dot{r}\hat{r} + r\dot{\theta}\hat{\theta}\right)}{\gamma}\right] = \frac{-e^2}{4\pi\varepsilon_o r^2}\left[\left(1-\frac{\lambda_e}{r}\right)e^{\frac{-\lambda_e}{r}} - \frac{1}{4}\left(1-\frac{\lambda_p}{(2r)}\right)e^{\frac{-\lambda_p}{2r}}\right] \qquad (2.63)$$

Equation (2.63) separates into two equations

$$\frac{d}{dt}\left[\frac{m_p\dot{r}}{\gamma}\right] = \frac{-e^2}{4\pi\varepsilon_o r^2}\left[\left(1-\frac{\lambda_e}{r}\right)e^{\frac{-\lambda_e}{r}} - \frac{1}{4}\left(1-\frac{\lambda_p}{(2r)}\right)e^{\frac{-\lambda_p}{2r}}\right] \qquad (2.64)$$

and

$$\frac{d}{dt}\left[\frac{m_p r\dot{\theta}}{\gamma}\right] = 0.$$

The second of these equations says that the angular momentum is given by

$$\frac{m_p r^2 \dot{\theta}}{\gamma} = L_p = n\hbar'_p, \tag{2.65}$$

where $\hbar'_p$ indicates that whereas the unit of angular momentum will be a constant for a given orbit, it may be different for different orbits.

The first of the equations, Equation (2.64) is

$$\frac{d}{dt}\left[\frac{m_p \dot{r}}{\gamma}\right] = \frac{\left(m_p \ddot{r} - m_p r\dot{\theta}^2\right)}{\gamma} - \frac{\left(m_p \dot{r}\frac{d\gamma}{dt}\right)}{\gamma}$$

$$= \frac{-e^2}{4\pi\varepsilon_o r^2}\left[\left(1 - \frac{\lambda_e}{r}\right)e^{\frac{-\lambda_e}{r}} - \frac{1}{4}\left(1 - \frac{\lambda_p}{(2r)}\right)e^{\frac{-\lambda_p}{2r}}\right] \tag{2.66}$$

but for circular motion $\dot{r} = 0$, therefore,

$$\frac{m_p r\dot{\theta}^2}{\gamma} = \frac{e^2}{4\pi\varepsilon_o r^2}\left[\left(1 - \frac{\lambda_e}{r}\right)e^{\frac{-\lambda_e}{r}} - \frac{1}{4}\left(1 - \frac{\lambda_p}{(2r)}\right)e^{\frac{-\lambda_p}{2r}}\right] \tag{2.67}$$

Substituting from Equation (2.65) into Equation (2.67) we have

$$\frac{\left(n\hbar'_p\right)^2 \gamma}{m_p r^3} = \frac{e^2}{4\pi\varepsilon_o r^2}\left[\left(1 - \frac{\lambda_e}{r}\right)e^{\frac{-\lambda_e}{r}} - \frac{1}{4}\left(1 - \frac{\lambda_p}{(2r)}\right)e^{\frac{-\lambda_p}{2r}}\right] \tag{2.68}$$

The potential energy for one of the protons can be found by integrating the force and is

$$V(r) = -\int F(r)\, dr$$

$$= \int \frac{e^2}{4\pi\varepsilon_o r^2}\left[\left(1-\frac{\lambda_e}{r}\right)e^{\frac{-\lambda_e}{r}} - \frac{1}{4}\left(1-\frac{\lambda_p}{(2r)}\right)e^{\frac{-\lambda_p}{2r}}\right]dr \,. \quad (2.69)$$

$$= \frac{e^2}{4\pi\varepsilon_o r}\left[\frac{1}{4}e^{\frac{-\lambda_p}{2r}} - e^{\frac{-\lambda_e}{r}}\right]$$

Then the total energy of the three body system, including rest energy, would be

$$E_T = \frac{2e^2}{4\pi\varepsilon_o r}\left[\frac{1}{4}e^{\frac{-\lambda_p}{2r}} - e^{\frac{-\lambda_e}{r}}\right] + \frac{2m_p c^2}{\gamma} + m_e c^2 \quad .(2.70)$$

However, by substituting Equation (2.65) into Equation (2.62) and solving for γ, we find

$$\gamma = \frac{1}{\sqrt{1+\left(\dfrac{n\hbar'_p}{m_p rc}\right)^2}} \,. \quad (2.71)$$

Thus, substituting Equation (2.71) into Equation (2.68) produces a transcendental equation whose solution gives *r(n)*, which may then be used in Equation (2.70) to obtain the total energy of the system. The mass of the system should then be found from

$$M = \frac{E_T}{c^2} \,. \quad (2.72)$$

Because this system has one electron and two protons, it has a total electric charge of +1 and would have a mass of approximately 2 amu. This is the same characteristic exhibited by the deuterium nucleus. If this is the structure of the H_2 nucleus, then the mass given by Equation (2.72) should correspond to the mass of the ground state nuclear mass for $n = 1$. If the $1e^-$, $2p^+$ case existing where $n = 1$ is the ground state H_2 nucleus, then is the excited state represented by two protons in the $n = 2$ state or can it be represented by one proton in an $n = 1$ orbit and one in an n

= 2 orbit? The equations developed here consider only the case when both protons are in the same orbit. Any consideration of the protons being in different orbits introduces an asymmetry in the forces and a similar difficulty faced in the neutron case. Therefore, for the moment we will consider only the simpler cases, where symmetry reduces the complexity of the solution. Notice, though, that even in the simpler symmetric case, no analytical solution exists of Equation (2.68) for *r(n)* because the force contains a transcendental function.

In my Los Alamos National Laboratory report titled, "The Unifying Effect of the Dynamic Theory" I explored the potential of modeling the nuclei using a model of a sub nuclear core made up of electrons or electrons and positrons which had an overall excess of negative electrical charge. This negative sub nuclear core was surrounded by shells of orbiting protons much like a reversal of the atomic structure of electrons orbiting around excess positive charge on the nucleus. Without exploring the sub nuclear core construction we can look at a structure for the nucleus by denoting the excess electron charge of the core by the integer Y, by which we mean the total number of core electrons less the number or core positrons, then by denoting the number of shell protons in orbit around this nuclear core with the currently used mass number, A, we find that the charge on the nucleus, Z, is given by

$$Z = A - Y. \tag{2.73}$$

Equation (2.73) indicates that the excess core electron number behaves identically with the neutron numbers in current nuclear theory, although there are no neutrons as such in this nuclear model. Indeed, the neutron, in this picture, is simply another state, namely Y = 1 and A = 1.

This suggests a picture of the nucleus in which there are protons in orbits about a nuclear core. The number of protons are given by the current mass number, A. The radii of the proton shell orbits are approximately the value of λ_p;

that is, about 1 F ermi. The core may be made up of electrons in orbit about positrons and is sized approximately the same as λ_{e-}, which is much, much less than λ_p. This view of the nucleus is similar to that of the atomic view, but here the nuclear core plays the role of the atomic nucleus. The force law for the shell proton orbits might then be given for the two proton symmetric case by

$$F = \frac{-e^2}{4\pi\varepsilon_o r^2}\left[\left(1-\frac{\lambda_e}{r}\right)e^{\frac{-\lambda_e}{r}} - \frac{1}{4}\left(1-\frac{\lambda_p}{(2r)}\right)e^{\frac{-\lambda_p}{2r}}\right] \qquad (2.74)$$

The equations specifying the proton shell orbits are given by Equation (2.68) and the total energy of the nuclei would be given by

$$E(A,Y) = \sum_n \left\{ \frac{A(n)}{R(n)}\left[\frac{1}{4}e^{-\left[\frac{\lambda_p}{R(n)}\right]} - Ye^{-\left[\frac{\lambda_e}{R(n)}\right]}\right] + \frac{A(n)m_p^2}{\gamma(n,R(n))}\right\} + E_c(Y) \qquad (2.75)$$

where $A(n)$ is the number of protons with the quantum number, n; $R(n)$ is the radius of the proton orbit with the number n; $E_c(Y)$ is the energy of the nuclear core for which Y is the excess electron charge; and $\gamma(n,R(n))$ is the relativistic γ evaluated for n and $R(n)$. The mass of the nuclei with energies given by Equation (2.75) would then be

$$M(A,Y) = \frac{E(A,Y)}{c^2}. \qquad (2.76)$$

This approach has a simple look to it. For instance, if $Y = 1$ then $E_c=0.511$ MeV, the rest energy of the electron. Then the ground state for ^{2}H would be

$$E(2,1) = 2E_1 + E_c \qquad (2.77)$$

whereas the excited state is

$$E(2,1)^* = 2E_2 + E_c. \qquad (2.78)$$

Using $E_c=0.511$ MeV we find that the energy of a single proton in the n=2 orbit would be

$$E_2 = \frac{\left[E(2,1)^* - E_c\right]}{2}. \qquad (2.79)$$

Non-Singular Potential and Nuclear Physics

Thus, the ^{3}He nuclei energy would be

$$E(3,1) = E(2,1) + \frac{\left[E(2,1)^{*} - E_{c}\right]}{2}. \qquad (2.80)$$

Using this approach I put together the binding energies for nuclei up t o an A=10 by determining the average energy of each level from the experimental data. The resulting predicted nuclei masses had a 2.9 MeV RMS error. The binding energy per mass number is plotted versus mass number in Figure 2.7.

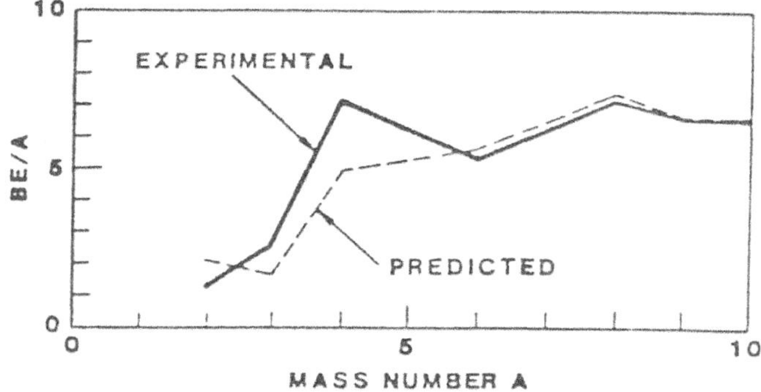

Figure 2.7. Binding Energy per mass number versus mass number.

Chapter 3 Red Shifts, Cosmology and Dark Stuff

The Weyl unity scale factor leads to a non-singular gauge potential that also depends upon t ime. These two features play large roles in gravitational phenomenon. In particular, the non-singular aspect of the gravitational field leads to a prediction of the advance of the planetary orbits that matches the prediction of Einstein, while the time dependence leads to red shifts, dark matter and dark energy effects.

The geometrical basis for Weyl's scale factor introduces a new interpretation to Heisenberg's uncertainty principle. This is most easily seen when one considers that the differentiation required in the Poison brackets involves the geometry in which the differentiation is done. What has not been discussed much in the literature is that the value of the uncertainty depends upon the gauge function in its role in geometry. The result of this is that the unit of action depends upon the gauge function. This leads to the frequency of light being both emitted and received being dependent upon the gravitational field strength and the time the light is emitted and when and where it is received. This dependence upon the gauge function predicts red shifts from distant stellar objects more in keeping with experimental evidence than Einstein's general theory of relativity.

The non-singular gauge function leads to a cosmology that supports an old universe, but does not support a big bang beginning. This cosmology does have an initial inflationary period that causes an extremely rapid expansion early and then slows the expansion to the current expansion rate.

Red Shifts, Cosmology and Dark Stuff

The time dependence of the gauge function shows up in stellar objects such as spiral galaxies in the sense that stars in the arms of the spiral are responding to the field strength of the center of the galaxy when the gravitational field was stronger in the past. Any change in the field strength cannot travel faster than the speed of light and this means that the outer stars in the arms of the spiral galaxy are responding to a greater gravitational field than are the stars in the inner part of the arms. This leads to the tangential velocities that currently are interpreted as the result of dark matter.

The time dependence of the gauge function influences the light we see from Type Ia supernovas in two ways. One way is similar to the influence the time dependence has upon the stars in the arms of spiral galaxies. However, the other way the time dependence influences light from the Type Ia supernovas stems from the influence the time dependence has upon the Chandrasekhar mass limit of supernovas. As time weakens the gravitational field the limiting mass is reduced. This means that newer supernovas have less mass than do older supernovas. This means that newer supernovas are less luminous than older supernovas.

Red Shifts

One of the predictions of Einstein's General Theory of Relativity concerns the tendency of light from stars and other objects in the heavens to be shifted towards the red color end of the spectrum. Looking at the emission and reception of light within the framework of the proposed theory one finds that the unit of action, which establishes the energy of any quantum state, depends upon both the relative time and the gravitational field at the time and place of the emission and reception. This is so because the theory holds the gravitational field to be a gauge field and it is the gauge function that determines the applicable unit of action. It has been shown that the quantum Poison bracket is given by

$$[x^j, p^k]\psi = i\hbar g^{kl}\left[\delta_{jl} + \binom{j}{sl}x^s\right]\psi. \qquad (3.1)$$

Thus, for a metric with only a gauge function the effective unit of action would be given by

$$\hbar' = \hbar \exp[2f_{,t}f_{,r}f_{,\gamma}]. \qquad (3.2)$$

By recalling the gauge gravitational field of Equation (2.15), one may use Equation (3.2) to find the expression for the unit of action for emission of a photon to be

$$\hbar_e = \hbar \exp\left[\frac{W_e(1+bt_e)}{R_e}e^{-\left(\frac{\lambda_e}{R_e}\right)}\right] \qquad (3.3)$$

where the subscript, e, denotes emission and the first order approximation has been used for the exponential time dependence. Similarly, the unit of action for the reception of a photon can be found to be

$$\hbar_r = \hbar \exp\left[\frac{W_r(1+bt_r)}{R_r}e^{-\left(\frac{\lambda_r}{R_r}\right)}\right]. \qquad (3.4)$$

If photon energy is conserved between emission and reception then

$$\hbar_e\nu_e = \hbar_r\nu_r. \qquad (3.5)$$

By setting $t_e = 0$, $t_r = L/c$, $W = (-GM/c^2)$, and $b = -H$, then it may be seen that the shift in frequency is given by

$$z = \frac{\Delta\lambda}{\lambda_e} = \exp\left\{\left(\frac{-G}{c^2}\right)\left[\frac{M_r e^{\frac{-\lambda_r}{R_r}}}{R_r} - \frac{M_e e^{\frac{-\lambda_e}{R_e}}}{R_e}\right] + \left(\frac{HL}{c}\right)\left(\frac{\frac{M_r}{R_r}}{\frac{M}{R}}\right)\right\} - 1. \qquad (3.6)$$

By looking at the first order approximations of this prediction one finds that the time dependence of the gravitational field produces the linear dependence and is given by Hubble's constant while the gravitational potential produces the same prediction that comes from Einstein's theory.

Looking a little closer one finds that the time dependence of the red shift produces an experimental number, $H^{-1} = (5.6+0.6) \times 10^{17}$ sec. $(1.61 \times 10^{-18}$ sec$^{-1} < H < 2.0 \times 10^{-18}$ sec$^{-1})$, that corresponds to the same time dependence that has been measured and reported for the moon's orbit $(b=1.9 \times 10^{-18}$ sec$^{-1})$, well within experimental error. It is somewhat pleasing that a prediction coming from the same time dependence originating in the gauge function leads to a comparison of phenomena involving cosmological distances agrees with phenomena involving the much shorter distance involved in the moon's orbit. Another possible plus to this prediction is that, because the prediction involves an exponential dependence upon time and gravitational potential between the emission and reception of the light, then the distances that are currently ascribed to distant bodies by their red shifts may be much greater than the actual distances. Also, the possible red shifts from dense gravitating bodies may be much greater than is now believed possible thereby removing the mystery from many objects.

Cosmology

The hot big bang model of the Universe is the model that is in vogue now. Virtually all the journals print numerous articles relating to some aspect of the hot big bang model. The model is based upon t he Newtonian gravitational force and the notion of a scale of the universe that is changing with time. This notion is borrowed from Einstein's general theory of relativity; however, Einstein's theory is not used in the hot big bang model itself. A new, non-singular, gravitational potential such as presented above would have an impact upon cosmology that may be compared to the hot big bang model.

The Dynamic Theory

The development of the standard big bang model begins by considering a spherical piece of the universe with an observer at the center. This sphere is considered to be filled with "dust" of density $\rho(t)$ with a galaxy of interest placed at the outer boundary of the sphere which has a radius denoted by x.

When Gauss's law and Newton's laws of motion and gravitation are used one arrives at

$$m_g \frac{d^2x}{dt^2} = \frac{4\pi}{3}\frac{x^3\rho(t)Gm_g}{x^2} = \frac{4\pi}{3}G\rho(t)m_g x. \qquad (3.7)$$

But consider what happens if one wishes to compare this with the cosmology produced by the non-singular, gravitational gauge potential. Then Equation (3.7) becomes

$$m_g \frac{d^2x}{dt^2} = \frac{4\pi}{3}\frac{x^3\rho(t)Gm_g}{x^2}\left(1-\frac{\lambda}{x}\right)e^{-\frac{\lambda}{x}}$$

$$= \frac{4\pi}{3}G\rho(t)xm_g\left(1-\frac{\lambda}{x}\right)e^{-\frac{\lambda}{x}}. \qquad (3.8)$$

In Equation (3.8) the time dependence of the gravitational field has been left out in order to be able to compare the effect of the non-singular potential more directly with the hot big bang model.

Now replace x with the co-moving coordinate $x=R(t)r$ where $R(t)$ is the scale factor of the universe and r is the co-moving distance coordinate as is done in the standard model. When we also normalize the density to its value at the present epoch, ρ_o , by $\rho(t)=\rho_o R^{-3}(t)$ we obtain

$$\frac{d^2R}{dt^2} = \frac{4\pi G\rho_o}{3}\left(1-\frac{\lambda}{Rr}\right)\frac{e^{-\frac{\lambda}{Rr}}}{R^2}. \qquad (3.9)$$

A trend may now be seen in the universe expected from Equation (3.9) by noting that should we look back in time to the point when $R=\lambda/r$ then we would have a point in time, say T_1, when the acceleration of the universe would have been zero. At times before T_1 there would have been an acceleration outward while for times after T_1, such as the current epoch, the rate of the expansion of the universe is slowing down. This is a very different story than is told by the standard model. But how is it different? It is the same as the standard model in that from Equation (3.9) one sees that the universe was forced into expansion at early times and is now slowing down its rate of expansion. One big difference between the story to be told by Equation (3.9) and the standard model is that Equation (3.9) gives the reason for the initial expansion and it denies that the universe was ever collapsed to a singular point as supposed by the standard model. To better see the first contention we should proceed a little further.

A plot of the dynamics of the universe might be helpful in understanding the predictions of Equation (3.9). First, Figure 3.1 s hows the deceleration expected in the past. It shows that at times prior to roughly 10 billion years ago the universe was in a period of very rapid expansion. Around 10 bi llion years ago there arrived a point of transition from an accelerated expansion in the universe to an expanding universe wherein the expansion was slowing down, or decelerating.

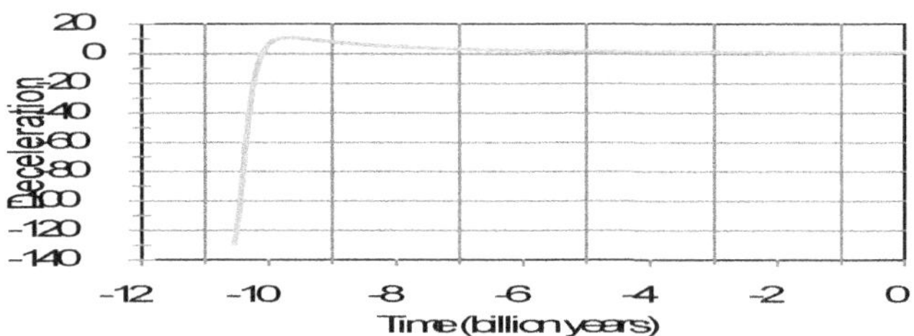

Figure 3.1 Cosmological Deceleration

Next Figure 3.2 s hows the velocity of expansion that accompanies the cosmological deceleration of Figure 3.1. The early rise in expansion velocity echo's the early inflation shown in Figure 3.1.

The Dynamic Theory

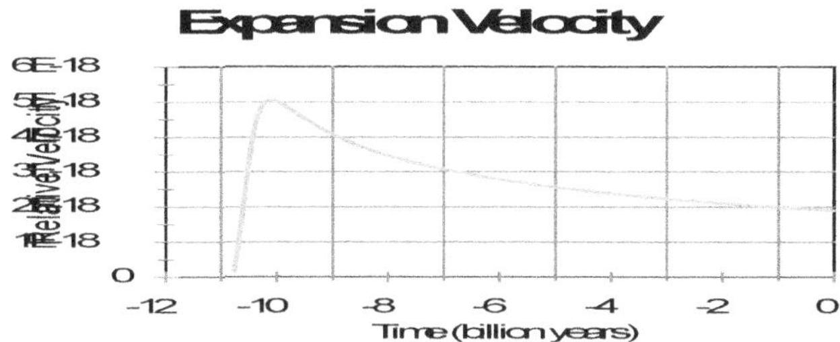

Figure 3.2 Cosmological Expansion Velocity

In Figure 3.3 the evolution of the universe is displayed, however, the evolution does not show the early inflation as well as the deceleration and expansion velocity graphs do.

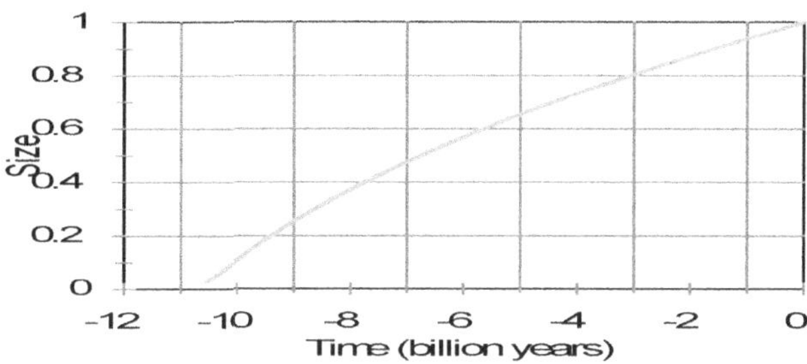

Figure 3.3 Cosmological Evolution

Multiply Equation (3.9) by *dR/dt* and integrating with respect to time obtains

$$\dot{R}^2 = \frac{8\pi G}{3}\left[\rho(t)R^2 e^{\frac{\lambda}{Rr}} + \frac{\varepsilon(t)R^2}{2c^2} \right] kc^2. \qquad (3.10)$$

In Equation (3.10) the radiation term has been included for completeness. Now the constant of integration, k, may be evaluated by setting the values of R, dR/dt, ρ, and ε at their present day values of *1*, H_o, ρ_o, and ε_o. Then Equation (3.10) becomes

65

Red Shifts, Cosmology and Dark Stuff

$$H_o^2 = \frac{8\pi \rho_o G}{3} e^{\frac{\lambda}{r}} + \frac{4\pi G \varepsilon_o}{3c^2}\, kc^2. \qquad (3.11)$$

By making the definitions

$$\rho_c \equiv \frac{3H_o^2}{8\pi G} \quad and \quad \Omega \equiv \frac{\rho_o e^{\frac{\lambda}{r}}}{\rho_c}, \qquad (3.12)$$

Equation (3.12) may be put into Equation (3.10) to obtain

$$\dot{R}^2 = H_o^2 \left(\frac{\Omega e^{\frac{\lambda}{Rr}}}{R e^{\frac{\lambda}{r}}} + 1 - \Omega \right) + \frac{H_o^2 \varepsilon_o}{2c^2 \rho_c}\left(\frac{1}{R^2} \right). \qquad (3.13)$$

We may now take a look at some of the implied dynamics from Equation (3.13). First look at the dynamics as R tends to infinity and there is no radiation. For this case we would have

$$\dot{R}_\infty^2 = H_o^2(1-\Omega), \qquad (3.14)$$

which is the same as in the standard model.

Now suppose we look backward in time to the time when dR/dt was zero? Then Equation (3.13) becomes

$$0 = \rho_o R e^{\frac{\lambda}{Rr}}\, \rho_o R^2 e^{\frac{\lambda}{r}} + \rho_c R^2 + \frac{\varepsilon_o}{2c^2}\left(1 - R^2\right). \qquad (3.15)$$

This is a transcendental equation that could be solved for R if we knew λ, r and the density of the dust and radiation at the current epoch. It may be seen from Equation (3.15) that if there is no radiation and R does not equal zero then

$$R = \frac{e^{\frac{\lambda}{Rr}}}{\left[e^{\frac{\lambda}{r}}\, \dfrac{\rho_c}{\rho_o} \right]}. \qquad (3.16)$$

There is also a trivial solution at $R=0$ in Equation (3.15), but for this case the acceleration is also zero and, therefore, no dynamics are allowed.

Perhaps this is sufficient to point out that the overall picture of cosmology given by the non-singular gravitational gauge potential is very different from the hot big bang model of the universe. Will it a llow for high

66

temperatures needed for accounting for the abundances of the elements? Since it allows for the universe to be much smaller in the past it would have the associated high temperatures. Yet it should not have the infinite temperatures associated with a singular universe.

Dark Matter

Data wherein the tangential velocities of stars in the arms of a spiral galaxy differed from Newtonian predictions were first reported nearly seventy years ago. A fundamental theory supporting these data has not heretofore been given, though empirical theories have been presented. The best of these empirical theories may be the Modified Newtonian Dynamics (MOND). The theory presented here had its beginning in 1974 and only recently has it been applied to the dynamics of spiral galaxies.

Newtonian uniform circular motion equates the gravitational acceleration to the centripetal acceleration so that

$$\frac{GMm}{r^2} = \frac{mv^2}{r}.$$ (3.17)

A time-dependent, non-singular gravitational field, such as the Dynamic Theory predicts, alters Equation (3.17) to

$$\frac{GMm\left(1 - H_0\tau\right)}{r^2}\left(1 - \frac{\lambda}{r}\right)e^{-\frac{\lambda}{r}} = \frac{mv^2}{r}$$ (3.18)

where H_0 is Hubble's constant, only the first order in time is considered, and $\lambda \equiv GM/c^2$ as determined by planetary orbits. For this time dependent gravitational field the gravitational acceleration acting on an arm of a galaxy is due to the gravitational field of the mass of the galaxy, M, at a previous time. This previous time is given by the time that it takes for the field to travel from the site of the gravitational field to the point on the arm under consideration. This means that when all the mass is considered to be at the center of the galaxy the time that enters into Equation (3.18) is $\tau = -r/c$ so that, when $r>>\lambda$ the velocity of the arm of the galaxy would be given by

$$v = \sqrt{\frac{GM\left(1-H_o\tau\right)}{r}} = \sqrt{GM\left(\frac{1}{r}+\frac{H_0}{c}\right)}. \qquad (3.19)$$

Equation (3.19) shows a very different character than the expression for the velocity for the time independent gravitational field. This expression shows that the velocity of the galaxy arms should not be expected to drop off as the time independent Newtonian gravitational field does.

We may look at the 5-dimensional approach of the Dynamic Theory by looking at the Lagrangian

$$L = \frac{1}{2}mc^2\left(\dot{\tau}\right)^2 + \frac{1}{2}m\dot{r}^2 + \frac{1}{2}m\left(r\dot{\theta}\right)^2 + GMm\left(1-H_o\tau\right)\frac{e^{-\frac{\lambda}{r}}}{r} \qquad (3.20)$$

where the universe time, τ, is treated as another variable and t is the local time. The universe time, τ, becomes a geometrical coordinate that makes the problem local-time independent in five dimensions.

The time Lagrange equation may then be written as

$$\frac{d}{ds}\left[\frac{\partial L}{\partial \dot{\tau}}\right] - \frac{\partial L}{\partial \tau} = 0 = \frac{d}{dt}\left[m\dot{\tau}\right] + H_o\lambda m\frac{e^{-\frac{\lambda}{r}}}{r}. \qquad (3.21)$$

For a spherically symmetric field the radial equation is

$$\frac{d}{dt}\left[\frac{\partial L}{\partial \dot{r}}\right] - \frac{\partial L}{\partial r} = 0 = \frac{d}{dt}\left[m\dot{r}\right] - mr\dot{\theta}^2$$
$$+GMm\left(1-H_o\tau\right)\left(1-\frac{\lambda}{r}\right)\frac{e^{-\frac{\lambda}{r}}}{r^2} \qquad (3.22)$$

The third Lagrange equation becomes

$$\frac{d}{dt}\left[\frac{\partial E}{\partial \dot{\theta}}\right] - \frac{\partial E}{\partial \theta} = 0 = \frac{d}{dt}\left[mr^2\dot{\theta}\right]. \qquad (3.23)$$

For the problem of spiral galaxy behaviour we may assume the $\lambda<<r$ and write the equations of motion as

The Dynamic Theory

$$\ddot{\tau} = \frac{-H_o \lambda}{r},$$

$$\ddot{r} - r\dot{\theta}^2 = -\left(1 - H_o \tau\right)\frac{GM}{r^2}\left(1 - \frac{\lambda}{r}\right), \text{ and} \quad (3.24)$$

$$\ddot{\theta} + \frac{2}{r}\dot{r}\dot{\theta} = 0$$

If we now look at uniform circular motion we find that the first of Equations (3.24) becomes

$$\ddot{\tau} = \frac{-H_o \lambda}{r} = \text{constant} \Rightarrow \frac{d\dot{\tau}}{dt} = \frac{-H_o \lambda}{r} \quad (3.25)$$

and Equation (3.25) may be integrated again to get

$$\tau = \tau_o - \frac{H_o \lambda}{2r}t^2 + \left(\dot{\tau}_o + \frac{H_o \lambda}{r}t_o\right)t \quad (3.26)$$

Also for the assumed uniform circular motion the second of Equations (3.24) may be written as

$$v^2 = \left(1 - H_o \tau\right)\frac{GM}{r}\left(1 - \frac{\lambda}{r}\right) \quad (3.27)$$

where v is the tangential velocity of the uniform circular motion. Putting Equation (3.26) into Equation (3.27) obtains

$$v^2 = \frac{GM}{r}\left\{ \begin{array}{c} 1 - H_o \tau_o + \dfrac{H_o^2 \lambda}{2r}t^2 \\ -\left(H_o \dot{\tau}_o + \dfrac{H_o^2 \lambda}{r}t_o\right)t \end{array} \right\}\left(1 - \frac{\lambda}{r}\right). \quad (3.28)$$

We must keep in mind there are two times to be considered. First there is the time it takes for the gravitational change to travel from the center of the galaxy to the point of measurement in the galaxy arm. The second time is for the light signal to travel from the galaxy to the Earth.

If $\tau_o = 0$ and $t_o = 0$ at the point in time when the light left the star on its way toward Earth, then Equation (3.28) becomes

$$v^2 \cong \frac{GM}{r}\left\{1+\frac{H_o^2\lambda}{2r}t^2-H_o\dot{\tau}_o t\right\}\left(1-\frac{\lambda}{r}\right). \qquad (3.29)$$

Time runs from the time the gravitational signal left the center of the galaxy at $t=-r/c$ where r is the distance from the center of the galaxy to the star. Then Equation (3.29) becomes

$$v^2 \cong \frac{GM}{r}\left\{1+\frac{H_o\dot{\tau}_o r}{c}\left(1-\frac{H_o^2\lambda}{2c^2}\right)\right\}$$

$$\cong \frac{GM}{r}\left\{1+\frac{H_o\dot{\tau}_o r}{c}\right\} \qquad (3.30)$$

To establish a value for $\dot{\tau}_o$ look at the energy at time $t=0$ and $\tau=0$ with $r>>\lambda$, or

$$E_o = \frac{1}{2}mc^2\left(\dot{\tau}_o\right)^2 + \frac{1}{2}mv^2 - \frac{GMm}{r_o} \qquad (3.31)$$

This may be rewritten as

$$\frac{2E_o}{mc^2} = \left(\dot{\tau}_o\right)^2 + \frac{v^2}{c^2} - \frac{\lambda}{r_o}. \qquad (3.32)$$

Since the tangential velocities are non-relativistic this requires that

$$\dot{\tau}_o = \sqrt{\frac{2E_o}{mc^2} + \frac{\lambda}{r_o}}. \qquad (3.33)$$

Equation (3.33) shows that the initial conditions establish the point at which the tangential velocities begin to differ from those predicted by Newtonian gravity. In the absence of a means of evaluating the initial conditions experimental results may be used. First, write the acceleration in the arms of the galaxy as

$$a = a_N\left\{1-\frac{H_o^2\lambda}{2r}t^2-H_o\dot{\tau}_o t\right\}$$

$$\cong a_N\left\{1+H_o\dot{\tau}_o\frac{r}{c}\right\} \qquad (3.34)$$

Now use the data that shows the acceleration begins to deviate from Newtonian when the acceleration drops to a value of 1.2×10^{-10} m/sec^2 so that

$$a_N = \frac{GM}{r_c^2} \cong 1.2 \times 10^{-10}$$

$$\Rightarrow r_c \cong \sqrt{\frac{GM}{1.2 \times 10^{-10}}} \quad . \tag{3.35}$$

Then requiring $\dot{\tau}_o = c/r_c H_o$ sets a value of $\dot{\tau}_o$ in keeping with the data. Equation (3.35) becomes

$$a \cong a_N \left\{ 1 + \frac{r}{r_c} \right\} \tag{3.36}$$

where the short range Newtonian acceleration may be seen and the long range acceleration is as predicted by MOND.

It should be noted that the approximate linearity of the tangential velocity with respect to time displays an independence of the time it takes for light to travel from the galaxy to Earth. This apparent independence of time masks the fact that the gravitational strength of the galaxy, relative to the current epoch, depends upon the time of light travel to Earth.

Dark Energy

Data displaying evidence that provided the beginning of the hypothesized dark energy was first presented in 1998. To date no fundamental theory has had success in explaining these data. However, the time dependence of the gravitational field provides an explanation for the data. The explanation comes in three parts. First, the time dependence of the gravitational field satisfies the expansion relation without the need of a cosmological constant. Secondly, the new relations for the red shift must be included. However, it is the third part that makes the most important impact upon the explanation of the data. The time dependence of the gravitational field changes the Chandrasekhar mass limit of the type Ia supernovas. The time dependence of the gravitational field causes gravity to become weaker in time. Below it is shown that less mass is needed for newer supernovas. This makes newer supernova dimmer than older ones. The assumption of a constant standard candle then would argue incorrectly for acceleration in the expansion of universe rather than the actual deceleration of the expansion.

The universe expansion factor is currently taken from general relativity and is

$$\frac{\ddot{a}}{a} = -\frac{4}{3}\pi G\left(\rho + 3\frac{p}{c^2}\right). \tag{3.37}$$

The mean density and pressure are currently taken to include dark energy and are taken to obey the local energy conservation of energy relation

$$\dot{\rho} = -3\frac{\dot{a}}{a}\left(\rho + \frac{p}{c^2}\right). \tag{3.38}$$

The first integral of Equations (3.37) and (3.38) is the Friedman equation

$$\dot{a}^2 = \frac{8}{3}\pi G\rho a^2 + \text{constant}. \tag{3.39}$$

But consider what happens if one wishes to compare this with the cosmology produced by the non-singular, time dependent, gravitational gauge potential. Then Equation (3.8) becomes

$$m_g\frac{d^2x}{dt^2} = \frac{4\pi}{3}G\rho(t)x\left(1 - \frac{\lambda}{x}\right)(1 - H_o\tau)e^{-\frac{\lambda}{x}} \tag{3.40}$$

where τ is the universe time.

Now let us replace x with the co-moving coordinate $x=R(t)r$ where $R(t)$ is the scale factor of the universe and r is the co-moving distance coordinate as is done in the standard model. When we also normalize the density to its value at the present epoch, ρ_o, by $\rho(t)=\rho_oR^{-3}(t)$ we obtain

The Dynamic Theory

$$\frac{d^2 R}{dt^2} = \frac{4\pi G \rho R}{3}\left(1 - \frac{\lambda/_r}{R}\right)\left(1 - H_o \tau\right) e^{-\lambda/_r / R} . \qquad (3.41)$$

Multiplying Equation (3.41) by dR/dt and integrating with respect to time obtains

$$\frac{\dot{R}^2}{2} - \frac{\dot{R}_o^2}{2} = \frac{4\pi G \rho}{3} \int \left(1 - H_o \tau\right) R\, dR . \qquad (3.42)$$

We now need to know how to integrate the right hand side of Equation (3.42). Suppose we consider the time it takes for light to travel from the distant star to Earth, or $t = -R_s/c$, where R_s is the distance from the star to Earth and the minus sign comes from looking backwards in time. The radius of the universe now has two parts. The first part is the radius of the universe when the light left the star on its journey to the Earth. Let this radius be R_0. Thus we see that

$$R = R_o + R_s \qquad (3.43)$$

and $\dot{R} = \dot{R}_s$. Further, from considerations of the dark matter it was determined that the world time was given by

$$\tau = \tau_o - \frac{H_o GM}{2c^2 R} t^2 + \left(\dot{\tau}_o + \frac{H_o GM}{c^2 R} t_o\right) t . \qquad (3.44)$$

When we set both initial times to zero and use the value of $\lambda_U \equiv GM/c^2$, Equation(3.44) becomes

$$\tau = -\frac{H_o \lambda_U}{2R} t^2 + \dot{\tau}_o t . \qquad (3.45)$$

Now we find that Equation (3.42) may be written as

$$\dot{a}^2 = \frac{8\pi G \rho}{6c^2}\left\{ \begin{array}{l} 2R_o c^2 a \\ + \left(c^2 + H_o \dot{\tau}_o c R_o\right) a^2 \\ + \dfrac{1}{3}\left(H_o^2 \lambda_U + H_o \dot{\tau}_o 2c\right) a^3 \end{array} \right\} + K . \qquad (3.46)$$

If we set the constant of integration, K, to zero, then Equation (3.46) becomes

$$\dot{a}^2 = H_o^2 \Omega_M' \left\{ \begin{array}{l} R_o a \\[6pt] +\dfrac{1}{2}\left(1+\dfrac{H_o \dot{\tau}_o R_o}{c}\right)a^2 \\[12pt] +\dfrac{1}{6}\left(\dfrac{H_o^2 \lambda_U}{c^2}+\dfrac{2H_o \dot{\tau}_o}{c}\right)a^3 \end{array} \right\} \qquad (3.47)$$

where we have used the definitions

$$\rho_c \equiv \frac{3H_o^2}{8\pi G}, \quad and \quad \Omega_M' \equiv \frac{\rho}{\rho_c}. \qquad (3.48)$$

In Equation (3.47) we find that the mass density term splits into three terms for a time-dependent gravitational field. For a time-independent gravitational field there was only one term.

An interesting aspect of Equation (3.47) is that the two new mass terms both involve the same time dependence factor as the one that causes the tangential velocity of the arms of spiral galaxies to differ from Newtonian behaviour. That is to say that should the two new terms provide a basis for the current experimental evidence for dark energy it comes from the same source as the basis for dark matter. The time dependence of the gravitational field explains both phenomena.

Consider Equation (3.47) again and add the usual term for radiation so that we find

$$\left(\frac{\dot{a}}{a}\right)^2 = H_o^2 \left\{ \Omega_M' \left[\begin{array}{l} \dfrac{R_o}{a} \\[12pt] +\dfrac{1}{2}\left(1+\dfrac{H_o \dot{\tau}_o R_o}{c}\right) \\[12pt] +\dfrac{1}{6}\left(\dfrac{H_o^2 \lambda_U}{c^2}+\dfrac{2H_o \dot{\tau}_o}{c}\right)a \end{array} \right] +\Omega_{RO} \right\} \qquad (3.49)$$

where we did not add a term for the cosmological constant.

The Dynamic Theory

If the relationship between the red shift and a is used along with the assumption $\Omega_{RO} = 0.25\,\Omega_M$, it has been shown that Equation (3.49) becomes

$$\left(\frac{\dot{a}}{a}\right)^2 = H_o^2 \left\{ \left[\begin{array}{c} \Omega_M z \\[2mm] +\Omega_M \dfrac{12-2z}{(3+2z)} \\[4mm] +\Omega_M \dfrac{z(7-2z)}{(3+2z)} \end{array} \right] + \Omega_{RO} \right\} \qquad (3.50)$$

where Ω_M varies as $(1+z)^3$.

The expansion of the universe taken with the red shift of Equation (3.6) where the red shifts are measured at the Earth's surface allows Equation (3.50) to be written as

$$\left(\frac{\dot{a}}{a}\right)^2 = H_o^2 \left\{ \left[\begin{array}{c} 0.25\left(1+\log\left(1+z_{exp}\right)\right)^3 \log\left(1+z_{exp}\right) \\[3mm] +0.25\left(1+\log\left(1+z_{exp}\right)\right)^3 \dfrac{12-2\log\left(1+z_{exp}\right)}{\left(3+2\log\left(1+z_{exp}\right)\right)} \\[4mm] +0.25\left(1+\log\left(1+z_{exp}\right)\right)^3 \dfrac{z\left(7-2\log\left(1+z_{exp}\right)\right)}{\left(3+2\log\left(1+z_{exp}\right)\right)} \end{array} \right] + \Omega_{RO} \right\}. \qquad (3.51)$$

Standard Candles

One reason for choosing the Type Ia supernova in the universe expansion research is the assumption that the mass of this type supernova are all the same; roughly the Chandrasekhar Limit mass of 1.39 solar masses. However, a time dependent gravitational field changes this limit. This may be seen by considering the Newtonian equation of hydrostatic equilibrium known as the Tolman-Oppenheimer-Volkov [TOV] equation, or

Red Shifts, Cosmology and Dark Stuff

$$\frac{dp}{dr} = \frac{-GM(r)\left(1-H_o\tau\right)\rho}{r^2}.$$

(3.52)

The time dependence of the gravitational field that is holding the star together against the internal pressure is diminishing in time. This means that the mass of the Chandrasekhar Limit diminishes in time. Supernova found closer to Earth will have less mass, and therefore less luminosity, than supernova at greater distances. A reduction in luminosity from the assumed constancy would show up in an analysis by making the supernova appear further away than it really is. The natural conclusion, based on the time-independent gravitational field that produces the constant Chandrasekhar limiting mass, would be that the expansion of the universe is accelerating.

Using the Virial Theorem development by Collins who arrives at the Chandrasekhar limiting mass with the equation

$$\frac{R_o}{\left(\dfrac{2GM}{c^2}\right)} > 228\left(\frac{M_\odot}{M}\right)^{\frac{4}{9}} \approx 200$$

(3.53)

we find that the time dependent gravitational field requires that this relation become

$$\frac{R_o}{\left(\dfrac{2GM}{c^2}\right)\left(1-H_o\tau\right)} > 228\left(\frac{M_\odot}{M}\right)^{\frac{4}{9}} \approx 200\,.$$

(3.54)

This gives the limiting mass as

$$M > M_{Ch}\left(1-H_o\tau\right)^{\frac{9}{4}},$$

(3.55)

where M_{Ch} is the Chandrasekhar limiting mass. By differentiating Equation (3.55) with respect to universe time we find the limiting mass for the type Ia supernovae to change according to

$$\frac{dM}{d\tau} > -\frac{9H_o}{4}M_{Ch}\left(1-H_o\tau\right)^{\frac{5}{4}}.$$

(3.56)

This displays the reduction to be expected with time.

Conclusions

A time-dependent gravitational field, that gets weaker in time, shows the physical effects of this past, stronger field in the dynamics of spiral galaxies. This weakening gravitational field also shows up in the analysis of the distances to, and red shift of light from, supernovas. Here it adds terms to the universe expansion velocity relations that are not present in the analysis of time-independent fields. It also changes the luminosity of the supernovas that were assumed to have constant luminosity. These effects of the time dependent gravitational field remove the need for hypothesizing new matter or energy to explain these effects.

There have been many attempts in the past to find different solutions to Einstein's field equations and to show how an expanding universe may be viewed in different ways. Portions of the above may be reminders of prior approaches. Therefore, it may prove useful to point out what is new in this approach.

Fundamentally there are three things that are new in this theory. First, the fifth dimension is considered to be a real physical entity. All five dimensional theories that I know of in the past, whether by Kulsa-Klein, Einstein with his many collaborators, and others, did not consider the fifth dimension to be real and, therefore, required several terms in the resulting gauge field equations to be zero. Here these terms are non-zero and require that the gravitational potential and field be time-dependent. Second, this theory uses the Weyl Gauge Principle as its basis for quantum theory and this requires that the gravitational potential be a non-singular potential. These two things require the gravitational field to be a time-dependent, non-singular, gauge field not seen previously. The third aspect of the approach is that the Weyl Gauge Principle requires that the unit of action be dependent upon the gauge function. This requires the red-shift from distant objects to have an exponential dependence upon bot h the time and distance between emission and reception. The new red-shift relation becomes important in both dark matter and dark energy predictions because both phenomena are witnessed by red-shifted light. The time dependence, or weakening, of the gravitational field is the major factor in predicting effects interpreted as dark matter. The time dependence of the gravitational field also provides the major factor in predictions with respect to dark energy as it is responsible for the diminishing of the luminosity of the supernovas used as standard candles and the expression for the expansion of the universe.

Chapter 4 The Foundation of Physics

In the scientific community's search for a fundamental theory the laws of classical thermodynamics have been overlooked. Though these laws have not yet been seen to be violated they do not seem to be a good basis from which to obtain mechanical motion, field theories and quantum mechanics. However, they do j ust that.

The presentation in the sections above shows that the classical laws of thermodynamics require relativity as a subset of the entirety of the universe they describe (Relativity–Special and General). Another section showed that thermodynamics leads to quantization of paths and electric charge as well as a n on-singular gauge potential that produces a nuclear model more accurate than the standard model (Non-Singular Potential and Nuclear Physics). The third section showed how thermodynamics, through the time and mass dependence of the gauge function predicts a cosmology that includes red shifts and predictions of effects currently ascribed to dark matter/energy, but excludes a big bang beginning (Red Shifts, Cosmology and Dark Stuff).

This section weaves the preceding sections together to display the unifying properties of classical thermodynamic laws. It will show how the first and second laws combine to require both geometry and a set of equations of motion for all dynamics. Actually thermodynamics produces two sets of equations of motion. The First Law gives one set and the Second Law produces another set. It will show how the combination of stability conditions and the entropy principle for isolated systems produce the space-time structure of relativity which is required of a system with the speed of light as a limiting velocity. This space-time set of equations given by the Second Law place restrictions upon the equations of motion from the First Law. The presentation will show how the entropy principle requires two geometrical manifolds for every motion. One manifold is a non-integrable energy manifold with a Weyl geometry. The second manifold is an integrable Riemannian manifold obtained from the first manifold by the use of a gauge function as a geometrical integrating factor. It is the five dimensional gauge function stemming from the five dimensional first law that field theories and the connection between classical, relativistic and quantum mechanics.

Weyl showed how to use the gauge function to produce field equations and the five dimensional first law demands five dimensional fields that include inductively coupled electric, magnetic and gravitational fields and forces. The imposition of conservation of mass has been shown to require a four dimensional hyper surface, with an Einstein curvature, to be embedded into the five dimensional manifold. A very stable isentropic system has been shown to require the Weyl scale factor to take on a value of unity which means that the length of a vector must return to its original length after completing a closed path. This unity scale factor quantizes the path integral of the gauge potentials. For given gauge potentials the isentropic paths must satisfy quantum mechanics. The gauge function for fundamental particles whose gauge characteristics do not change in space or time must be quantized, non-singular with respect to space, depend upon time and depend upon mass.

Thus, the classical thermodynamic laws, through the stability conditions, entropy principle, and restrictive assumptions such as the isentropic assumption, require the fields, forces and dynamics currently seen in nuclear, atomic, and cosmology. It may, therefore, be seen that thermodynamic laws can and do provide the foundation of physics as we now know it and is a viable starting point for a search for a fundamental theory.

Fundamental Laws

There are at least two reasons that classical thermodynamics would not be expected to provide a foundation for the current dynamic theories. First, thermodynamics, as currently studied, does not provide a description of motion like the mechanistic theories do. Secondly, texts teach, as Einstein appeared to believe, that classical thermodynamics might be obtained from statistical procedures applied to Newtonian mechanics. A third reason to discount thermodynamics is that it is typically thought to apply only to equilibrium conditions rather than dynamic ones. Studies in non-equilibrium thermodynamics attempted to address dynamic conditions.

The thermodynamic laws are seen to apply to dynamic conditions when it is recognized that dynamics, as started by Newton, requires a freely chosen geometry, a variational principle and a force law. The statement of the First Law allows for a chosen geometry. The Second Law requires variational principles in the Maximum Entropy Principle for isolated systems and the Minimum Free Energy Principle for non-isolated systems. What may not be so readily apparent is that the stability conditions, which tell what direction a system will go when displaced from equilibrium, provides several metrics whose geometries are specified by the two laws.

This leaves a force law remaining to be found. It may be recalled that in classical mechanics the principle of Maupertuis provided a means of transforming from motion given by a force law in a chosen geometry to a geometrical space in which motion was along geodesics without forces. However, when it was noticed that the line element included a multiplicative function it was thought that the function's appearance precluded geometrization of the dynamics as it implied different geometries with different energy levels. The recognition that a multiplicative function in a l ine element was a feature of Weyl geometry that included the multiplicative gauge function allowed the means to derive the gauge forces from the gauge function.

The remaining step was then to see if the two laws required a gauge function in the metric of the stability conditions. The fact that the laws do r equire a gauge function in the metric of an isolated system fills the last requirement for dynamics. Then through the transformations of the principle of Maupertuis, dynamics may be determined either as geodesics in a manifold whose geometry is specified by the laws or in a geometry freely chosen wherein the dynamics is force driven.

Figure 4.1 is a logic flow chart that starts with the adopted laws of thermodynamics and proceeds by restrictive assumptions to show how these restrictive assumptions lead to classical theories. The use of restrictive assumptions displays the fact that these classical theories are subsets of the thermodynamics laws. This chart may be referred to in order to see which assumptions are required to arrive at which feature of classical physics.

The key insight needed to understand the fundamental nature of the laws of thermodynamics is to note that the First Law is a Pfaff differential equation and to apply the Second Law of thermodynamics as Carathéodory did in 1909. C arathéodory's principle guarantees the existence of a property called entropy along with the energy statement of the First Law. The form of these laws are such that they may be expressed, without preference, in any coordinate system and of any dimension, as Einstein stated should be required of a fundamental set of laws. Though the necessary, complimentary existence of energy and entropy appears to complicate any mechanistic description of nature, it is their simultaneous existence that provides a logical description.

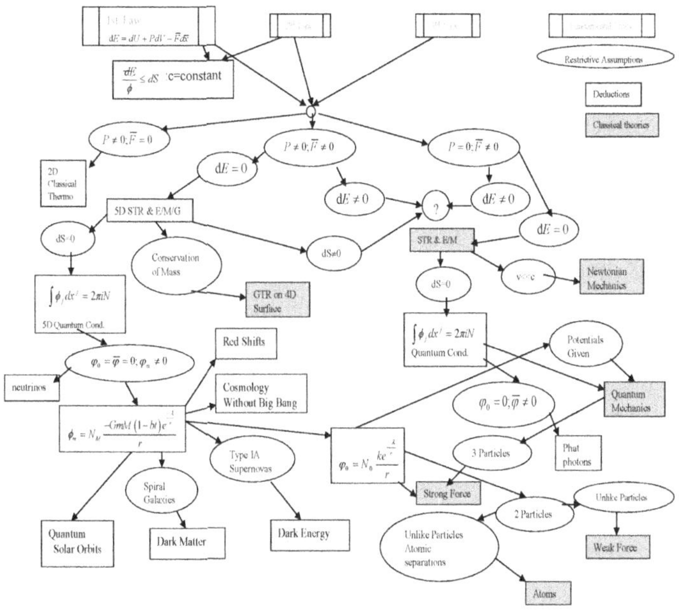

Figure 4.1 Logic Flow

Today, the concept of entropy is almost universally related to order or information. However, the concept demanded by the Second Law is best thought of as 'energy that becomes unavailable' as the thermal engineers have been known to call it. In this form, it is easier to connect the second law with the denial of perpetual motion. The more you do the greater the amount of energy that becomes unavailable. This becomes the entropy principle for isolated systems. For all other systems, it requires the minimum free energy principle. This provides variational principles that may be used to determine motion should a geometric metric also be given.

Now there is a surprising, but encouraging, immediate result from these laws. Using the same logic Carathéodory used to prove that the connection between energy (heat) and entropy is strictly a f unction of temperature and that this function leads to an absolute limit in temperature, these laws provide a function of velocity as the connecting function between energy and entropy for mechanical systems. More importantly, these laws provide a universal (i.e. independent of the type of force) limiting velocity for the same inertial systems that Einstein used in his special theory of relativity. There is now no need to assume that this as a fundamental postulate.

Bolstered by the fact that the laws seem to support the special theory of relativity by providing Einstein's postulate concerning the constancy of the speed of light, there is more reason to believe that mechanics may indeed come from these laws. Again following the lead of thermodynamics, in which there is a choice of the variables one may choose with which to write the second order differential equations that give the stability conditions, there is a choice provided for the description of mechanical systems' stability. These differential equations are natural metrics for use in determining motion. Choosing the stability conditions in a manifold of space and entropy seems very unlike mechanics. Yet, when this metric is scaled using local time as the arc length, a r ewarding requirement occurs when an isolated system is considered. These fundamental laws do not provide a variational principle in time. Therefore, the scaled metric must be solved for the element of entropy before the principle of increasing entropy may be applied. This gives equations of motion in a Riemannian manifold as Einstein used in his relativistic theories.

The two most remarkable features of this result are that the fundamental laws specify the geometry of the Second Law equations of motion and that the laws require *two* metrics. The first metric is the relativistic metric whose arc length is the entropy playing the role of Einstein's proper time. The second metric is a somewhat similar

84

relativistic metric with the energy as the arc length. This is the same requirement that occurs in thermodynamics where the change of entropy does not depend upon how the system changes. It depends only upon the end points while the heat depends upon how the change occurs. Here the distance between two points in the entropy manifold does not depend upon t he path, but the distance between the same two points in the energy manifold does depend upon the path. The laws require the energy metric to have a geometry that was first developed by Weyl when he proposed this geometry as a means to unify the electromagnetic and the gravitational fields in terms of the gauge fields that appeared within the geometry. Einstein argued that the path dependence of Weyl's geometric differed from experience with the result that Weyl's thoughts were abandoned so far as they went toward any unification.

In 1922, however, Schrödinger noticed that, should one require a unity scale in a Weyl space, then Bohr's quantized paths were allowed. Schrödinger went on t o develop his wave equations of quantum mechanics in 1926. In 1927, London showed that the requirement of unity scale in a Weyl space could only be satisfied by paths that obeyed Schrödinger's wave equations. Further, London showed that Schrödinger's wave function was proportional to Weyl's scale factor. Weyl seized upon t his result and raised London's result to the level of a principle, referred to as Weyl's Quantum Principle. This, together with Weyl's display that the gauge potentials formed the scale factor in his geometry, leads to the electromagnetic gauge fields and provided the basis for all the subsequent gauge field work that has followed in the search for a description of the weak and the strong nuclear forces.

This historical review is of importance when very stable mechanical systems are sought. These would be represented by systems with constant entropy, called isentropic systems. Isentropic mechanical systems must have the unity scale factor of Weyl's Quantum Principle. Therefore, stable isentropic mechanical systems must satisfy quantum mechanics! Quantum mechanics is required by the fundamental laws only for the isentropic subset of all the mechanical systems in nature. Thus, quantum mechanics is required by these laws, but may never be considered by them to be fundamental to the description of all motions in nature. Furthermore, London's result that Schrödinger's wave function must be proportional to the Weyl scale factor, limits the statistical interpretation of quantum mechanics. Rather, London showed that the wave function contained information about the tendency of the scale factor to vary around unity. This requires the same mathematical operations as are currently used, yet with an entirely different interpretation.

One more result, from the fact that the Entropy Principle provides the variational principle for descriptions of motions for an isolated system, is one Einstein suggested was needed for any proper foundation of physics. The need for an explanation of inertia is satisfied by the fact that the entropy is an extensive property of the system requiring it to be proportional to the mass. This is the origin of the multiplicative mass term in the equations of motion.

The unsatisfactory separation between the theories of relativity and quantum mechanics, referred to by Einstein, is also hereby resolved. All isolated systems must obey relativity theory through the Second Law. Isentropic systems must also obey the laws of quantum mechanics.

Weyl showed that Maxwellian electromagnetism can be derived from his gauge potentials. The isentropic condition requires these potentials and the subsequent electro-magneto-gravitic fields to be quantized. This requires the electrostatic potential to be quantized in integer values as is seen by experiment. Existing theories do not require this quantization of electric charge. When the quantized electrostatic field is forced to satisfy the extended Maxwell equations, the dependence of the electrostatic potential upon s pace is determined to be a non-singular potential $[(1/r^2)\exp(-\lambda/r)]$ with the familiar $1/r^2$ long range dependence. The point, λ, at which the potential begins to deviate significantly from Coulomb's $1/r^2$ potential, is different for different particles! This produces a violation of Newton's action and reaction law for unlike particles that get very close to each other. Yet this has been shown to provide the basis of the nuclear force currently described as the weak force. A lso, the equations of the Yang-Mills theory have been derived from this requirement for non-singular interactions between unlike particles that do not obey Newton's law of action and reaction. On the other hand, the non-singular potential requires that like particles change their character as they approach each other. This means that like particles, such as protons, that have a repulsive long-range interaction will have an attractive interaction as they are forced into the near proximity of each other. The equations for non-singular interactions between like particles have the SU(3) group characteristics.

How is gravitation required of the fundamental laws? Suppose one looks into the equations that describe a four dimensional hyper-surface that is embedded into a five dimensional Weyl manifold that have the four dimensions of space-time and an unknown, but physically real, fifth dimension. Without knowing what the fifth dimension might be, restricting it to be conserved similar to the statement of the conservation of mass provides a restriction that may quickly be explored. The equations resulting from the conservation of the fifth dimension are seen to be identical in form to those chosen by Einstein as his gravitational field equations in his general theory. Now the determination of the fifth dimension may be seen for the only physically real property that could give Einstein's equations is gravitating mass!

The Foundation of Physics

Substantiation for this conclusion may be had when one considers the First Law of thermodynamics and includes three spatial work terms plus the thermodynamic work term. The usual manner of writing the thermodynamic work term involves the specific volume. However, the reciprocal of the specific volume is the mass density. Therefore, the fundamental laws allow the mass as a fifth dimension. When this property is conserved, which is the only condition Einstein considered, there are at least three ways these laws describe gravitational phenomena. First, they may be described using the five dimensional gauge field descriptions, wherein the electromagnetic and gravitational fields form a single, inductively coupled, electro-magneto-gravitic field. Secondly, either of the two fundamental metrics for a surface embedded into a manifold may be used. The fundamental metric of the second type produces the Einstein equations. Further, Einstein's principle of the equivalence of inertial and gravitational mass is a further requirement of the fundamental laws and need not be made separately.

The fundamental laws require the quantization of gravitational phenomena for isentropic systems as well as a non-singular, time dependent, gravitational potential. The appearance of the non-singular, time dependent, gravitational potential changes the interpretation of black holes, the big bang, red shifts of cosmological objects, and dark matter/energy. Now the tie between gravitation and quantum mechanics has been established.

One last feature of these fundamental laws should be mentioned. It concerns Einstein's position that two separate theoretical descriptions of light, on the one hand as a particle and on the other hand a wave was intolerable. Electromagnetic waves follow from Weyl's gauge fields; that is, from the Maxwell equations. Isentropic propagation of electromagnetic energy must also satisfy Weyl's quantum condition and hence, must simultaneously satisfy the wave equations and be quantized. Further, the fundamental laws require that the quantized, isentropic propagation of electromagnetic energy must satisfy Plank's blackbody radiation law. The wave and the particle nature of light are, therefore, both required by these fundamental laws.

Einstein stated that there appears to be two choices for a foundation for physics; statistical or deterministic. Here we see a foundation that is fundamentally deterministic. Non-isolated systems and systems with variable entropy must be deterministic while isentropic systems must be quantized and, therefore, may have a statistical nature even though the probabilistic interpretation of Schrödinger's waves was shown by London to be in error. Einstein's desire for a logically simple foundation for physics is also satisfied; for these laws have been shown to produce the foundations of each of the various branches of physics without yet coming upon a measured difference from experiment.

Part 2 Theoretical Development

In the previous part an over view of the theory was presented. In this part of the book a detailed development of the Dynamic Theory will be presented. This is the background for all the material previously published in the 2009 book t itled "Physics – Against the Odds." The fundamental laws were seen to be a starting point for a theory that included equations of motion in 1975 and the initial findings were put into my master's thesis when I graduated from the US Naval Postgraduate School in 1976. The first venture into five dimensions and five dimensional quantum mechanics was reported in a technical report while I was teaching at the US Naval Academy in 1978. The derivation of the non-singular potential was done while I was a m ilitary research associate at the Los Alamos National Laboratory in 1981 a nd this led to the nuclear model. Also, while I was at the laboratory in Los Alamos I was involved in shock physics and that prompted the research into the five dimensional hydrodynamics. The research resulting in the details of the fusion of deuterium nuclei into a helium nucleus was started while I was at New Mexico Tech. Most of the remainder included here was completed after I retired from the university in 2001.

The material presented herein was not developed temporally as it is presented here for it was developed rather piecemeal as my interest was piqued by one thing or another. However, the presentation herein is grouped together in a somewhat more logical progression of applications. It is interesting to look at this part of the book and think of how differently the research turned out than what I had anticipated when I started the first steps back in 1974. I had initially found it hard to accept the notion of the speed of light being a limiting velocity for gravitational phenomena based only upon experiments with electromagnetic fields and forces. My discovery of an absolute velocity changed my view of the speed of light drastically and permanently. Initially I had felt that something was very wrong with quantum mechanics even though I could understand that a restriction could lead to quantization. For example, fixing the ends of a guitar string restricted the motion of the string to only a limited set of standing waves. I did not think that quantum mechanics could be used as a basis for all theories of dynamics. I thus felt rewarded when I found that quantum mechanics was demanded by restricting one's attention to isentropic systems.

The fundamental laws and some of their immediate consequences are in Chapter 5. I put several immediate results of the extension to five dimensions into Chapter 6. Chapter 7 looks at the gauge fields allowed for particles which satisfied the fundamental assumption that their fields are independent of where and when they existed. Chapter 8 concentrates on electromagnetic fields and the wave/particle duality aspect of light required by the fundamental laws. Chapter 9 turns to nuclear physics and the resulting nuclear model. Gravitational phenomena and quantum gravity are addressed in Chapter 10. Chapter 11 looks at several general aspects of the five dimensional fields and a new communication concept. While it seems a little out of place following several chapters discussing field concepts, Chapter 12 presents the five dimensional hydrodynamics that is then reduced to the four dimensional, non-relativistic Navier-Stokes equations to show that the conservation of mass imposes a natural viscosity upon all hydrodynamic flows.

Chapter 5 Fundamental Laws

Though I had often asked "Why?" when confronted with some new assumption or adopted postulate, the first really puzzling facet of current physics I encountered was the concept of relativistic kinetic energy from Einstein's Special Theory of Relativity. The puzzling part was that it depended upon the speed of light independent of the mechanism by which this energy might be transferred. To better illustrate what puzzled me, consider the transfer of energy between two charged particles on collision courses. If the particles have near miss trajectories, then the energy is primarily transferred by the electrical forces between the charges. From the view of retarded potentials, or the concept of a limiting speed of electromagnetic signal transmission, it is rather easy to accept the energy transferred being dependent upon this limiting velocity. But suppose the particles are uncharged and the interaction is strictly a gravitational one. Again the concept of a limiting signal speed would imply that the energy exchanged between the particles depend upon this limiting velocity. But is the gravitational limiting velocity the same as the limiting signal velocity for the electromagnetic case? Do gravitational waves travel at the same speed as electromagnetic waves?

Einstein, in the Special Theory of Relativity, adopted the position that the constancy of the speed of light forces a modification of Newton's dynamic law. This modification implies that all forces have the same limiting velocity, namely, the speed of light. There exists an abundance of theoretical and experimental evidence that the speed of light becomes the limiting velocity whenever electromagnetic forces are involved. The point that bothered me was whether other forces, such as gravitational, should also have the same limiting velocity. Though we have had reports of the detection of gravitational waves, we have no experimental determination of the speed of a gravitational wave. Therefore, I object to the viewpoint that the modification to Newton's law should be applied to all forces without some additional justification.

Let me describe an analogy which may not hold in the strictest sense yet may serve to illustrate my point of view. A river, flowing toward the sea, carries energy with it. The speed with which this energy can move from one point to another is the velocity of the river's current. The river produces a force on a boat tied up to a pier on the river. When the boat is set adrift, this force accelerates the boat. However, the maximum velocity to which the river can accelerate the boat is the current velocity; this is the velocity with which the energy of the river can propagate.

From this point of view the speed of light, being the propagation velocity of electromagnetic energy must be the limiting velocity associated with electromagnetic forces. Certainly nature would be much simpler if all forces have the same limiting velocity. Yet without some experimental evidence of the propagation of gravitational energy, I find it difficult to feel comfortable with Einstein's modification of Newton's law justified by electromagnetic experimental evidence and arguments of simplicity.

The fundamental philosophical viewpoint that the force depends upon velocity and vanishes as the velocity approaches the limiting velocity raises another question concerning Einstein's modification of classical mechanics. Under Einstein's modification Hamilton's principle is written with a relativistic mass which depends upon the velocity and a velocity independent force. Does these represent a different philosophy or are both views equivalent? More specifically, are the "real" concepts to be taken as a mass independent of velocity together with a velocity dependent force or should we associate the velocity dependent relativistic mass and velocity independent forces with "real" world? Or does it make any difference which we chose?

At this point I faced the first major decision. If I adopted Einstein's postulates, then it appeared that I would be required to change my intuitive beliefs concerning certain physical phenomena. I found this extremely difficult to do. On the other hand, if I did not embrace these postulates, I would have to replace them with something that would say essentially the same thing in all cases where the Special Theory of Relativity has been found to be very accurate. Not only this but if a new point of view were adopted, then virtually the entire sphere of physics may need to be reviewed in order to ensure that the new point of view did not conflict with currently used theories where they have experimental verification.

History records the advancements in physics which came from the efforts of people new to the field. Therefore my lack of training in physics might be turned into an advantage if I sought to determine a philosophical basis unhampered by the directed philosophy that comes from a study of physics as currently taught. This is in contradistinction with current practices and procedures of academicism where mastery of current theories generally precedes the development of a new one. To deliberately choose this deviation risks accusations of arrogance and naiveté. On the other hand such a choice seemed the best way of avoiding the danger of becoming so familiar with current ways of thinking as to make it improbable of giving due attention to other ways.

Having decided to look for a new foundation for physics I was faced with the question of how to begin. I recalled some Ozark hill philosophy I overheard as a youngster. A native Ozarkian was giving directions to a stranger who was trying to find a certain fishing hole. The directions went something like this: "See yonder road going down that holler? Well, go down thar 'bout five mile and you'll come to a fork in the road. Take the right hand fork. Now that's the wrong one but you take it anyways. After you've gone a piece, you'll come to a log across the road. Now you know you're on the wrong road. Go back and take the left hand fork. You can't miss it."

A quick review of physics reveals that there are different branches with different sets of fundamental laws or postulates. Though it is easy to see how the distinction between these branches came about, it was difficult to believe that nature shared the same divisions. I felt that all natural phenomena should be explained by a single set of fundamental laws. This belief is somewhat like a grove of redwood trees or bamboo forest. Above the ground each tree appears as a distinct plant. Yet we know that below the ground they may be found to grow from the same root system. Thus, I felt that a more fundamental approach might display the unity in nature and that prior attempts at unification in the search for a unified field theory could be likened to attempts to tie the trees together at the tree top level rather than down at the root level.

Is nature symmetrical in time? Does everything run backward in time as well as forward? Obviously, not every process in nature will run backwards, yet the equations of motion in Newtonian and relativistic mechanics are time symmetrical. I believe in an asymmetrical nature and this belief played a role in the eventual selection of fundamental laws.

How then did I use this philosophy to determine a set of generalized laws on which to base an attempt to construct a new approach to physics?

Before proceeding let me offer a word of caution. During any theorization the philosophy of the theorist plays such an important role that an attempt to understand the theory is aided by knowledge of this philosophy. Therefore the following includes not only the philosophical basis upon which the theory is based and the mathematical development but also ideas and beliefs which played a part in the various decisions. Because of the individualistic nature of philosophy the following will deviate occasionally from a strict third person presentation, risking a loss of professional appearance, to the clearly personal first person.

Newtonian mechanics fails to describe events involving high velocities, relativistic mechanics fails to describe the atom, and gravitational effects have resisted quantization. If these are viewed as logs and the Ozarkian's directions are followed, then we must retrace our steps and seek another approach rather than attempting to chop up the log and continue to push forward up one of these roads.

The branch of thermodynamics, however, does not appear to have a l og anywhere along the way. Here the classical thermodynamic laws are very general, particularly Carathéodory's statement of the second law. Thus the thermodynamic laws appeared to be the fork in the road where a new route might be chosen.

However, in mechanics we talk of equations of motion, field equations, and geometry while in thermodynamics we speak of equations of state and equilibrium. If a generalization of the classical thermodynamic laws is adopted, how might we obtain the equations with which we are familiar in mechanics? More particularly, how could this type of general law yield geometry and a variational principle? The second law of thermodynamics can produce a variational principle through principles such as increasing entropy and minimizing free energy, but can it also produce geometry?

This seemed to be a crucial point. If the laws could not produce geometry, then a geometry would have to be assumed, thus necessitating an additional assumption. The belief that a simple fundamental set of laws should lead to the fundamental principles of the different branches of physics made the thought of additional assumptions abhorrent. The notion that the adopted laws should specify the type of geometry that must be used seemed very satisfying. Newton found that the absolute nature of Euclidean geometry brought undesirable features. Einstein, in his General Theory, displayed the benefits that might be gained by going to a more general geometry. He showed that physical phenomena might be displayed as elements determined by certain physical laws. This is essentially the question here. Can a set of laws, which are generalizations of the classical thermodynamic laws, determine the metric elements and hence the geometry?

By appealing to the mathematics of functions of more than one variable we find that a quadratic form becomes involved when a maximum or minimum is sought. Further, this quadratic form generates a n atural geometry for that function. In thermodynamics the stability conditions provide a similar quadratic form and therefore the quadratic form which specifies the stability conditions should form a natural geometry for a physical system governed by laws such as the thermodynamic laws.

Thus the foundations of the theory have been outlined, namely the belief that all physical phenomena should be derivable from a single set of physical laws which are generalizations of the classical thermodynamic laws. Such a theory should be capable of describing all the dynamic events in nature. Therefore it seems appropriate to call it the "Dynamic Theory". Obviously, for such a theory to be tenable it must reproduce, or be consistent with, the various fundamental postulates and/or laws currently used in the various branches of physics. Indeed it should do e ven more. It should also reduce the number of necessary assumptions and provide an unprecedented unification of physics. Further, there is the possibility that the theory might produce an experimentally verifiable prediction.

The first requirement that should be placed upon the Dynamic Theory is that it reproduces, or be consistent with, current theories. In order to show that the Dynamic Theory satisfies this requirement, this chapter states the adopted laws and then the remainder of this part shows how appropriate restrictions upon the system do yield the fundamental principles for the various current theories.

Though a theory which has the capability of displaying a unification of physical theories might have significant value based solely upon this capability, it would become more attractive if it could explain phenomena for which no explanation exists or make some new prediction which might lead to an experimental test of the theory. Since restrictions will be placed upon the system in order to show how current theories may be obtained, the easiest way to see the expanded coverage of the theory is to relax one or more of the restrictions and consider a more general system.

A theory, such as the Dynamic Theory, immediately poses several problems which are not associated with its validity or applicability. First, there is a new point of view to be dealt with. Initially it would appear to be inconsistent with all past concepts of system energy or relativistic concepts. Yet in the end it is completely consistent with current theories and sheds an entirely new light upon physical phenomena.

Another imposing difficulty with the Dynamic Theory stems from its generality. The scope of the theory includes all physical phenomena while in the past half century the vast amount of scientific knowledge that has been accumulated has demanded specialists. The increasing expansion of mankind's knowledge demands further specialization. Such a progression produces no demand for a generalist. The result is that the greater portion of this theory will be outside the field of many readers.

Closely associated with this problem is another. Throughout science symbols and words are used to denote concepts and quantities. The limited number of available symbols and words together with the expanded scope of scientific knowledge requires duplication. For the specialists this duplication can be somewhat minimized. However, in the case of a general theory touching virtually all areas of specialization the problem becomes very significant. In particular, if a certain symbol or set of words is used, a particular notion or concept may be associated with them by the reader. This association will likely depend upon the reader's specialty and therefore will vary with the reader. Any attempt to choose symbology or word usage aimed at a particular specialty risks increased confusion for readers in other fields. Therefore, the reader is cautioned to keep in mind that conceptualizations and symbology familiar because of its use in one branch of physics may now take on an entirely new meaning.

Fundamental Laws

Einstein used his postulate concerning the constancy of the speed of light in showing that there was a limiting velocity for material objects. This limiting velocity is similar to the limiting aspect of the absolute zero temperature that appears in classical thermodynamics. This gives rise to the question as to whether or not there might be a fundamental connection between the two limiting concepts, one in mechanics and the other in thermodynamics. This question was first addressed by the author in 1976 and provides the basis for later work, yet the author later learned that he was not the only one to have considered such a question. The presentation below will show that the laws of thermodynamics require Einstein's postulate.

First Law (Conservation of Energy)

The concept of conservation of energy is fundamental to all branches of physics and is the beginning of thermodynamics and mechanics. In terms of generalized coordinates or independent variables, the notion of work, or mechanical energy, is considered linear forms of the type

$$\text{đ}W = F_i(q^1,...,q^n,u^1,...,u^n)\,dq^i \quad (i=1,2,...,n), \quad (5.1)$$

where the forces F_i may be functions of the velocities $(dq^i/dt=u^i)$ as well as the coordinates q^i and the summation convention is used. The line integral $\int_C F_i\,dq^i$ then represents the work done along the path C by the generalized forces.

A system may acquire energy by other means in addition to the work terms; such energy acquisition is denoted $\text{đ}E$. The system energy, which represents the energy possessed by the system, is considered to be $U(q^1,...,q^n,u^1,...,u^n)$.

With these concepts, then the First Law, which is the generalized Law of Conservation of Energy, has the form

$$\text{đ}E = dU - \text{đ}W = dU - F_i\,dq^i \quad (i=1,...,n). \quad (5.2)$$

Positive $\text{đ}E$ is taken as energy added to the system by means other than through the work terms and F_i is taken as the component of the generalized force acting on t he system which caused displacement dq^i.

In the First Law the dimensionality is n+1 and is determined by the system considered. There is no limitation on the quantity or type of variables that may be used. A system with only one work term which is the *pdv* expansion work of classical thermodynamics will be called a "thermodynamic" system and the dimensionality will be two. A system with three or less $F_i dq^i$ work terms will be called a "mechanical" system with the appropriate dimensionality. Obviously, if there are three mechanical work terms, the dimensionality will be four.

In an infinitesimal transformation, the First Law is equivalent to the statement that the differential

$$dU = \text{đ}E + F_i\,dq^i \quad\quad (5.3)$$

is exact. That is, there exists a f unction U whose differential is dU; or the integral $\int dU$ is independent of the path of the integration and depends only on t he limits of integration. This condition is not shared by $\int đE$ or $\int đW$.

The path dependence of $\int đW$ is another reason that the generalized forces are assumed to be functions of velocity as well as position for if the forces were only functions of position the work done by such forces would be independent of the path. This is the concept of conservative forces in mechanics. In Newtonian mechanics forces are usually assumed to be dependent on position only so that the simplicity of path independence may be used. Though even in Newtonian mechanics certain forces are taken as velocity dependent. Friction forces are an example.

This statement of the generalized First Law is consistent with the first law of thermodynamics in that if there is only one generalized force, which is taken to be the pressure, and one generalized coordinate, the volume, then Equation (5.2) becomes

$$đQ = đE = dU + Pdv$$

where $F=-P$ with the convention that work of expansion is work done by the system on i ts surroundings. Here the system energy, U, is the thermodynamic internal energy. There should then be no c onfusion when Caratheódoryy's statement of the second law is applied to this thermodynamic system. However, when considering the application of generalizations of the classical thermodynamic laws to mechanical systems some confusion may be expected. In this chapter, it is desired to demonstrate the applicability of the generalized laws to mechanical systems. Therefore, it may help avoid confusion to think of the generalized coordinates of a mechanical system as the space coordinates of a m ass point. Obviously, there exist systems in nature that may be considered to consist of a continuous distribution of mass points. Such a system may be thought of as a composite system of an infinite number of subsystems and, therefore, involve an infinite number of "generalized coordinates," or "degrees of freedom." However, just as in classical mechanics, we may later make the transition from mass points to matter in bulk; then the generalized coordinates, q^i, used here may better be termed independent variables.

There are a couple of points that should be made here before we move on to the Second Law. Take another look at the first law, or

$$đE = dU - đW = dU - F_i \, dq^i \quad (i = 1,...,n). \quad (5.4)$$

If in Equation (5.4) we consider only isolated systems with three work terms and let them be the terms associated with the space dimensions, then we would write

$$dU = F_i \, dq^i \quad (i = 1,2,3). \quad (5.5)$$

Now suppose the system's energy is given only as a function of the velocity somewhat like the usual expression of kinetic energy. Now Equation (5.5) becomes

$$\frac{\partial U}{\partial \bar{v}} \frac{d\bar{v}}{dt} \equiv \bar{p} \bullet \bar{a} \equiv m\bar{v} \bullet \bar{a} = F_i \frac{dq^i}{dt} = \bar{F} \bullet \bar{v} \; . \quad (5.6)$$

Now Equation (5.6) looks a lot like Newton's second law, or his equation of motion. However, we must remember that, first, the force here must be path dependent and the force usually used in Newton's second law is path independent; second, Newton's second law doesn't care whether time runs forward or backward. These two aspects of Equation (5.6) make it v ery different from Newton's second law. This will become even more evident as we look into what the Second Law of thermodynamics has to offer.

Second Law

There are processes, or motions, that satisfy the First Law but are not observed in nature. The purpose of the Second Law is to incorporate such experimental facts into the model of dynamics. We will also see how the Second Law places restrictions on the First Law that makes the First Law differ greatly from Newton's equation of motion.

The statement of the Second Law is made using the axiomatic statement provided by the Greek mathematician Carathéodory[4], who presented an axiomatic development of the second law of thermodynamics that may be applied to a system of any number of variables. The Second Law may then be stated as follows:

> In the neighborhood (however close) of any equilibrium state of a system of any number of dynamic coordinates, there exist states that cannot be reached by reversible E-conservative $\left(dE = 0 \right)$ processes or motions.

When the variables are thermodynamic variables, the E-conservative processes are known as adiabatic processes.

A reversible process, or motion, is one that is performed in such a way that, at the conclusion of the process, both the system and the local surroundings may be restored to their initial states without producing any change in the rest of the universe.

Consider a system whose independent coordinates are a generalized displacement denoted q, a generalized velocity u (with $u' \equiv \frac{du}{dt}$), and a generalized force F. It can be shown that the E-conservative curve comprising all equilibrium states accessible from the initial state, i, may be expressed by $\sigma(u,q)$ = constant, where σ represents some as yet undetermined function. Curves corresponding to other initial states would be represented by different values of the constant.

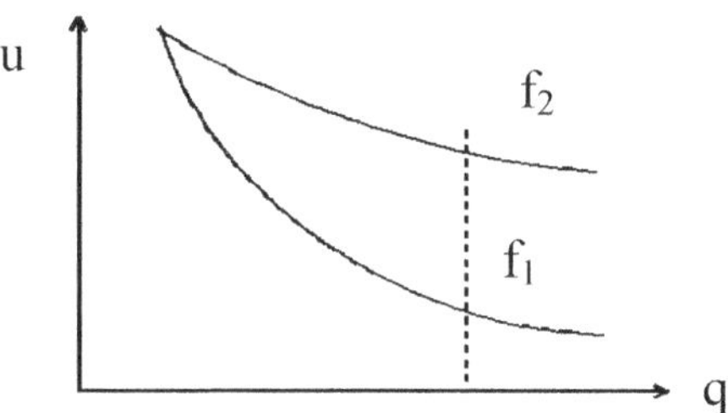

Figure 5.1. Reversible E-conservative curves cannot intersect.

Reversible E-conservative curves cannot intersect, for if they did, it would be possible, as shown in Figure 5.1, to proceed from an initial equilibrium state i, at the point of intersection, to two different final states f_1 and f_2, having the same q, along reversible E-conservative paths, which is not allowed by the Second Law. This is an important requirement of the Second Law and we will now show this contention.

When the system can be described with only two independent variables, such as on the E-conservative curve, then if these variables are the space variable q and the velocity variable u and F is a generalized force,

$$\text{đ}E = dU - F dq.$$

Regarding $U = U(q,u)$, then

$$\text{đ}E = \left[\frac{\partial U}{\partial u}\right]_q du + \left[\left[\frac{\partial U}{\partial q}\right]_u - F\right] dq$$

where all quantities on the right hand side are functions of both u and q.

An E-conservative motion for this system requires that

$$\left[\frac{\partial U}{\partial u}\right]_q du + \left[\left[\frac{\partial U}{\partial q}\right]_u - F\right] dq = 0.$$

Solving for du/dq yields

$$\frac{du}{dq} = \frac{-\left[\left[\dfrac{\partial U}{\partial q}\right]_u - F\right]}{\left[\dfrac{\partial U}{\partial u}\right]_q}. \tag{5.7}$$

The right hand member is a function of u and q, and therefore, the derivative du/dq, representing the slope of an E-conservative curve on a (u,q) diagram, is known at all points. The (u,q) diagram is better known in classical physics as the phase diagram. Equation (5.7) has therefore a solution consisting of a family of curves, see Figure 2, and the curve through any one point may be written $\sigma = \sigma(u,q) = constant$.

A set of curves is obtained when different values are assigned to the constant. The existence of the family of curves $\sigma = \sigma(u,q) = constant$, generated by Equation (5.7) representing reversible E-conservative processes follows from the fact that there are only two independent variables and not from any law of physics. Thus, it can be seen that the First Law may be satisfied by any of these $\sigma = constant$ curves. The Second Law requires that these curves do not intersect. Therefore the Second Law, together with the First Law, leads to the conclusion that through any arbitrary initial state point, all reversible E-conservative processes lie on a curve, and E-conservative curves through other initial states determine a family of non-intersecting curves.

To see the results of this conclusion consider a system whose coordinates are the generalized velocity u, the generalized displacement q and the generalized force F. The First Law is

$$đE = dU - Fdq \tag{5.8}$$

where U and F are functions of u and q. Since the phase diagram, or (u,q) surface, is subdivided into a family of non-intersecting E-conservative curves $\sigma(u,q) = constant$ where the constant can take on various values σ_1, σ_2, ..., and points on the surface may be determined by specifying the value of σ along with q, in all regions where the Jacobian of the transformation does not vanish, so that U, as well as F, may be regarded as functions of σ and q. Then

$$dU = \left[\frac{\partial U}{\partial \sigma}\right]_q d\sigma + \left[\frac{\partial U}{\partial q}\right]_\sigma dq \qquad (5.9)$$

and

$$\text{đ}E = \left[\frac{\partial U}{\partial \sigma}\right]_q d\sigma + \left[\left[\frac{\partial U}{\partial q}\right]_\sigma - F\right]dq. \qquad (5.10)$$

Since σ and q are independent variables Equation (5.10) must be true for all values of $d\sigma$ and dq.

Suppose $d\sigma = 0$ and $dq \neq 0$. The provision that $d\sigma = 0$ is the provision for an E-conservative process in which $\text{đ}E = 0$. Therefore, the coefficient of dq must vanish. Then, in order for σ and q to be independent and for $\text{đ}E$ to be zero when $d\sigma$ is zero, the equation for $\text{đ}E$ must reduce to

$$\text{đ}E = \left[\frac{\partial U}{\partial \sigma}\right]_q d\sigma,$$

Defining a function λ by $\lambda \equiv \left[\frac{\partial U}{\partial \sigma}\right]_q$ then $\text{đ}E = \lambda d\sigma$ where $\lambda = \lambda(\sigma,q)$.

Integrating Factor

Now, in general, an infinitesimal of the type

$$Pdx + Qdy + Rdz + ... = 0, \qquad (5.11)$$

known as a linear differential form, or a Pfaffian expression, when it involves three or more independent variables, does not admit of an integrating factor. This equation is integrable if, and only if, in the neighborhood of any arbitrary point G_o there are points G which are inaccessible from G_o along solution curves Equation (5.11).

It is only because of the existence of the Second Law that the differential form for dE referring to a physical system of any number of independent coordinates possesses an integrating factor.

Two infinitesimally neighboring reversible E-conservative curves may be considered. One curve is characterized by a constant value of the function σ_A, and the other by a slightly different value $\sigma_A + d\sigma = \sigma_B$. In any motion represented by a displacement along either of the two E-conservative curves $dE = 0$. When a r eversible motion connects the two E-conservative curves, energy $dE = \lambda d\sigma$ is transferred. The various infinitesimal motions that may be chosen to connect the two neighboring reversible E-conservative curves that involve the same change of σ but take place at different λ. In general λ is a function of u and q. However, it is obvious that λ may be expressed as a function of σ and u.

To find the velocity dependence of λ consider two systems, one and two, such that in the first system there are two independent coordinates u and q and the E-conservative curves are specified by different values of the functions σ of u and q. When dE is transferred, σ changes by $d\sigma$ and $dE = \lambda d\sigma$ where λ is a function of σ and q.

The second system has two independent coordinates u, and q' and the E-conservative curves are specified by different values of the function σ' of u and q'. When dE is transferred, σ' changes by $d\sigma'$ and $dE = \lambda' d\sigma'$ where λ' is a function of σ' and u.

The two systems are related through the coordinate u in that both systems make up a composite system in which there are three independent coordinates u, q, and q' and the E-conservative curves are specified by different values of the function σ_c of these independent variables. To help visualize the situation it ma y be noted that the composite system might be thought of as a particle which

108

has a velocity u and whose position is specified in terms of two independent space coordinates such as x and y. Then we should more properly refer to the variable u as the speed. Obviously, in one dimension the velocity and the speed become the same thing. However, in spaces with two or more dimensions the speed and velocity are very different things. For the remainder of this chapter we will be careful to call a variable by its proper name.

Since $\sigma=\sigma(u,q)$ and $\sigma' = \sigma'(u,q')$, using the equations for σ and σ', σ_c may be regarded as a function of u, σ and σ'.

For an infinitesimal motion between two neighboring E-conservative surfaces specified by σ_c and $\sigma_c+d\sigma_c$, the energy transferred is $dE = \lambda_c d\sigma_c$ where λ_c is also a function of u, σ and σ'. Then

$$d\sigma_c = \left[\frac{\partial \sigma_c}{\partial u}\right]du + \left[\frac{\partial \sigma_c}{\partial \sigma}\right]d\sigma + \left[\frac{\partial \sigma_c}{\partial \sigma'}\right]d\sigma'. \qquad (5.12)$$

Now suppose that in a motion there is a transfer of energy dE_c between the composite system and an external reservoir with energies dE and dE' being transferred, respectively, to the first and second systems, then

$$dE_c = dE + dE'$$

and

$$\lambda_c d\sigma_c = \lambda d\sigma + \lambda' d\sigma',$$

or

$$d\sigma_c = \left[\frac{\lambda}{\lambda_c}\right]d\sigma + \left[\frac{\lambda'}{\lambda_c}\right]d\sigma'. \qquad (5.13)$$

Comparing Equations (5.12) and (5.13) for $d\sigma_c$ then

$$\frac{\partial \sigma_c}{\partial u} = 0.$$

Therefore σ_c does not depend on u, but only on σ and σ'. That is $\sigma_c=\sigma_c(\sigma,\sigma')$. Again comparing the two expressions for $d\sigma_c$ we find

$$\frac{\lambda}{\lambda_c} = \frac{\partial \sigma_c}{\partial \sigma} \quad \textit{also} \quad \frac{\lambda'}{\lambda_c} = \frac{\partial \sigma_c}{\partial \sigma'}.$$

Therefore the two ratios λ/λ_c and λ'/λ_c are also independent of u, q and q'. These two ratios depend only on the σ's, but each separate λ must depend on t he speed as well (for example, if λ depended only on σ and on nothing else, the $đE = \lambda d\sigma$ would equal $f(\sigma)d\sigma$ which is an exact differential). In order for each λ to depend on the speed and at the same time for the ratios of the λ's to depend only on the σ's, the λ's must have the following structure:

$$\lambda = \phi(u)\, f(\sigma),$$

$$\lambda' = \phi(u)\, f'(\sigma'), \quad with \tag{5.14}$$

$$\lambda_c = \phi(u)\, g(\sigma, \sigma').$$

(The quantity λ cannot contain q, nor can λ' contain q', since λ/λ_c and λ'/λ_c must be functions of the σ's only.)

Referring now only to the first system as representative of any system of any number of independent coordinates, the transferred energy is, from Equations (5.10),

$$đE = \phi(u)\, f(\sigma)\, d\sigma. \tag{5.15}$$

Since $f(\sigma)d\sigma$ is an exact differential, the quantity $1/\phi(u)$ is an integrating factor for $đE$. It is an extraordinary circumstance that not only does an integrating factor exist for the $đE$ of any system, but this integrating factor is a function of speed only and is the same function for all systems. Since the integrating factor is the same form for all systems it does not depend upon t he type of force involved and is, therefore, unique.

It would be nice if there were a simple way of deriving the functional form of $\phi(u)$. In thermodynamics we opted to take the easy way out by assuming that the integrating factor was simply the reciprocal of the temperature as was determined from Carnot's work.

110

However, for mechanical systems we will find the functional form of the integrating factor when we determine the equations of motion.

The fact that a system of two independent variables has a dE that always admits an integrating factor regardless of the axiom is interesting, but its importance in physics is not established until it is shown that the integrating factor is a function of speed only and that it is the same function for all systems.

Absolute Speed and Einstein's Postulate.

The universal character of $\phi(u)$ makes it possible to define an absolute speed.

Through the years I have had the privilege of talking to many good physicists and mathematicians. The person who most readily saw the necessity and importance of showing that the fundamental laws produced an absolute velocity was Dr. Edward Teller. He was visiting Los Alamos National Laboratory (LASL then) around 1980 and I had sent him a copy of some of my work upon the advice of my group leader. Dr. Teller called me and stated "Come over. Let's talk." I went over to his office and he quickly asked, "How do you prove the laws require Einstein's postulate?" I told him there were three steps to the proof and, before I could state what they were, he invited me to put them on the blackboard. The first step is to show that the second law requires that solution curves of the first law cannot cross. I put this on the board. The second step is to show that an integrating factor must exist for the first law. I put this on the blackboard. Dr. Teller then jumped up and stated that he could finish the proof and did so by showing that the integrating factor must become infinite at a unique velocity. Alas, he became disinterested when he learned that neutrinos were not required in neutron disintegration and could not withhold his judgment until I could find the logical source of neutrinos.

The first two steps of the proof of Einstein's postulate of the speed of light have been completed. We need now to finish the proof as Dr. Teller also did.

Consider a system of two independent variables q and u, for which two constant speed curves and E-conservative curves are shown in Figure 4. Suppose there is a constant speed transfer of energy E between the system and its surroundings at the speed u, from a state b, on a n E-conservative curve characterized by the value σ_1, to another state c, on another E-conservative curve specified by σ_2. Then since it is seen that

$$đE = \phi(u) \int_{\sigma_1}^{\sigma_2} f(\sigma)d\sigma \quad \text{at constant } u. \tag{5.16}$$

For any constant speed process between two other points a to d, at a speed u_3 between the same E-conservative curves the energy transferred is

$$đE(u_3) = đE_3 = \phi(u_3) \int_{\sigma_1}^{\sigma_2} f(\sigma)d\sigma \quad \text{at constant } u_3. \tag{5.17}$$

Taking the ratio of

$$\frac{đE}{đE_3} = \frac{\phi(u)\int_{\sigma_1}^{\sigma_2} f(\sigma)d\sigma}{\phi(u_3)\int_{\sigma_1}^{\sigma_2} f(\sigma)d\sigma} = \frac{\phi(u)}{\phi(u_3)} \frac{a\,fct\,of\,v\,at\,which\,đE\,is\,transferred}{a\,fct\,of\,v\,at\,which\,đE_3\,is\,transferred}\,.$$

(5.18)

Then the ratio of these two functions is defined by

$$\frac{\phi(u)}{\phi(u_3)} = \frac{đE\,(between\,\sigma_1\,and\,\sigma_2\,at\,u)}{đE_3\,(between\,\sigma_1\,and\,\sigma_2\,at\,u_3)} \tag{5.19}$$

or

$$đE = \left[\frac{đE_3}{\phi(u_3)}\right]\phi(u). \tag{5.20}$$

By choosing some appropriate speed u_3 it follows that the energy transferred at constant speed between two given E-conservative curves decreases as $\phi(u)$ decreases, or the smaller the value of $đE$ the lower the corresponding value of $\phi(u)$. When dE is zero $\phi(u)$ is also zero. The corresponding velocity u_0 such that $\phi(u_o)$ is zero is the "absolute speed". Therefore, if a s ystem undergoes a constant speed motion between two E-conservative curves without an exchange of energy, the speed at which this takes place is called the absolute speed.

The definition of the absolute speed was obtained using constant speed motions. These constant speed motions allowed us to quickly see that the absolute velocity exits, but do not limit the absolute velocity to only constant speed motions. Of course, all Galilean frames of reference will display these constant speed processes as processes of constant speed. Further, if all reference frames are to be of equal status then observers in all Galilean reference frames must share the $dE = 0$ constant speed motion equivalently. Furthermore, each Galilean observer will have the same value for the absolute speed or else one of the frames will enjoy a privileged nature. This is Einstein's postulate concerning the speed of light. Here, though, there is much more to the picture. First, the absolute velocity is not limited to constant speed motions and, second, the propagation velocity of light was never needed in showing the absolute velocity existed. This means the absolute velocity is more fundamental than Einstein's constancy of the speed of light and Einstein's postulate now denotes a subset of the physical phenomena contained within the first two laws of thermodynamics.

Just as the absolute temperature in classical thermo-dynamics is a limiting quantity we may suspect that the absolute speed will also turn out to be a limiting quantity. Because of our experimental evidence that the speed of light behaves as a limiting speed when electromagnetic forces are involved and the absolute speed is independent of the force or type of system and is therefore unique, it must be the speed of light. Thus, the first two laws require Einstein's postulate concerning the speed of light.

To be more specific, the absolute speed is unique for all Galilean frames of reference. There is one such speed already known and that speed is the speed of light, c. Therefore, the absolute speed must be the speed of light and the same for all Galilean observers. This is Einstein's postulate.

Mechanical Entropy.

In a system of two independent variables, all states accessible from a given initial state by reversible E-conservative motions lie on a $\sigma(u,q)$ curve. The entire (u,q) space may be thought of as being filled by many non-intersecting curves of this kind, each corresponding to a different value of σ. In a reversible non-E-conservative motion involving a transfer of energy $đE$, a system in a state represented by a point lying on a surface σ will change until its state point lies on a nother surface $\sigma+d\sigma$. Then $đE = \lambda d\sigma$, where $1/\lambda$, the integrating factor of $đE$, is given by $\lambda = \phi(u) f(\sigma)$, and therefore $đE = \phi(u) f(\sigma) d\sigma$ or

$$\frac{đE}{\phi(u)} = f(\sigma) d\sigma. \tag{5.21}$$

Since σ is an actual function of u and q, the right-hand member is an exact differential, which may be denoted by dS; and

$$dS = \frac{đE}{\phi(u)} \tag{5.22}$$

where S is the mechanical entropy of the system and the motion is a reversible one.

The Dynamic Theory's Second Law may be used to prove the equivalent of Clausius' theorem, which is stated here.

114

Theorem: In any cyclic transformation throughout which the speed is defined, the following inequality holds: $\int \frac{\text{đ}E}{\phi(u)} \leq 0$, where the integral extends over one cycle of the transformation. The equality holds if the cyclic transformation is reversible. Then for an arbitrary transformation $\int_A^B \frac{\text{đ}E}{\phi(u)} \leq S(B) - S(A)$, with the equality holding if the transformation is reversible. The proof of this theorem will now be given.

The proof of this statement may be seen by letting R and I denote respectively any reversible and any irreversible path joining A to B, as shown in Figure 5. For path R the assertion holds by definition of S. Now consider the cyclic transformation made up of I plus the reverse of R. From Clausius' theorem

$$\int_I \frac{\text{đ}E}{\phi} - \int_R \frac{\text{đ}E}{\phi} \leq 0, \qquad (5.23)$$

or

$$\int_I \frac{\text{đ}E}{\phi} - \int_R \frac{\text{đ}E}{\phi} \equiv S(B) - S(A). \qquad (5.24)$$

Another result of the Second Law is that the mechanical entropy of an isolated ($\text{đ}E = 0$) system never decreases. This can be seen since an isolated system cannot exchange energy with the external world because $\text{đ}E = 0$ for any transformation. Then by the previous property of the entropy,

$$S(B) - S(A) \leq 0 \qquad (5.25)$$

where the equality holds if the transformation is reversible.

Thus, the Second Law provides an answer to the question that is not contained within the scope of the First Law: "In what direction does a process take place?" The answer is that a process always takes place in such a direction as to cause an increase of the mechanical entropy in the universe. In the case of an isolated system, it is the entropy of the system that tends to increase.

Three Dimensional Space

It may thought that since the above arguments were made in the simple case of considering only one space dimension that the arguments concerning the integrating factor do not carry over into the three dimensional world. The proof that they do is given below.

With one work term the differential of the entropy was written as

$$dS = \frac{đE}{\phi} = f_i d\sigma_i. \tag{5.26}$$

Then if for each dimension the exchange of energy is denoted by $đE_j$, then

$$dS_i = \frac{đE_i}{\phi_i} = f_i d\sigma_i, \tag{5.27}$$

where there is no summation intended for $f_i d\sigma_i$. Since each dS_i is a p erfect differential, then the total change in mechanical entropy may be written as

$$dS = \Sigma_i dS_i = \Sigma_i \frac{đE_i}{\phi_i} = \Sigma_i f_i d\sigma_i. \tag{5.28}$$

However, the question which arises is whether there exists a single integrating factor ϕ such that

$$dS = \frac{đE}{\phi} = \Sigma_i \frac{đE_i}{\phi_i} = \Sigma_i f_i d\sigma_i. \tag{5.29}$$

To see this consider the element of work considered before as

$$đW = \Sigma_i F_i dq^i ; \quad i = 1,...,n. \tag{5.30}$$

Since each dU_i is in itself a perfect differential, then $dU = \Sigma_i dU_i$ so that

$$đE = \Sigma_i dU_i - \Sigma_i F_i dq^i = \Sigma_i (dU_i - F_i dq^i) \tag{5.31}$$

The Dynamic Theory

Or $\text{đ}E = \sum_i \text{đ}E_i$. If the system is total E-conservative in the sense that $\text{đ}E = \sum_i \text{đ}E_i = 0$, then $\text{đ}E = 0$ is a Pfaffian differential equation. This equation is integrable and has an integrating factor ϕ. The integrability is guaranteed by the Second Law since it is impossible to go from one initial state to <u>any</u> neighboring state. Then, just as in the one dimensional case, the perfect differential follows:

$$dS = \frac{\text{đ}E}{\phi} = \sum_i \frac{\text{đ}E_i}{\phi}. \tag{5.32}$$

But since

$$\text{đ}E = \sum_i \phi_i f_i d\sigma_i, \tag{5.33}$$

then

$$dS = \sum_i \phi_i \frac{f_i}{\phi} d\sigma_i. \tag{5.34}$$

Now following the same argument presented above concerning the composite system, $\text{đ}E = \lambda d\sigma$ where σ is a function of all the σ_i and the u_i. Therefore, since $\text{đ}E_i = \lambda_i d\sigma_i$, then

$$\text{đ}E = \sum_i \lambda_i \left[\frac{\partial \phi_i}{\partial \sigma} d\sigma + d\sigma_i \right]. \tag{5.35}$$

Now

$$d\sigma = \sum_i \left[\frac{\partial \phi}{\partial u^i} du^i + \frac{\partial \lambda}{\partial \sigma_i} d\sigma_i \right] \tag{5.36}$$

so that $\text{đ}E = \sum_i \text{đ}E_i$ or $\lambda d\sigma = \sum_i \lambda_i d\sigma_i$ and

$$d\sigma = \sum_i \left[\frac{\lambda_i}{\lambda} \right] d\sigma_i. \tag{5.37}$$

It follows that the $\partial\sigma / \partial u^i = 0$ and that the ratios λ_i/λ are also independent of the q^i. Therefore the λ's have the form $\lambda_i = \phi f_i$ and $\lambda = \phi F(\sigma_1, \sigma_2, ..., \sigma_n)$ and also

$$dS = f d\sigma = \Sigma_i F \left[\frac{\lambda_i}{\lambda} \right] d\sigma_i$$

$$= \Sigma_i \left[\frac{\lambda_i}{\phi} d\sigma_i \right] = \Sigma_i f_i d\sigma_i.$$

(5.38)

The right hand side is a perfect differential and therefore so is the left.

Since each λ_i/f_i is an integrating factor and λ/F is also an integrating factor, it follows that $\phi(u_1, u_2, ..., u_n)$ is an integrating factor for the $đE$ as well as for $đE = \sum_i đE_i$. Therefore

$$dS = \frac{đE}{\phi} = \Sigma_i \frac{đE_i}{\phi}.$$

(5.39)

This shows that the form of the integrating factor is independent of dimensionality.

The thermodynamic basis for the constancy of the speed of light provides the starting point for a thermodynamic basis for Einstein's special theory of relativity with its space-time metric and, ultimately, quantum mechanics. The starting point in this chain of logic is the adoption of the classical thermodynamic laws and their use to define the mechanical entropy and arrive at the maximum entropy principal for isolated mechanical systems. The ability of these laws to require the constancy of the speed of light displays the connection to relativistic theories. However, we now need to display the logic through to the development of mechanics. There will be numerous points in this logic where one could branch off on interesting and useful topics. This chapter will stay strictly on the track of presenting the logic from the adopted laws through relativistic mechanics to quantum mechanics.

118

Entropy and Geometry

It was previously shown that the Second Law guarantees the existence of an integrating factor for the First Law where the integrating factor is a function of only the speed. The Second Law also requires that the entropy must be maximized for an isolated system for which $dE = 0$. This principle becomes a variational principle for isolated systems. Now we need to look into how these requirements impact the equation of motion we saw in the First Law. Plus, we need to establish how the integrating factor depends upon the velocity.

In thermodynamics when one seeks to learn about the stability of a system it is the stability conditions to which one may turn. The stability conditions basically look around an equilibrium point to see how the system may go if it is displaced from this equilibrium. The stability conditions may be stated in terms of several sets of variables. Each set of variables used in the stability conditions are equally valid, but lead us to different looking conditions that have differing degrees of difficulty in assessing the stability of the equilibrium.

To provide an example of how using different variable in the stability conditions may offer different degrees of difficulty suppose one of the variables was the velocity. The velocity is the time differential of the change of position. The stability conditions are second order differentials. This means that a metric obtained using the velocity as one of the variables would result in third order differential equations with respect to time. Third order differential equations are difficult, if not impossible, to solve.

After looking at various sets of variables to use in the stability conditions I found that the choice of space and entropy lead to a metric that produced equations of motion which would provide an expression for the velocity dependence of the integrating factor. Though it may seem that using space and entropy in the stability conditions is odd, the results will justify their use.

A metric may be obtained by considering the stability conditions when q and S are taken as the independent variables.

For example, consider the terms of second order in small displacements beginning with the general condition

$$\delta U - F\delta q - \phi\delta S > 0. \tag{5.40}$$

Choose $U = U(q,S)$, which, because of the identity

$$\phi dS = dU - Fdq$$

is a natural choice for the independent variables, and expand δU in powers of the δq and δS

$$\delta U = \phi\delta S + F\delta q$$

$$+ \frac{1}{2}\left[\frac{\partial^2 U}{\partial q^2}\right]\delta q^2 + 2\left[\frac{\partial^2 U}{\partial q\partial S}\right]\delta q\delta S + \left[\frac{\partial^2 U}{\partial S^2}\right]\delta S^2 \tag{5.41}$$

$$+\ terms\ of\ third\ order...$$

The inequality (5.40) then shows that in Equation (5.41),

second order terms + third order terms + $\ldots > 0.$

Retaining only the second order terms, the criterion of stability is that a quadratic differential form be positive definite;

$$\frac{\partial^2 U}{\partial S^2}(dS)^2 + \frac{\partial^2 U}{\partial S\partial q^\alpha}(dS)(dq^\alpha)$$

$$+ \frac{\partial^2 U}{\partial q^\alpha \partial q^\beta}(dq^\alpha)(dq^\beta) > 0; \tag{5.42}$$

$$\alpha,\beta = 1,2,3.$$

Adopting this quadratic form as the metric of a general system whose thermodynamic variables are held fixed, we may then write this metric as

$$(ds)^2 = h_{ij}dq^i dq^j ; \quad i,j = 0,1,2,3, \tag{5.43}$$

where the summation convention is used and

$$h_{ij} = \frac{\partial^2 U}{\partial q^i \partial q^j}, \tag{5.44}$$

with $q^0 \equiv S/F_0$, the scaled mechanical entropy for dimensional correctness.

Thus, the stability conditions provide a metric in the four-dimensional manifold of space-mechanical entropy. The arc length s in the space-mechanical entropy manifold may be parameterized by choosing $ds \equiv u_0 dt \equiv cdt$, where $u_0 = c$ is the unique velocity appearing in the integrating factor of the second postulate. There are two reasons for choosing the unique velocity. First, it is the only well-defined velocity we have thus far. Secondly, we may look ahead to the metric of the Special Theory of Relativity. The metric may now be written as

$$c^2 (dt)^2 = h_{ij} dq^i dq^j ; \quad i,j = 0,1,2,3. \tag{5.45}$$

Now suppose the systems considered are restricted to only E-conservative systems. Then the principle of increasing mechanical entropy may be imposed in the form of the variational principle

$$\delta \int \sqrt{(dq^0)^2} = 0. \tag{5.46}$$

In order to use this variation principle, Equation (5.45) may be expanded, solved for (dq^0) and squared to arrive at the quadratic form

$$(dq^0)^2 = \frac{1}{h_{00}} \left[c^2 (dt)^2 + 2 h_{0\alpha} A dt dq^\alpha - h_{\alpha\beta} dq^\alpha dq^\beta \right], \tag{5.47}$$

where

$$A = \frac{h_{0\alpha}}{h_{00}} u^\alpha \pm \sqrt{ \frac{c^2}{h_{00}} + \frac{h_{\alpha\beta}}{h_{00}} u^\alpha u^\beta + \frac{h_{0\alpha}}{h_{00}} (u^\alpha)^2 }$$

with $u^\alpha \equiv dq^\alpha / dt$.

By defining $x^0 \equiv ct$, $x^\alpha \equiv q^\alpha$, $\alpha = 1,2,3$, then Equation (5.47) may be written as

$$(dq^0)^2 = \frac{1}{f} g'_{ij} dx^i dx^j ; \quad i,j = 0,1,2,3 \tag{5.48}$$

where $f \equiv h_{00}$. This metric obviously reduces, in the Euclidean limit o f constant coefficients, to the metric of Minkowski's space-time manifold of special relativity. It is interesting to note that in the metric of Equation (5.47) the difference in the sign on the time and space elements of the metric comes from the fact that the stability conditions are given in terms of space and mechanical entropy while the variational principle was taken to be the Entropy Principle. In this fashion the Second Law guarantees the limiting aspect found in Einstein's Special Theory of Relativity.

In his General Theory of Relativity, Einstein assumed the space-time manifold to be Riemannian. However, this assumption involves the a priori assumption that the scalar product be invariant. This assumption was later questioned by Weyl in his generalization of geometry. From the viewpoint that the adopted postulates should contain the other theories within them then it becomes desirable to determine whether or not these postulates specify the geometry of the $(dq^0)^2$ space-time manifold. More particularly do t he adopted postulates lead to a geometry that includes the geometry of current theories? To arrive at a more general geometry would not be a limitation for it would certainly include the others.

Recalling Equation (5.48), we can define

$$(dq^0)^2 = \frac{1}{f} g'_{ij} dx^i dx^j$$

$$= \frac{1}{f} (d\sigma)^2 \equiv g_{ij} dx^i dx^j. \tag{5.49}$$

Now the Second Law guarantees the existence of the function mechanical entropy and that dq^0 be a p erfect differential; therefore $dq^0 = q_i^0 dx^i$, where

$$q_i^0 \equiv \frac{\partial q^0}{\partial x^i}. \tag{5.50}$$

Then the exactness of dq^0 is stated by

The Dynamic Theory

$$q^0_{i|j} - q^0_{j|i} = 0. \qquad (5.51)$$

By defining the parallel displacement of a vector to be

$$d\zeta_i = \Gamma^v_{is} dx^s \zeta_v \qquad (5.52)$$

and using Equations (5.51) and (5.52) it may be seen that the connections must be symmetrical, or

$$\Gamma^v_{ik} = \Gamma^v_{ki}. \qquad (5.53)$$

This result should not be taken to mean that only symmetric connections need to be considered. R ather it means that given the g_{ij}'s that maximize $(dq^0)^2$, then the connections are symmetrical. However, since a v ariational principle must be used to determine the g_{ij} 's, then both symmetric and antisymmetrical connections will have to be considered.

In Weyl's generalization of geometry he found it necessary to assume the symmetry of the connections. He proved a theorem showing that the symmetry of the connections guaranteed the existence of a local Euclidean limiting manifold and used this theorem in support of the symmetry assumption. Here we find that the Second Law requires that the connections formed by the solution coefficients must be symmetrical thus guaranteeing, through Weyl's theorem, the existence of a local Euclidean geometry within the Dynamic Theory.

Suppose now we consider whether the order of differentiating the change in entropy makes any difference. This means that we must use symmetric connections since the actual change in entropy will be determined by the metric coefficients that generate a maximum. Therefore, consider the difference

$$\Delta (dq^0)^2 \equiv \frac{\partial^2 (dq^0)^2}{\partial x^i \partial x^j} - \frac{\partial^2 (dq^0)^2}{\partial x^j \partial x^i}. \qquad (5.54)$$

Since $\left(dq^0\right)^2 = q_i^0 q_j^0 dx^i dx^j$ from Equation (5.50), using Equation (5.49) we find $q_i^0 q_j^0 = g_{ij}$. Then

$$\frac{\partial(dq^0)^2}{\partial x^k} = [q^0_{j|k}q^0_{,i} + q^0_{i|k}q^0_{,j}]dx^i dx^j + (dq^0_{,k})^2. \quad (5.55)$$

Thus

$$\frac{\partial^2(dq^0)^2}{\partial x^k \partial x^l} = [q^0_{j|k|l}q^0_{,i} + q^0_{j|k}q^0_{,i|l}$$
$$+ q^0_{i|k|l}q^0_{,j} + q^0_{i|k}q^0_{,j|l}]dx^i dx^j \quad (5.56)$$
$$+ 2q^0_{l|k}q^0_{,i} + 2q^0_{k|l}q^0_{,k}.$$

Likewise

$$\frac{\partial^2(dq^0)^2}{\partial x^l \partial x^k} = [q^0_{j|l|k}q^0_{,i} + q^0_{j|l}q^0_{,i|k}$$
$$+ q^0_{i|l|k}q^0_{,j} + q^0_{i|l}q^0_{,j|k}]dx^i dx^j \quad (5.57)$$
$$+ 2q^0_{k|l}q^0_{,k} + q^0_{l|k}q^0_{,r}.$$

Therefore the difference must be

$$\Delta(dq^0)^2 = [(q^0_{j|k|l} - q^0_{j|l|k})q^0_{,i}$$
$$+ (q^0_{i|k|l} - q^0_{i|l|k})q^0_{,j}]dx^i dx^j. \quad (5.58)$$

Using the definition Equation (5.52) we see that

$$dq^0_{,i} = \Gamma^r_{is}dx^s q^0_{,r},$$
$$q^0_{i|k} = q^0_{,r}\Gamma^r_{ik}, \, also \quad (5.59)$$
$$q^0_{k|i} = q^0_{,r}\Gamma^r_{ki}.$$

Now

$$q^0_{\;i|k|l} = \frac{\partial}{\partial x^l}[\,q^0_{\;r}\Gamma^r_{\;ik}\,]$$

$$= q^0_{\;r|l}\Gamma^r_{\;ik} + q^0_{\;r}\frac{\partial \Gamma^r_{\;ik}}{\partial x^l}$$

$$= q^0_{\;s}\Gamma^s_{\;rl}\Gamma^r_{\;ik} + q^0_{\;r}\frac{\partial \Gamma^r_{\;ik}}{\partial x^l} \tag{5.60}$$

$$= q^0_{\;r}\left[\Gamma^r_{\;sk}\Gamma^s_{\;ik} + \frac{\partial \Gamma^r_{\;ik}}{\partial x^l}\right].$$

Similarly

$$q^0_{\;i|l|k} = q^0_{\;r}\left[\Gamma^r_{\;sk}\Gamma^s_{\;il} + \frac{\partial \Gamma^r_{\;il}}{\partial x^k}\right]. \tag{5.61}$$

Therefore

$$q^0_{\;j}[\,q^0_{\;i|k|l} - q^0_{\;i|l|k}\,] = q^0_{\;i}q^0_{\;r}\left[\begin{array}{c}\dfrac{\partial \Gamma^r_{\;ik}}{\partial x^l} - \dfrac{\partial \Gamma^r_{\;il}}{\partial x^k} \\ +\Gamma^r_{\;sl}\Gamma^s_{\;ik} - \Gamma^r_{\;sk}\Gamma^s_{\;il}\end{array}\right]. \tag{5.62}$$

Then defining the vector curvature as

$$R^r_{\;ilk} \equiv \frac{\partial \Gamma^r_{\;ik}}{\partial x^l} - \frac{\partial \Gamma^r_{\;il}}{\partial x^k} + \Gamma^r_{\;sl}\Gamma^s_{\;ik} - \Gamma^r_{\;sk}\Gamma^s_{\;il}, \tag{5.63}$$

the difference may be written as

$$\Delta(dq^0)^2 = [\,q^0_{\;j}q^0_{\;r}R^r_{\;ilk} + q^0_{\;i}q^0_{\;r}R^r_{\;jlk}\,]dx^i dx^j. \tag{5.64}$$

However, recall that $q^0_{\;i}q^0_{\;j} = g_{ij}$; then

$$\Delta(dq^0)^2 = [\,g_{jr}R^r_{\;ilk} + g_{ir}R^r_{\;jlk}\,]dx^i dx^j. \tag{5.65}$$

But $g_{ri} = g_{ir}$ and $R_{ijkl} = g_{ir}R^r_{\;jkl}$, so that

$$\Delta(dq^0)^2 = [\,R_{jilk} + R_{ijlk}\,]dx^i dx^j. \tag{5.66}$$

So the difference will vanish if $R_{jilk} = -R_{ijlk}$. Now since

$$(dq^0)^2 = q^0_{\;i}q^0_{\;j}dx^i dx^j = g_{ij}dx^i dx^j, \tag{5.67}$$

differentiation will result in

$$d(dq^0)^2 = d(q^0_{\;i}q^0_{\;j}dx^i dx^j) = d(g_{ij}dx^i dx^j) \tag{5.68}$$

or

$$d\,q^0{}_{,i}q^0{}_{,j}\,dx^i\,dx^j + q^0{}_{,i}\,dq^0{}_{,j}\,dx^i\,dx^j + q^0{}_{,i}q^0{}_{,j}\,d(\,dx^i\,dx^j\,)$$
$$= dg_{,ij}\,dx^i\,dx^j + g_{,ij}\,d(\,dx^i\,dx^j\,)$$
$$,(5.69)$$

which can be written as

$$\Gamma^r{}_{is}\,dx^s q^0{}_{,r}q^0{}_{,j}\,dx^i\,dx^j + q^0{}_{,i}\Gamma^r{}_{js}\,dx^s q^0{}_{,r}\,dx^i\,dx^j$$
$$+ q^0{}_{,i}q^0{}_{,j}\,d(\,dx^i\,dx^j\,) \tag{5.70}$$
$$= dg_{,ij}\,dx^i\,dx^j + g_{,ij}\,d(\,dx^i\,dx^j\,).$$

But $g_{ij} = q^0{}_{,i}q^0{}_{,j}$. Therefore

$$\Gamma^r{}_{is}\,dx^s g_{,rj} + \Gamma^r{}_{js}\,dx^s g_{,ir} = dg_{,ij} \tag{5.71}$$

or

$$g_{,rj}\Gamma^r{}_{is} + g_{,ri}\Gamma^r{}_{js} = \frac{\partial g_{,ij}}{\partial x^s} \tag{5.72}$$

and

$$\Gamma_{,jis} + \Gamma_{,ijs} = \frac{\partial g_{,ij}}{\partial x^s}. \tag{5.73}$$

Now interchange *jis* to *sij* to get

$$\Gamma_{,sij} + \Gamma_{,jsi} = \frac{\partial g_{,is}}{\partial x^j}. \tag{5.74}$$

Then interchange *jis* to *isj* so that

$$\Gamma_{,isj} + \Gamma_{,sij} = \frac{\partial g_{,si}}{\partial x^j}. \tag{5.75}$$

Add Equations (5.73) and (5.74) then subtract Equation (5.75) to get

$$\Gamma_{,sij} + \Gamma_{,jsi} + \Gamma_{,isj} + \Gamma_{,sij} - \Gamma_{,jis} - \Gamma_{,ijs} = \frac{\partial g_{,si}}{\partial x^j} + \frac{\partial g_{,js}}{\partial x^i} - \frac{\partial g_{,ij}}{\partial x^s}$$

or

$$\Gamma_{,sij} = \frac{1}{2}\left[\frac{\partial g_{,si}}{\partial x^j} + \frac{\partial g_{,sj}}{\partial x^i} + \frac{\partial g_{,ij}}{\partial x^s} \right] \tag{5.76}$$

and

$$\Gamma^r_{ij} = g^{rs}\Gamma_{sij} = \frac{g^{rs}}{2}\left[\frac{\partial g_{si}}{\partial x^j} + \frac{\partial g_{sj}}{\partial x^i} + \frac{\partial g_{ij}}{\partial x^s}\right]. \qquad (5.77)$$

Now by using the symmetry of g_{ij} it can be shown that

$$R_{jilk} = -R_{ijlk} \qquad (5.78)$$

and therefore $\Delta(dq^0)^2 = 0$.

Equation (5.78) states the necessary and sufficient condition that the differential entropy change may be transferred from an initial point to all points of the space in a manner that is independent of the path.

The distinguishing feature of Riemannian geometry is the invariance of the scalar product under vector transplantation. Therefore to determine whether the $(dq^0)^2$ space is a Riemannian space, consider the vector ξ_i and η_i. Now since $\xi_i = g_{ij}\xi^j$ and

$$d\xi_i = \Gamma^r_{is}dx^s\xi_r = \Gamma^r_{is}dx^s g_{rk}\xi^k$$
$$= \frac{\partial g_{ij}}{\partial x^s}\xi^j dx^s + g_{ij}d\xi^j, \qquad (5.79)$$

then

$$g_{ij}d\xi^j = \Gamma^r_{is}dx^s g_{rk}\xi^k - \frac{\partial g_{ij}}{\partial x^s}\xi^j dx^s. \qquad (5.80)$$

Or, since $g^{ij}g_{ij} = \delta^i_i = 1$ and

$$\frac{\partial g_{ij}}{\partial x^s} = \Gamma_{jis} + \Gamma_{ijs}, \qquad (5.81)$$

then

$$d\xi^j = g^{ij}[\Gamma_{kis}\xi^k - (\Gamma_{jis} + \Gamma_{ijs})\xi^j]dx^s$$
$$= g^{ij}(-\Gamma_{iks})\xi^k dx^s \qquad (5.82)$$
$$= -\Gamma^j_{ks}dx^s\xi^k.$$

Thus the change in the covariant and the contravariant vectors are given by

$$d\xi_i = \Gamma^r_{is}dx^s\xi_r$$
$$d\xi^i = -\Gamma^i_{rs}dx^s\xi^r. \qquad (5.83)$$

Now consider the change in the scalar product $\xi_i\eta^i$. Then

$$d(\xi_i\eta^i) = d\xi_i\eta^i + \xi^i d\eta_i$$

$$= \Gamma^r_{is}dx^s\xi_r\eta^i + \xi_i(-\Gamma^i_{rs}dx^s\eta^r) \qquad (5.84)$$

$$= \Gamma^r_{is}dx^s\xi_r\eta^i - \Gamma^i_{rs}dx^s\xi_i\eta^r.$$

Renaming the indices in the second term yields

$$d(\xi_i\eta^i) = (\Gamma^r_{is}\xi_r\eta^i - \Gamma^r_{is}\xi_r\eta^i)dx^s. \qquad (5.85)$$

Thus the geometry of the $(dq^0)^2$ manifold is Riemannian.

Next consider the question of what is the geometry of the $(d\sigma)^2$ space?

Equation (5.49) shows that we may write $(d\sigma)^2 = f(dq^0)^2$, which is reminiscent of Weyl's generalized geometry. Further we have

$$g'_{ij} = f\,g_{ij}. \qquad (5.86)$$

Then in the sigma space an arbitrary vector ξ^i would have a length l' given by the self-scalar product

$$l' = ||\xi||^2 = g'_{ij}\xi^i\xi^j - fg_{ij}\xi^i\xi^j = fl^2, \qquad (5.87)$$

where l is the length of the vector in the entropy space.

If we differentiate Equation (5.87), we have

$$2l'd'l = l'^2\frac{\partial f}{\partial x^i}dx^i + 2fldl. \qquad (5.88)$$

However, in the entropy space the length of the vector is unchanged under parallel displacement so that

$$dl' = \frac{1}{2}\frac{\partial f}{\partial x^i}dx^i\frac{l'}{f} = \frac{1}{2}\frac{\partial lnf}{\partial x^i}dx^i l'. \qquad (5.89)$$

Comparing Equation (5.89) with the definition of the parallel displacement of a vector, Equation (5.52), we find that

$$\phi_i = \frac{\partial lnf}{\partial x^i} \qquad (5.90)$$

plays a role similar to that of the connections Γ^i_{jk} in the definition of parallel displacement of a vector. Therefore

The Dynamic Theory

we shall define the change in the length of a vector under displacement to be

$$dl = (\phi_i dx^i)l. \qquad (5.91)$$

This is the same definition Weyl made in his generalization of geometry. However, there is a difference in the way it was obtained. Weyl chose this definition in analogy with the connections Γ and the definition then led to the second more general metric. In this theory the fundamental laws lead us to two metrics and Equation (5.91) for the change in the length of a vector under displacement. Therefore, we have no choice; the fundamental laws require it.

Here Equation (5.89) is a derived equation and Equation (5.90) only renames the logarithmic derivative.

Using Equation (5.91) we may obtain, in general,

$$dl^2 = 2l^2(\phi_i dx^i) = d(g_{ij}\xi^i\xi^j)$$
$$= g_{ij|k}\xi^i\xi^j dx^k + g_{ij}\Gamma^i_{lk}\xi^l\xi^j dx^k + g_{ij}\Gamma^j_{lk}\xi^i\xi^l dx^k. \qquad (5.92)$$

Renaming the various summation indices, rearranging terms, and using the length of a vector, we obtain

$$[g_{ij|k} + g_{lj}\Gamma^l_{ik} + g_{il}\Gamma^l_{jk}]\xi^i\xi^j dx^k = 2g_{ij}\phi_k\xi^i\xi^j dx^k. \qquad (5.93)$$

Since this must hold for arbitrary choice of ξ^i and dx^k, we conclude that

$$(g_{ij|k} - 2g_{ij}\phi_k) + g_{lj}\Gamma^l_{ik} + g_{li}\Gamma^l_{jk} = 0. \qquad (5.94)$$

This is the same system of linear equations for the connections Γ^i_{jk} as Equation (5.81) except that the inhomogeneous term $g_{ij|k}$ has now to be replaced by $g_{ij|k} - 2g_{ij}\phi_k$. Therefore the same linear algebra as before leads to

$$\Gamma^i_{jk} = -\begin{pmatrix} i \\ jk \end{pmatrix} + g^{li}[g_{lj}\phi_k + g_{lk}\phi_j - g_{jk}\phi_l] \qquad (5.95)$$

where $\binom{i}{jk}$ is the usual Christoffel symbol of the second kind.

Fundamental Laws

Now, since the entropy space is Riemannian, then in the entropy space we have $\phi_k = 0$ and $\Gamma''_{jk} = -\binom{i}{jk}$ and the length l of a vector is unchanged under parallel displacement. However, the same displacement law in the sigma space, with metric g'_{ij}, leads to the relation

$$dl' = \pm d\sqrt{g'_{ij}\xi^i\xi^j} = \pm d\sqrt{fg_{ij}\xi^i\xi^j}$$

$$= l\frac{\partial\sqrt{f}}{\partial x^k}dx^k \qquad (5.96)$$

$$= \pm\frac{l}{2}\frac{\partial lnf}{\partial x^k}dx^k l'.$$

Thus, $\pm\dfrac{1}{2}\dfrac{\partial \ln f}{\partial x^k}$ plays the role of ϕ_k in Equation (5.91). It follows then that the ordinary connections $-\binom{i}{jk}$ constructed from $\hat{g}_{ij}$ are equal to the more general connections Γ''_{jk} constructed according to Equation (5.96) from g_{ij} and $\phi_k = \dfrac{1}{2}\dfrac{\partial \ln f}{\partial x^k}$:

$$\Gamma''_{jk} = \Gamma'_{jk}. \qquad (5.97)$$

This can also be seen by direct computation from Equation (5.96) and

$$g'_{ij} = fg_{ij}. \qquad (5.98)$$

We may interpret the change of metric from g_{ij} to g'_{ij} by Equation (5.98) as a change of scale for the length at every point of the Riemannian manifold by the variable gauge factor f. This transformation is called a gauge transformation, and ϕ_k is called a gauge vector field.

The generalized geometry thus separates the problem of measurement of angles from that of measurement of length. For instance, the angle between the two vectors ξ^i and η^i at a given point of the space is measured by the ratio

The Dynamic Theory

$$\frac{\xi^i \eta_i}{\|\xi\|\,\|\eta\|} = \frac{g_{ij}\xi^i\xi^j}{\sqrt{(g_{ij}\xi^i\xi^j)(g_{ij}\eta^i\eta^j)}}. \tag{5.99}$$

This ratio does not change under the gauge transformation Equation (5.98). The gauge transformation is therefore an angle preserving, or conformal, change of metric. On the other hand, the length of vectors will change under Equation (5.98) according to Equation (5.91). Thus the metric tensor g'_{ij} determines angles, while one needs also the gauge vector ϕ_k to measure length.

Considering the sigma space, which is characterized by the tensor field g'_{ij} and gauge vector ϕ_k. The same argument as before shows that we may, without changing the intrinsic geometric properties of vector fields, replace the geometric quantities by use of a scalar field f as follows:

$$g'_{ij} = fg_{ij}, \quad \phi'_k = \phi_k + \frac{1}{2}\frac{\partial \ln f}{\partial x^k}, \quad \Gamma''^{i}_{\ jk} = \Gamma'^{i}_{\ jk}. \tag{5.100}$$

That is, in the new metric, vectors will have the same law of affine transplantation and the angle between different vectors at the same point of the manifold will be preserved, but the local lengths of a vector will be changed according to

$$l'^2 = fl^2. \tag{5.101}$$

Thus the general Weyl geometry of the sigma space admits also a conformal gauge transformation.

Entropy and Equations of Motion

Classical mechanics describes the motion of a system, which could be a particle, for which the energy of the system is a constant. The equations of motion yield trajectories resulting from the action of forces; they may also be obtained from the Principle of Least Action. When the action integral is treated as a variational problem with variable end points, the method of Lagrangian multipliers yields the same equations as does Hamilton's Principle. However, if the variational problem is transformed to a new space in which the new variational problem has fixed end points, then the metric for this space is displayed, and the equations of motion are geodesics in this space.

In classical mechanics the Principle of Least Action as formulated by Lagrange has the integral form

$$A = \int_{P_1}^{P_2} m\bar{v} \bullet d\bar{s}. \tag{5.102}$$

In curvilinear coordinates the integral assumes the form

$$A = \int_{P_1}^{P_2} mg_{\alpha\beta} \frac{dx^\alpha}{dt} dx^\beta = \int_{t(P_1)}^{t(P_2)} mg_{\alpha\beta} \frac{dx^\alpha}{dt} \frac{dx^\beta}{dt} dt,$$

where $\alpha,\beta = 1, 2, 3$.

Defining

$$T = \frac{m}{2} g_{\alpha\beta} \frac{dx^\alpha}{dt} \frac{dx^\beta}{dt},$$

the integral becomes

$$A = \int_{t(P_1)}^{t(P_2)} 2T dt. \tag{5.103}$$

Then the principle of least action may be stated as:

Of all curves C' passing through P_1 and P_2 in the neighborhood of the trajectory C, which are traversed at a rate such that, for each C', for every value of t, $T+V=F$, that one for which the action integral A is stationary is the trajectory of the particle.

The transformation of variables may be carried out to display the metric

$$(ds)^2 = h_{\alpha\beta} dx^\alpha dx^\beta \tag{5.104}$$

where $h_{\alpha\beta} = 2m(E_0-V)g_{\alpha\beta}$. Here different particles in the same field and with different energies E_0 would appear to have different geometries, a s ituation which has been previously taken to be impossible and therefore precluded thc gcometrization of dynamics(see page 6 of ref. 46). However, in view of Weyl's generalization of geometry,

132

treating the variational problem in the Principle of Least Action as transformed to a new space in which the variational problem has fixed end points, in effect, is a transformation into a space with Weyl geometry where the gauge function is $2m(E_0-V)$. Thus changing the energies does not change the geometry since it will still be a Weyl space.

Suppose now that the concepts of classical mechanics are compared with the concepts from the point of view of the Dynamic Theory. The energy of the system in classical mechanics is a constant of the motion and therefore the change in kinetic energy is the negative of the change in potential energy, which may be written as

$$dH = dT + dV = 0.$$

However, for classically conservative forces dH is a perfect differential. Therefore for this system with only one work term the force is a function of position only.

This suggests the association of the classical energy of the system, H, with the system energy, U, which is also a perfect differential. Now if the system is isolated, or E-conservative, then

$$0 = đE = dU - Fdq.$$

But if $dU=dH=0$ then F must be zero. This situation points out an important difference between classical physics and the Dynamic Theory. A classically conservative system is one for which the system's energy is a constant of the motion. However, the E-conservative system, within the Dynamic Theory, is one for which $đE = 0$. Thus an E-conservative system which is also conservative in the classical sense must have no forces. This means that the First Law requires all forces to be velocity dependent forces.

Suppose we now turn our attention to the mechanics of special relativity. In the special theory of relativity Einstein sought to put Newtonian mechanics into a form which would leave the speed of light invariant. The resulting dynamics exhibits the notion of a unique velocity in a similar sense to the previously defined absolute velocity.

Within the Dynamic Theory we may display the appearance of the special theory's foundations by using the generalized entropy principle rather than being required to assume the existence of Newton's equations of motion on an <u>a priori</u> basis.

Newtonian mechanics is displayed in its simplest form for particles, so we shall make the restrictive assumption that the mass density, γ_0, such that

$$\int \gamma_0 d(vol) = m_0$$

We will also assume that the g_{ij} are constants, thus

$$(d\sigma)^2 = c^2(dt)^2 - g_{\alpha\beta} dx^\alpha dx^\beta, \quad \alpha,\beta = 1,2,3,$$

and $g_{\alpha\beta} = \delta_{\alpha\beta}$. Our variational problem depends upon the integral

$$S = \int_0^S dS = \int_0^S \sqrt{(dS)^2} = \int_0^{mq^0} \sqrt{m_0{}^2 (dq^0)^2} = \int_{\sigma_1}^{\sigma_2} \sqrt{m_0^2 f g_{jk} u^j u^k} \, d\sigma,$$

where we have used the definition $u^j = dx^j / d\sigma$.

Because we have assumed that the mass, m_0, is a constant we can write our integral as

$$q^0 = \frac{S}{m_0} = \int_{\sigma_1}^{\sigma_2} \sqrt{f g_{jk} u^j u^k} \, d\sigma.$$

We can make a change of variables by letting

$$ds = \sqrt{g_k u^j u^k} \, d\sigma, \qquad (5.105)$$

so that the integral becomes

$$q^0 = \int_{S_1}^{S_2} \sqrt{f} \, dS.$$

We can now define a new function

$$T = \frac{f}{2} m_0 c^2,$$

and then consider a further change in variables such that

$$dS = c\sqrt{f}, \quad d\tau = \sqrt{\frac{2T}{m_0}} \, d\tau.$$

If we substitute this new variable into our integral we find

$$m_0 c q^0 = \int_{\tau(P_1)}^{\tau(P_2)} 2T \, d\tau, \qquad (5.106)$$

with the auxiliary selection that $2T - m_0 c^2 f = 0$.

The problem of determining geodesics has now been converted into a statement of the principle of maximum generalized entropy:

> Of all curves C', passing through P_1 and P_2 in the neighborhood of the trajectory C, which are traversed at a r ate such that, for each C', and for every value of τ, $2T = m_0 c^2 f$, that one for which the generalized entropy integral, q^0, is maximum (stationary) is the trajectory of the particle.

When stated in the form of a variational equation, this principle reads

$$\delta \int_{\tau(P_1)}^{\tau(P_2)} 2T \, d\tau = 0, \qquad (5.107)$$

with the auxiliary condition $2T - m_0 c^2 f = 0$, on C'.

The dynamics is unaffected by the addition of a constant to the gauge function; therefore, let

$$V = h - m_0 c^2 \frac{f}{2},$$

where h is a constant. The auxiliary condition now reads

$$T + V - h = 0, \text{ on } C'.$$

We can solve this variational problem by making use of the Lagrangian multiplier method for a problem with constraints. We construct a function $G = 2T + \lambda\varphi$, where $\varphi = T + V - h = 0$, and determine the solution of the system of equations

$$\frac{\partial G}{\partial x^j} - \frac{d}{d\tau}\left[\frac{\partial G}{\partial u^j}\right] = 0, \quad j = 0, 1, 2, ..., n, \qquad (5.108)$$

$$with \, T + V - h = 0.$$

Fundamental Laws

This system has a solution for which $\lambda=-1$, and it follows that the trajectory C is determined by the solution of the system

$$\frac{d}{dt}\left[\frac{\partial T}{\partial u^j}\right]-\frac{\partial T}{\partial x^j} = -\frac{\partial V}{\partial x^j}, \quad j=0,1,2,...,n. \quad (5.109)$$

We assumed that the g_{jk} were constants; therefore, if we make the definition

$$F_j = -\frac{\partial V}{\partial x^j},$$

then the equations in Equations (5.109) become

$$\frac{d}{d\sigma}\left[m_0 c^2 g_{jk} u^k\right] = F_j,$$

because $d\sigma=cd\tau$ and $\tau=m_0 g_{jk} u^j u^k$. If we multiply these equations by g^{li} and sum them, we obtain

$$\frac{d}{d\sigma}\left[m_0 u^l\right] = g^{lj} F_j = F^l, \quad l=0,1,2,...,n. \quad (5.110)$$

From the metric with constant coefficients we get

$$\frac{d\sigma}{dt} = \sqrt{c^2 - g_{\alpha\beta}\dot{x}^\alpha \dot{x}^\beta}, \quad \alpha,\beta=1,2,3,$$

or

$$\frac{d\sigma}{dt} = \sqrt{c^2 - v^2}. \quad (5.111)$$

Substituting Equation (5.111) into Equation (5.110) we find that

$$F^l = \frac{1}{\sqrt{1-\beta^2}}\frac{d}{dt}\left[\frac{m_0 v^l}{\sqrt{1-\beta^2}}\right], \quad (5.112)$$

where $\beta=v/c$.

Because $\beta=0$ in a local coordinate system x,

$$F^l = m_0 \frac{d^2 x^l}{dt^2} = m_0 a^l, \quad (5.113)$$

where, in a local reference frame x, $a^l=(1/c^2)(d^2 x^l/dt^2)$. In Equation (5.113) we have the form of Newton's second law in classical mechanics.

The Dynamic Theory

We may rewrite Equation (5.112) in the form

$$\sqrt{1-\beta^2}\,F^l \equiv f^l = \frac{d}{dt}\left[\frac{m_0 v^l}{\sqrt{1-\beta^2}}\right]. \qquad (5.114)$$

These equations are the equations of motion of the special theory of relativity and come from the geodesic equations of the variational problem, in Equation (5.106), based upon the generalized entropy principle with the restrictive assumptions that the mass density, γ, be a constant and that the metric coefficients, g_{jk}, are independent of the mass distribution. Thus, we have shown the Second Law of thermodynamics requires a relativistic modification to the First Law in that the it s pecifies the integrating denominator that must be used to convert the path dependent First Law to a path independent change in entropy. The Second Law then specifies the functional form of the function of velocity that must exist within the force of the First Law. In addition we see that the special relativistic equations of motion are required by the generalized entropy principle, but that they represent a subset of the First Law.

What is real?

This point might be an appropriate one to introduce the question of what one considers real in terms of space, time and motion. When Einstein proposed his special theory the scientific community was forced to consider new concepts of time dilation and space contraction. Of course equations similar to Einstein's had been introduced by Lorentz, with the special theory of relativity came the concept of space and time behaving is almost inconceivable ways. The inconceivability of the behavior of space was worsened by Einstein's general theory of relativity. Now, not only was the scientific community wrestling with the concepts of time dilation and space contraction, but also they now had the concept that we lived in a curved space to deal with. That is, they were told that the influence of gravitating mass was to curve the space we lived in.

The concept of conservation of energy in the First Law allowed us to choose the space in which we were describing our motion. However, we then investigated how the Second Law modified and limited the First Law. We showed that for an isolated system the principle of maximizing entropy required us to use a manifold, or space, in which space was curved. This manifold was the Riemannian manifold of special relativity. We also introduced the transformation of the description of motion using the Principle of Least Action into a Weyl manifold where there exists a multiplicative gauge function which alters the gauge of the space and time variables. The gauge function is the $2m(E_0\text{-}V)$ seen in Equation (5.104). We further introduced the equivalent method of describing motion through the use of the Lagrange equations of Equation (5.109). If the reader became confused by the various requirements of the First and Second Laws in order for us to describe the motion allowed, it is understandable. A similar confusion may be expected when considering what is real. Where we started out with our choice of geometry in the First Law, this choice seemed to be twisted around by the Second Law which imposed the principle of maximizing the entropy for isolated systems.

Pausing here for the moment we may consider which statement of our problem of motion is real. The First Law is the statement of the conservation of energy and we may choose the space we use to describe this exchange of energy. Therefore, we should consider our choice of space as real. Or perhaps this space, which we are free to specify, might better be called the physical space. Turning to the principle of maximizing the entropy that comes from the Second Law we cannot claim that it says the same thing as the First Law or there would be no need of both laws. What the Second Law does through the entropy principle is to put the problem into a different manifold in order that we may find the solution. Is this manifold real? I would say the motion it helps us describe is real, but we should not expect the space in the First Law to be changed by this solution. Indeed it cannot be changed else the First Law would be changed. However, the First and Second Laws are telling us two different things even though at times they appear extremely close. In this case the physical space used in the First Law should be taken as the real space while the manifold demanded by the Second Law may be considered as the solution manifold. That is the solution manifold is one necessary in order to see the limitations and modifications that the Second Law makes upon the First Law.

What the two laws require is that we choose a space in which to state the First Law and the Second Law tells us what transformation we need to make in order to obtain a description of the motion. The laws do not require a real curved space, only that the solution space be curved. This is a very different point of view than that of Einstein's relativity where inertial mass in interpreted to increase with velocity and space is curved by gravity.

Perhaps it would be easier to understand the situation by saying that the First Law gives the equations of motion and the Second Law modifies them. Newton's equation of motion did not specify the force, but rather stated that given a force this is the motion that will occur. Here the First Law also says that given a force this is what the motion will be. However, here the Second Law jumps in and requires that whatever the force is it must be velocity dependent in that it must vanish at the absolute velocity. Further, the Second Law requires that the entropy must increase and this requirement means that time is not reversible within the equations of motions given by the First Law.

Energy Concepts

Newtonian and relativistic mechanics talk of potential and kinetic energy while classical thermodynamics, which forms the basis of the Dynamic Theory, contains concepts with units of energy such as entropy, enthalpy, and free energy. We may use these common fundamental principles within the Dynamic Theory to explore how the mechanical energy concepts fit among the general thermodynamic energy functions. It seems that this will be of more than a little benefit when trying to keep all of the energy based concepts in proper perspective.

First, let us recall the First Law, with n=1,

$$đE = dU - Fdx, \tag{5.115}$$

whereas the differential change in the generalized entropy is

$$dS = \frac{dU}{\phi} - \frac{Fdx}{\phi}, \tag{5.116}$$

where dS is a total differential. If we suppose that the system energy, U, may be a function of position, x, and the velocity, $v = dx/dt$, then we may write

$$dS = \frac{1}{\phi}\frac{\partial U}{\partial v}dv + \frac{1}{\phi}\frac{\partial U}{\partial x}dx - \frac{F}{\phi}dx. \tag{5.117}$$

Because dS is a total differential, then

$$\frac{\partial}{\partial v}\left\{\frac{1}{\phi}\left[\frac{\partial U}{\partial x} - F\right]\right\} = \frac{\partial}{\partial x}\left\{\frac{1}{\phi}\frac{\partial U}{\partial v}\right\}.$$

This requires that

$$-\frac{\frac{\partial \phi}{\partial v}}{\phi^2}\left[\frac{\partial U}{\partial x} - F\right] + \frac{1}{\phi}\frac{\partial}{\partial v}\left[\frac{\partial U}{\partial x} - F\right] = \frac{1}{\phi}\frac{\partial^2 U}{\partial x \partial v} \tag{5.118}$$

because $\phi = \phi(v)$ from the second law. Further, dU is a total differential so Equation (5.118) becomes

The Dynamic Theory

$$\frac{\dfrac{\partial F}{\partial v}}{\left[\dfrac{\partial U}{\partial x} - F\right]} = -\frac{\dfrac{\partial \phi}{\partial v}}{\phi} = \textit{function of velocity only.}$$

Now consider the functional form of the force from the equations of motion in Equation (5.114),

$$F = \sqrt{1 - \beta^2}\,\hat{F}(x) \equiv \Phi\hat{F}(x), \qquad (5.119)$$

where $F(x)$ is strictly a function of x because it came from the gauge function. Then

$$\frac{\partial F}{\partial v} = \frac{d\Phi}{dv}\hat{F}(x).$$

Thus, Equation (5.118) may be written as

$$-\frac{\dfrac{\partial \phi}{\partial v}}{\phi^2}\left[\frac{\partial U}{\partial x} - \Phi\hat{F}\right] + \frac{1}{\phi}\frac{\partial}{\partial v}\left[\frac{\partial U}{\partial x} - \Phi\hat{F}\right] = \frac{1}{\phi}\frac{\partial^2 U}{\partial x \partial v}. \qquad (5.120)$$

In order to satisfy Equation (5.120) we find $\Phi = \phi$ and $U = U(v)$.

By substituting these results into the differential expression for the entropy, Equation (5.116) we find

$$dS = \frac{\left(\dfrac{dU}{dv}\right)dv}{\sqrt{1 - \beta^2}} - \hat{F}(x)\,dx, \qquad (5.121)$$

which is a perfect differential whereby we have found that U is strictly a function of velocity.

Now consider the First Law for an isolated system, or

$$đE = 0 = dU - Fdx,$$

but, using Equation (5.119) this may be written as

$$dU = \sqrt{1 - \beta^2}\,F(x)\,dx.$$

Then by integrating we find that

$$U - U_o = \int_{P_0}^{P} F\,dx. \qquad (5.122)$$

Fundamental Laws

This is Einstein's energy integral which, because of the equations of motion, becomes

$$U = \frac{m_0 c^2}{\sqrt{1-\beta^2}} + constant. \qquad (5.123)$$

In his special theory of relativity Einstein interpreted the right-hand side of Equation (5.123) to be the kinetic energy; therefore, he chose the integration constant to be $-m_0c^2$ in order that $T=0$ when $v=0$. Here, Equation (5.123) is the energy of the system and, therefore, will not be zero when $v=0$. Thus, the constant of integration should be taken as zero, giving the energy by

$$U = \frac{m_0 c^2}{\sqrt{1-\beta^2}}. \qquad (5.124)$$

If we differentiate Equation (5.124) with respect to the velocity, we find

$$\frac{dU}{dv} = \frac{m_o v}{\left(1-\beta^2\right)^{3/2}}.$$

Substituting this result into Equation (5.121), the change in entropy becomes

$$dS = \frac{m_0 v dv}{(1-\beta^2)^2} - \hat{F}(x)dx. \qquad (5.125)$$

This expression may now be integrated to get the mechanical entropy as

$$S = \frac{\frac{1}{2}m_0 c^2}{(1-\beta^2)} + V(x) + constant.$$

By setting the constant of integration at $1/2\ m_0c^2$, we get

$$S = \frac{\frac{1}{2}m_0 c^2 \beta^2}{(1-\beta^2)} + V(x) \quad or \quad S = \frac{\frac{1}{2}m_0 v^2}{(1-\beta^2)} + V(x). \qquad (5.126)$$

Thus, the generalized entropy for a purely mechanical system has two parts. One, depending entirely

The Dynamic Theory

upon the velocity, and which we may call kinetic entropy, is given by

$$T = \frac{\frac{1}{2}m_0 v^2}{(1-\beta^2)}. \qquad (5.127)$$

The second term in the mechanical entropy is a function of position only and may be called the entropy potential, $V(x)$.

We may look at the kinetic entropy differently if we go back to the variable changes during the presentation of the maximum entropy principle, because there we had

$$ds = \sqrt{\hat{g}_{jk}u^j u^k}\,d\sigma = \sqrt{\frac{2T}{m_0}}\,d\tau,$$

but $d\sigma = cd\tau\sqrt{1-v^2/c^2}$ therefore,

$$T = \frac{1}{2}m_0 c^2 g_{jk}u^j u^k = \frac{1}{2}m_0 c^2 g_{jk}\frac{du^j}{d\sigma}\frac{du^k}{d\sigma}$$

$$= \frac{1}{2}m_0 c^2 g_{jk}\dot{x}^j \dot{x}^k \left(\frac{dt}{d\sigma}\right)^2.$$

but, by Equation (5.111) this becomes

$$T = \frac{\frac{1}{2}m_0 v^2}{(1-\beta^2)},$$

which is the kinetic entropy of Equation (5.127). Thus, we find that it is the mechanical entropy, S, that must have a constant value along any trajectory for an isolated system, because

$$T + V - h = 0$$

for the trajectory, and therefore, $S=h=$constant.

Thus we have established the following for the trajectories of an isolated system:

Mechanical Entropy : $\qquad S = h = $ constant

$$\text{Kinetic Entropy}: \quad T = \frac{\frac{1}{2}m_o v^2}{\left(1-\beta^2\right)}$$

$$\text{Potential Entropy}: \quad V(x) = -\int \hat{F}(x)\,dx$$

$$\text{Energy of the system}: \quad U = \frac{m_0 c^2}{\sqrt{1-\beta^2}}$$

$$\text{First Law}: \quad \mathrm{d}E = 0 = dU - F_j dx^j, \text{ or}$$

$$0 = \frac{m_0 v\,dv}{(1-\beta^2)^{\frac{3}{2}}} - \sqrt{1-\beta^2}\,\hat{F}_j dx^j$$

$$\text{Force}: \quad F_j = \sqrt{1-\beta^2}\,\hat{F}_j(x)$$

$$\text{Work done by the system}: \quad W = \int_{x_1}^{x_2} F_j dx^j$$

$$\text{Kinetic energy}: \quad T = m_0 c^2 \left[\frac{1}{\sqrt{1-\beta^2}} - 1\right]$$

$$\text{Rest energy of the system}: \quad U(v=0) = m_0 c^2$$

One of the benefits of the Dynamic Theory is the capability of using procedures currently used in one branch of physics in another where prior to the unification displayed here would have been thought impossible. A system in which this procedure should produce significant results is a non-equilibrium thermodynamic system. Thermodynamics tells us that we must minimize the free energy, but the ability to use this as a variational principle to obtain equations of motion is a procedure which the Dynamic Theory now makes possible for this thermodynamic system.

Non-Isolated System

Thus far we have consistently required the system to be isolated. Obviously there are a large number of physical phenomena for which this restriction may not be used, even as an approximation. Therefore, relaxation of this restriction should provide description of a large and important class of systems. The remainder of this book involves the investigation of the predictions of the Dynamic Theory assuming the system is isolated. This may give the implication the non-isolated system is less important or less interesting. This is not the intention of the presentation. Rather, the presentation is aimed at displaying the fact that existing theories are subsets of the Dynamic Theory. In order to do this we must stay with the assumption of the isolated system.

Conclusions

We have now shown how the first two laws of thermodynamics specify the geometry for our solution manifold of physical phenomena that obey these laws. The First Law gives the equations of motion wherein the physical geometry is free to be specified and the force must be given. The Second Law modified the form of the force that may be used in the First Law by requiring the force to be velocity dependent and vanishing as the velocity approaches the absolute velocity. The Second law also requires that the entropy never decrease for an isolated system. We found these requirements by considering the manifolds required by the stability conditions and the entropy principle. Indeed we found there were two types of geometrical solution manifolds required. The entropy manifold was shown to have a Riemannian geometry with the entropy playing the role of Einstein's proper time. This gives a very different interpretation of Arrow of Time for physical phenomena: the entropy principle requires that proper time never go backwards. The second geometrical solution manifold, the energy manifold, was shown to have Weyl's geometry, with a gauge function governing how the length of a vector is transported around in the space.

The following are among those conclusions that may be made from the forgoing:

1. The thermodynamic laws requires Einstein's postulate concerning the constancy of the speed of light as a subset of requiring an absolute, unique velocity,

2. The thermodynamic laws provide a metric through the stability conditions that leads to the space-time solution manifold of Einstein's special theory of relativity when the entropy principle is used,

3. The thermodynamic laws require not one, but two solution manifolds for a complete description of an event and the theory requires that one be a Riemannian and the other be a Weyl manifold,

4. The thermodynamic laws require that in the equations of motion of the First Law the force MUST be a function of velocity and that this function cause the force to vanish as the velocity approaches the absolute velocity,

5. The thermodynamic laws require that the interpretations be such that mass remains fixed and the physical space of the statement of conservation of energy First Law is not curved by gravitational mass, and

6. The thermodynamic laws require that the relativistic space-time be limited to the solution manifold and not to the real physical space of the First Law.

Chapter 6 Space-Time-Matter

In 1918 W eyl published his book t itled "Space-Time-Matter" in which he discussed Einstein's theories and his own unified theory of gravitation and electromagnetism. Yet, within the text, matter did not share the same role as space and time, though equivalence was implied by the title. Space and time were coordinates in the relativistic manifold while matter was not.

The theory proposed here does for mass what Einstein's Special Theory of Relativity did for time; mass also becomes a coordinate on an equal footing with space and time. Just as relativity created considerable conceptual difficulty, so also can the proposed theory be expected to create conceptual difficulty. However, if the unification provided by the theory is considered as justification for attempting to see what the theory might produce, then a fifth dimension with physical interpretation follows rather quickly.

The first restriction placed upon any system discussed thus far has been that of restricting the system to be either a thermodynamic system, with only a *pdv* work term, or a mechanical system with three *fdx* work terms. By removing this restriction, a system must be considered, which may be called a thermo-mechanical system that experiences four types of work. Thus, the first law includes four work terms and, therefore, involves five dimensions. Since the specific volume is the reciprocal of the mass density, the First Law may also be written in terms of the mass density. For such a system the coordinates become the specific entropy, mass density, and the three space variables.

If five dimensions, which include the mechanical and thermodynamic variables, seem odd c onsider how thermodynamics is taught. The First Law of Thermodynamics is written on the blackboard (well, it used to be a blackboard), equating the differential heat exchanged between the system and its surroundings with the differential change in the internal energy plus the differential work terms. In the work terms are the three mechanical work terms in addition to the thermodynamic work. It is then pointed out to the students that the right hand side of the equation has five independent variables and it is stated that five equations are needed which relate these variables in order to have a solvable system. Usually the first statement made at this point is that conservation of mass guarantees that the mass density may be written as a function of space and time and, therefore, only four additional equations are needed, which are stated as being the Equation of State and the three mechanical equations of motion, such as Newton's.

But what about the case when mass isn't conserved? Can mass density be written as a function of space and time for this case also? If it may not then the fundamental dimensionality of nature must be five dimensions. Where does this lead? It has already been shown that the stability conditions lead to metrics upon w hich the Entropy Principle works to provide equations of motion when the metric coefficients are assumed known and field equations for these coefficients when they are not known. Thus, it is necessary only to work out what the implications of the five fundamental dimensions would be, compare them to the existing theories in those regions of physical phenomena where the existing theories are known to work and see if there is some predictable critical experiment that may be conducted to test the new theory.

To begin the investigation of the implications of a five dimensional system first consider the system to be isolated. The principle of increasing entropy becomes effective and the equations of motion are given by the First

The Dynamic Theory

Law while the entropy principle requires the equations of geodesics in a five dimensional solution manifold of space, time, and mass density.

The First Law for five dimensions may be written as

$$\text{d}\tilde{E} \;=\; d\tilde{U} - \frac{P}{\gamma^2}d\gamma - F_\alpha dq^\alpha, \quad (\alpha = 1,2,3) \qquad (6.1)$$

where the tilde denotes specific quantities. The entropy variational principle becomes

$$\delta \int \sqrt{(dS)^2} = \delta \int \sqrt{(\gamma \, dq^0)^2} = \delta \int \gamma \sqrt{(dq^0)^2} = 0 \qquad (6.2)$$

where now q^0 is the specific entropy.

The system's specific energy is now given in terms of the five variables specific entropy, space, and mass density. The stability condition, and hence the metric, is then stated in terms of these same variables. The stability condition is stated as

$$h_{ii} dq^i dq^i = \frac{\partial^2 \tilde{U}}{\partial q^i \partial q^i} dq^i dq^i > 0 \; ; \; (i = 0,1,2,3,4) \qquad (6.3)$$

where $q^4 = \gamma/a_0$. The metric may then be written as

$$(ds)^2 \;=\; h_{ij} dq^i dq^j \; ; \quad i,j = 0,1,2,3,4. \qquad (6.4)$$

Equations of Motion

The equations of motion are obtained when it is assumed that the coefficients of the metric are given and one looks at the Euler equations giving the variations of the coordinates which satisfy the variational condition. By using the variational principle from Equation (6.2) one finds the force densities to be given by

$$F^i \;=\; \gamma f^i \qquad (6.5)$$

with

$$f^i = \frac{du^i}{dq^0} + \begin{pmatrix} i \\ lk \end{pmatrix} u^l u^k \qquad (6.6)$$

where u^i are the components of the five dimensional velocity vector, the $\begin{pmatrix} i \\ lk \end{pmatrix}$ are the Christoffel symbols, and f^i are the components of the five dimensional acceleration vector. Obviously, if the mass density is considered to be conserved such that $u^4=0$ and the system is near equilibrium so that a flat metric makes a good approximation, then the volume integral of Equation (6.5) becomes the force-mass-acceleration relation of Special Relativity. Therefore, Einstein's Special Theory is obtained within this theory by employing the restrictions of an isolated system near equilibrium, with conservation of mass.

It is interesting to note that the inertial mass density comes from the fact that the stability conditions are given in terms of specific quantities while the Entropy Principle is stated in terms of the entropy. This fact will take on an even more interesting character when we consider the comparison between inertial and gravitating mass.

Another interesting fact is that if the First Law is considered for an isolated system, one obtains

$$\text{đ}E = 0 = d\tilde{U} - \frac{p}{\gamma^2}d\gamma - \tilde{F}_\alpha dx^\alpha \;\; ; \;\; (\alpha = 1,2,3) \quad (6.7)$$

so that

$$d\tilde{U} = \tilde{F}_\alpha dx^\alpha \;\; ; \;\; (\alpha = 1,2,3,4) \quad (6.8)$$

The forces in Equation (6.8) are the velocity dependent forces found in Part 1. That is they are

$$\tilde{F} = \sqrt{1 - \beta^2} \; \hat{F}(x), \quad (6.9)$$

where β is the four dimensional version so that Equation (6.9) may be expanded to

$$\tilde{F} = \sqrt{1 - \frac{v^2}{c^2} - \frac{\dot{\gamma}^2}{a_o^2 c^2}} \; \hat{F}(x). \quad (6.10)$$

The Dynamic Theory

When the energy integral of Equation (6.8) is evaluated with the forces of Equation (6.10) one finds

$$\tilde{U} = \gamma c^2 + \frac{1}{2}\gamma v^2 + \frac{1}{2}\frac{\gamma}{(a_o)^2}(\dot{\gamma})^2 \qquad (6.11)$$

where it is assumed that the system is near rest. The energy density in Equation (6.11) includes the rest energy term because the integral requires it; not because of a constant of integration as in the special theory of relativity. Further, because the system was considered to be isolated, $đE = 0$, then the appearance of the rest energy term in the expression for the system specific energy brings with it some subtleties of interpretation not found in Einstein's special theory where energy and mass are equated one-to-one. For instance, the one-to-one correspondence between energy and mass exists only for resting mass when mass is conserved. Also notice that the special theory of relativity energy equivalence may exist only for isolated systems. Also, if we require the usual conservation of mass then $\dot{\gamma} \equiv d\gamma/dt = 0$ and Equation (6.11) reduces to the rest energy plus the classical kinetic energy.

It is interesting to see how the rest energy is required by the five dimensional energy integral. However, rest energy is a well worn concept and this discussion adds nothing substantially new. What is even more interesting in Equation (6.11) is the prediction of a new physical concept. The new concept is that of a limiting rate of mass conversion. For instance suppose the system under consideration is one that is setting still so that $v = 0$ and energy is neither added to nor removed from the system so that U remains constant. Then Equation (6.11) becomes

$$\tilde{U}_o = \gamma c^2 + \frac{1}{2}\frac{\gamma}{(a_o)^2}(\dot{\gamma})^2 .$$

But the limiting aspect of the forces in Equation (6.10) requires $\beta=1$ as a limit. In relativistic theories this means the limiting velocity is the speed of light. However, here our system is setting still so the limiting aspect of the Second Law requires that

$$\dot{\gamma} \leq a_o c. \tag{6.12}$$

This means that the Second Law requires a limiting rate of mass conversion and this limit is given by $a_o c$.

Weyl's Quantum Principle

Einstein's geometry

Stephen Hawking stated, in his book titled "A Brief History of Time," that it would be virtually impossible for someone to set down and come up w ith a Theory of Everything. By this he meant that the task of trying to come up with a simple set of fundamental principles that could be written on a envelope and from which one could then derive all the descriptions of the physical phenomena we observe in nature appears next to impossible. To do s o would mean being able to see what has not been seen before. How does one do that? However, there is an even more difficult task than the one discussed by Dr. Hawking. The more difficult task would be that of convincing the majority of today's physicists of the validity of a proposed Theory of Everything. One might offer several reasons why this might be so, however, it stems from the fact that all of these physicists are human. We humans have a difficult time embracing new thoughts, new ways of thinking or descriptions of things we have not seen before.

One of the best physicists at seeing things not seen before, Albert Einstein, also had difficulty seeing what he had not yet seen and this shows up in his refutation of Weyl's suggested extension of Einstein's general theory of relativity to include Maxwell's electromagnetism. It may be recalled that Einstein had proposed the expansion of the geometry in which we describe physical phenomena from Euclidean geometry to Riemannian geometry where

The Dynamic Theory

Einstein proposed that the curvature of space was the result of gravitation.

A clarification probably should be made here. Newton came up with laws of motion and forces that were to be used in whatever geometry one chose to use to describe the space in which they desired to work. This is to say that the forces were separate from the space and the space was free to be chosen. This space might properly be called the physical space because that is the space in which the forces are at work. What Einstein saw was quite different. First, he must have known that many scientists had shown that there existed the means of transferring the description of motion from one where there were equations of motion *in* a space, freely chosen, to a manifold, chosen by the Newtonian forces, and in which the forces bent, or curved, the manifold so that the path of minimum distance between two points was determined *by* the forces. These are two very different, but mathematically equivalent, concepts. It may be easier to 'see' forces when they are independent of the physical space we choose to use for a description of our world. Rather than to visualize the forces that warp a manifold that we don't even 'see.' Yet, it was this second concept that Einstein chose to work with.

This effort had taken Einstein several years of hard work and one can easily understand why it took this time simply by trying to understand, and thereby 'see', what he had achieved in his proposal. Hermann Weyl was one of those who readily understood one aspect of what Einstein saw. Weyl was a mathematician and knew that in Euclid's geometry a vector had two properties that were invariant. It was Euclidean geometry that was used as the basis for the description of the physical world before Einstein made his proposal of the general theory of relativity. Weyl must have seen that when Einstein relaxed the assumption of invariance of the one property of a vector, the self outer or cross product, the curvature of the resulting space gave

Einstein room in which to express the properties of gravity in his field equations. Keep in mind that Einstein's space is one in which the minimum distances between points describe the path a mass is expected to take when gravity is operating upon it. Weyl must have asked himself if the relaxation of the other invariant property, the self-inner, or dot, product, of the vector might be able to describe electromagnetism for that is what he did and proposed a unified theory based upon his work. Indeed he showed that the scale factor, or length of a vector, that results from the formation of the self inner product as it is transported around throughout space and time produced the equations that Maxwell had derived for the description of electromagnetism!

In order to lay the foundation of the argument and Einstein's comments to Weyl's work a brief discussion of the general theory of relativity is necessary. There are many good books that describe Einstein's work. However, suppose we approach the subject a little differently. As mentioned earlier Einstein looked at the removal of the invariance of the outer vector product in order to describe gravitational effects. So what is an outer product? Indeed, what is a vector?

Remembering that we are trying to see what Einstein saw, argues for visualization as much as possible. For the time being we shall talk of two-dimensional vectors that will allow them to be pictured on the two dimensional pages of this book. A vector is one of these

with a certain length and direction. If we wanted to write it out in terms of the two dimensions on this book page we might chose the up di rection to be given the name of y while the horizontal direction to the right might be given the name of x. Then the direction of the vector could also be given by up y units and right x units. Pythagoras told us centuries ago that for a right triangle the length,

154

The Dynamic Theory

hypotenuse, could be found by $l^2 = x^2 + y^2$, where l is the length of the vector. We might also write the length of the vector as

$$l = \sqrt{g_x x^2 + g_y y^2} \tag{6.13}$$

where the coefficients of the square of the two distances have been labeled with g_x and g_y to indicate that these coefficients may be constants or they may be considered to be functions of the positions x and y. To see how the vector changes as it moves around in the space, we need to differentiate its length, as

$$dl = \frac{\frac{1}{2}\left[\frac{\partial g_x}{\partial x} x^2 dx + \frac{\partial g_x}{\partial y} x^2 dy + \frac{\partial g_x}{\partial x} x^2 dx + \frac{\partial g_x}{\partial y} x^2 dy + 2 g_x x dx + 2 g_y y dy\right]}{\sqrt{g_x x^2 + g_y y^2}}. \tag{6.14}$$

The first four terms in the numerator of Equation (1.2) give the change in the vector as the coefficients change from position to position in the space. The last two terms give the change of the vector components. It may be noticed that in these two terms the change of the vector components are represented by a product of the position and the change of position. This product is called a linear form as the position only appears to the first power. This linear form becomes a guide in further development of geometry when the g's are allowed to vary around in the space.

Another way of looking at this change is to consider the vector as carried to a "neighboring" point without changing the values of its components: it is displaced *parallel to itself*. (As long as we use a coordinate system where the g_x and g_y are constants that statement has an invariant meaning, because the coefficients are the same at both positions.) Then this parallel displaced vector is compared with the value of the vector (as a function of the positions) at the same point. The difference is given a specific name, but contains the first four terms of Equation (6.14). As you might imagine, there is a whole field of

155

mathematics involved with this parallel displacement of vectors.

Einstein had developed his special theory of relativity by using constant coefficients, that is; constant g's, in a four-dimensional manifold consisting of time and the three dimensions of space. Following his presentation of his special theory, Einstein went on to consider whether or not variable coefficients might contain the information about the effects of gravity by forming a curvature in the four-dimensional manifold of space and time. He used the assumption of the equivalence of gravitating mass and inertial mass and proposed a certain set of equations whose solutions gave the functions that were to act as the coefficients.

Weyl looked at an additional consideration of the possible dependence of the coefficients upon the position of the vector in the space. He considered the possibility of a multiplicative functional term in all of the coefficients, such that the length of the vector might be written as

$$l' = \sqrt{fg_x x^2 + fg_y y^2} = \sqrt{f}\, l \qquad (6.15)$$

where the new vector has a length that is different from the original vector by the function $\sqrt{f}$. Weyl called this function the gauge function in reference to its control over the vector's length. If we now consider the change in the new vector's length, we find

$$dl' = \frac{\dfrac{1}{2}\left(\dfrac{\partial f}{\partial x} dx + \dfrac{\partial f}{\partial y} dy \right) l}{\sqrt{f}} + \sqrt{f}\, dl \qquad (6.16)$$

where the dl in the last term is the same total change in the vector length given in Equation (6.14). Suppose we consider the g's to be constants so we may limit o ur attention to the change in vector length caused by the gauge function f. Then, the last term vanishes and we are left with

The Dynamic Theory

only the first term and the change in vector length may now be written as

$$dl' = \frac{\frac{1}{2}\left(\frac{\partial f}{\partial x}dx + \frac{\partial f}{\partial y}dy\right)l}{\sqrt{f}}\frac{l'}{\sqrt{f}} = \frac{1}{2}\left(\frac{\partial f}{\partial x}dx + \frac{\partial f}{\partial y}dy\right)l'. \quad (6.17)$$

Now if we divide both sides by the vector length, and take notice of the total derivative on the right hand side, we obtain

$$\frac{dl'}{l'} = \frac{1}{2}\frac{\partial \ln f}{\partial x}dx + \frac{1}{2}\frac{\partial \ln f}{\partial y}dy \equiv \frac{\partial \phi}{\partial x}dx + \frac{\partial \phi}{\partial y}dy \quad (6.18)$$

where we have defined the logarithmic derivatives of the gauge function as the partial derivatives of the function ϕ just as Weyl did. We will also follow Weyl's lead and call the partial derivatives on the right hand side of the equation the gauge potentials to denote their role in the changing of the vector length around in the space. Now let's integrate Equation (6.18), to get

$$\int \frac{dl'}{l'} = \int \left(\frac{\partial \phi}{\partial x}dx + \frac{\partial \phi}{\partial y}dy\right) \quad (6.19)$$

and we come to the point where Einstein arrived at when looking over Weyl's proposal. The right hand side of Equation (6.19) is a well-known integral in the world of mathematics. It is called a line integral and may be thought of as having three parts; first it h as an integrand (in this case the integrand consists of the partial derivatives), second it must have a path along which to evaluate the integral (in this case the path would be given by specifying the dx and dy) and the third part is the answer. The path dependence of the answer stems from the need to specify a path in order to evaluate the integral. We can, however, integrate the left hand side of Equation (6.19) to get

$$l'_f = l'_o e^{\int \left(\frac{\partial \phi}{\partial x}dx + \frac{\partial \phi}{\partial y}dy\right)}. \quad (6.20)$$

In Equation (6.20) we see that the coefficient with the exponent which is the line integral of the gauge potentials acts as a scale factor for the space. If the exponent is zero the length of the vector does not change. This would mean that the scale of the space would be unity, or one. This scale factor will be shown later to be significant when we desire to consider atomic and nuclear phenomena.

Einstein thought he saw a problem within Weyl's variation of the scale factor that rested upon the feature of a path dependence of the scale factor. Einstein argued that if the electrostatic potential is derived from a path dependent scale factor wouldn't this path dependence also show up in the properties of things with an electric charge, even the simple hydrogen atom? Experimental evidence rested upon Einstein's side of this argument for all hydrogen atoms demonstrated the same features independent of what they had undergone.

Schrödinger's observation and London's proof

In 1922 Schrodinger noticed and reported what he felt was a remarkable property of quantum orbits of a single electron atom. What he noticed was that for systems satisfying the Bohr-Somerfield quantization conditions, the exponent of the non-integrable, path dependent, Weyl scale factor became quantized and the scale factor was unity. Schrödinger did not mention this remarkable property four years later when he developed and reported his wave equation as a foundation for the quantum mechanics stemming from his equations. London was aware of Schrödinger's 1922 paper because he wrote to Schrödinger in 1926 w ishing to know if this was the same Mr. Schrödinger who had written the 1922 pa per. Whether Schrödinger's paper was the reason London turned his attention to Weyl's scale factor is not known, but it did capture his attention along with Schrödinger's remarkable property. London established the relation between Weyl's scale factor and the gauge principle as it occurs in the Hamilton-Jacobi, de Broglie and Schrödinger equations. It is true that, as some writers have pointed out, London showed that applying the gauge principle to a free wave equation was essentially the same as multiplying the free wave function by the Weyl scale factor. Indeed London obtained the relation

$$\frac{\Psi}{l} = \frac{1}{l_o} = C$$

where C is a constant.

An interpretation of London's work has been that it has effectively shown that the physical object that behaves like the Weyl scale factor has been found: it is the complex amplitude of the de Broglie wave. This interpretation suggests that the fault with Weyl's original theory was not with the appearance of the non-integrable scale factor, but that it w as real and applied to the metric. Rather it w as proposed, the scale factor should be converted to a phase factor and applied to the wave function.

C. N. Yang used this interpretation in 1980 to revisit the Einstein-Weyl controversy. Where Einstein had argued that Weyl's scale factor would result in two rods having different scale if taken along different paths, Yang argued that the Weyl scale factor would be applied to two charged rods and that they would have different phases if taken along different paths. This same question could be raised about the phases of two electrons that have been taken along two different paths. This is the question that was independently, and without reference to Einstein's objection, asked by Aharonov and Bohm in 1959. T he experimental answer concerning the phases of two electrons is a decisive, "Yes, they are different."

Does the experimental confirmation of the path dependence of phase of two electrons finally answer the Einstein-Weyl controversy or solidify the phase interpretation of the Weyl scale factor?

It is no violation to this interpretation to point out that revisiting London's work shows that what he did was to ask, and answer, a different question. "What paths are possible given an electrostatic gauge potential?"

To see why this question might come up let us look back at the line integral in the exponent of Weyl's scale factor, namely,

$$l'_f = l'_o e^{\int \left(\frac{\partial \phi}{\partial x} dx + \frac{\partial \phi}{\partial y} dy \right)}. \tag{6.21}$$

We know that there are an infinite number of values of the exponent for l_f and l_o and that they are for the exponents to be given by the complex values $2\pi in$ where n may take on only the values of 0, 1, 2, 3,... Now our line integral from Equation (6.21) has an answer, or:

$$2\pi in = \int \left(\frac{\partial \phi}{\partial x} dx + \frac{\partial \phi}{\partial y} dy \right) \tag{6.22}$$

Recall that there are three parts to the line integral. By requiring a unity scale factor the value of the line integral is known to be one of these infinite number of values and only two parts are remaining unspecified. If we consider only an electrostatic potential, as both Schrödinger and London did, then only the path remains undetermined. London's work then provides an answer to the question, "What paths are possible given the requirements of a unity scale factor and an electrostatic potential?" The answer from London's work is, "Only those paths that are solutions of Schrödinger's wave equations."

Another way of stating this result is that London started with a unity Weyl scale factor and an electrostatic gauge potential and derived Schrödinger's wave equation. At the time London did his work the Schrödinger wave equation had not yet received general acceptance. It is not surprising then that no one looked at London's work in this light.

Weyl's Gauge Principle

Due to Einstein's objection Weyl's geometric proposal was dropped. Was Einstein right? Had Einstein seen what Weyl had seen? Or did either one see what lay within the scale factor? Keep in mind that this interchange occurred around 1918 before the development of the quantum description of the hydrogen atom so interpretations from quantum principles did not play a role in this discussion. Yet, quantum interpretations would soon follow Weyl's scale factor and one should revisit Einstein's argument with this in mind and, then, perhaps one should reconsider Einstein's argument in light of the new developments, even as C. N. Yang did after his proposed weak force theory had gained acceptance.

In 1929 after London reported his work on Weyl's unity scale factor, Weyl revisited his earlier geometric work and then elevated his scale factor to the level of a principle. This principle became the starting point of much, if not all, of the gauge field work that has been reported since 1929. Therefore, let us state this principle for our use in the following chapters.

Weyl's Gauge Principle states that gauge fields may be transformed into a space where the fields are given by the scale factor in the new space wherein the fields are given by Maxwell's electromagnetic field equations.

In this transformation the logarithmic derivatives of the gauge function are the gauge potentials as may be seen in Equation (6.18). Weyl showed that the electromagnetic fields are gauge fields by deriving them using this definition for gauge potentials and his gauge principle

Weyl's Quantum Principle

Weyl proposed his scale factor as a gauge principle from which electromagnetism may be derived. London's work showed that Schrödinger's equation might be derived from Weyl's scale factor given the requirement of a unity scale and an electrostatic potential. If we are given Weyl's Gauge Principle we may derive the electrostatic gauge potential from this principle and need not be given the electrostatic potential as stemming from any different principle. The only remaining condition that London used was the unity scale factor. In this case, since Weyl's Gauge Principle states no limitation on the scale factor in deriving electromagnetism, we now impose the restriction of a unity scale factor and propose the following as Weyl's Quantum Principle:

> A space that satisfies Weyl's Gauge Principle and for which Weyl's scale factor is unity, is said to obey Weyl's Quantum Principle.

The objective of this section is now before us. With the statement of a single principle, Weyl's Quantum Principle, we have the foundation from, and the guidance by, which we may derive both Maxwell's electromagnetism and Schrödinger's quantum mechanics. Therefore, Weyl's Quantum Principle is the only fundamental principle we need to derive both Maxwell's electromagnetism and Schrödinger's quantum mechanics.

Five Dimensional Quantum Mechanics Derived

The strength of the quantum theoretical structure is such that it has swept aside virtually every attack upon it. However, using classical definitions of commutivity it may be shown that the anti-commutivity of the position and momentum is dependent upon the metric approximating a flat metric. If a realm of conditions exists that does not allow a flat metric approximation then the commutators must depend upon the geometry. The quantum Poisson brackets are given by

The Dynamic Theory

$$[x^j, p^k]\Psi = i\hbar g^{kl}\left[\delta_{jl} + \begin{pmatrix} j \\ sl \end{pmatrix} x^s\right]\psi \qquad (6.23)$$

where the $\begin{pmatrix} j \\ sl \end{pmatrix}$ are the Christoffel symbols. This much does not depend upon any theory whatsoever, but only upon the mathematics of differentiation. Since the quantum Poisson brackets must correspond to the classical Poisson brackets, they must also depend upon t he geometry in the same fashion. In the past it has been possible to argue that if the only physical field that affects the geometry is Einstein's gravitational field, then it is possible to ignore this geometrical effect upon the commutivity of space and momentum in nuclear phenomena. If, however, the gravitational field is described by a gauge field then this argument is nullified because the gauge fields do pl ay a large role in the realm of nuclear physics.

The German physicist London produced a quantization of Weyl's theory in 1927. In his work, London showed that if the arc length of the metric was required to return to its original value, a quantization was produced and that the wave function was proportional to this arc length. However, there was a difficulty with his work; it required an imaginary distance.

The proposed theory not only removes the difficulty of the imaginary distance but further, logically produces the quantization conditions when the system is placed under an additional restriction. The quantum condition, as stated before, comes from restricting one's attention to systems which are isentropic. The requirement that the system have constant entropy is the simplest restriction that produces London's quantization. The imaginary distance appearing in London's work also appears here in the entropy manifold. However, the attractive electromagnetic force comes from a negative gauge function which couples the "distance" in the manifold with the Weyl geometry to the entropy manifold.

In the entropy manifold the change in entropy is the distance and, therefore, this distance must always be real and non-negative for an isolated system because of the principle of increasing entropy.

The proposed theory then logically produces London's assumption and removes the difficulty with imaginary distances. Further, it is found that the quantization conditions are limited to a system with a distance curvature, or gauge function. Thus, the interpretation of universal application of a non-varying, least unit of action coming from Heisenberg's Uncertainty Principle rests with the existence, or lack, of a distance curvature and not with the existence of a vector curvature. Equivalently, only forces that may be expressed in terms of a gauge function, or distance curvature, may exhibit quantization, while forces describable by only a vector curvature cannot be quantized. If the above interpretation of the new field components as gravitational field components holds up as gauge field components then gravitational effects may be quantized as well as the electromagnetic effects.

This description of the derivation of quantum mechanics from generalizations of the classical thermodynamics runs counter to the commonly held belief that one may derive classical thermodynamics using statistical methods and a v ariety of force laws. This contention is, however, without rigorous support, as may be seen when one considers the development of statistical thermodynamics. For instance, in order to talk of a statistical temperature one must start by assuming Newtonian physics (this constitutes three fundamental assumptions). Given Newtonian, or other physics, one can talk of an energy distribution, canonical ensembles and statistical temperature; however, one must make an additional fundamental assumption (the Equipartition Law) before the statistical heat capacities may be obtained.

The Dynamic Theory

In order to obtain thermodynamics two more assumptions are required. It was pointed out by Peter G. Bergmann that using the statistical approach one may obtain an expression for the difference in the heat exchanged between the system and the surroundings and the element of work done. In classical thermodynamics this difference is the change in internal energy which is path independent. In the statistical approach the difference is obtained without reference to the internal energy. To claim that the statistically derived expression is an exact differential is a logically new assertion; it constitutes the First Law of Thermodynamics. In addition, the assumptions of statistical thermodynamics allow the derivation of the fact that the differential of heat exchanged must be greater than, or equal to, the multiplication of the statistical temperature by the differential change in the statistical entropy. This product of statistical thermodynamic properties is similar to an identical product of thermodynamic properties. In statistical thermodynamics it is asserted that the ratio of the statistical temperature and the classical temperature is Boltzmann's constant. Once this assertion is made, the statistical entropy may be related to the classical entropy. However, there is no logical necessity that the ratio of temperatures be a constant from the statistical approach; only if it is a constant can there be a one-to-one correspondence between the statistical entropy and the classical entropy.

The above quantum condition establishes the conditions assumed by London and, therefore, one may follow his work in deriving the Schrödinger quantum mechanics. London's work establishes how quantum mechanics may be derived within the framework of a larger theory and will not be repeated here. Rather, a sketch of the five dimensional quantum mechanics will be presented.

The variational principle required by the entropy principle is given by maximizing the entropy. Because multiplication by a constant does not change the problem, one may write

$$\delta \int \gamma c^2 \sqrt{(dq^0)^2} = 0. \tag{6.24}$$

By defining the velocity vector as $u^j = dx^j/dq^0$ and the momentum as $p_j = \gamma c g_{jk} u^k$, where the fact that $g_{jk} u^j u^k = 1$ has been used, one may show that $p_j p^j = \gamma^2 c^2$, which is the five dimensional "momentum-energy" relation.

Because of the benefits of a first order differential wave equation, Dirac sought to find a first order operator equation that also satisfied the second order Klein-Gordon equation (the operator equivalent to the momentum-energy relation). This can also be done in five dimensions by taking the specific Hamiltonian operator to be

$$h = i\left(\alpha_1 \frac{\partial}{\partial x^1} + \alpha_2 \frac{\partial}{\partial x^2} + \alpha_3 \frac{\partial}{\partial x^3} + \alpha_4 \frac{\partial}{\partial x^4}\right) - \beta. \tag{6.25}$$

By taking the four partial derivatives in Equation (6.25) as the components of the four-vector specific momentum operator, one may write

$$h = -(\overline{\alpha} \bullet \overline{P} + \beta) \tag{6.26}$$

where natural units, $h = c = 1$, have been used.

If one takes the $p^0 |> = h|>$ and requires that the alphas and beta are to be chosen such that solutions of this equation are also solutions of Equation (6.26), one finds the restrictions imposed upon the choice of the alphas and beta to be

$$(\overline{\alpha} \bullet \overline{P})^2 = P^2,$$

$$\beta^2 = 1, \tag{6.27}$$

$$\overline{\alpha}\beta + \beta\overline{\alpha} = 0.$$

The set of 4 x 4 matrices satisfying the requirements of Equation (6.27) is given as

$$\gamma^0 = \beta \; ; \; \gamma^j = -\beta\alpha_j \; ; \; j = 1,2,3,4 \tag{6.28}$$

where I is the 2 by 2 identity matrix and the sigma's are the 2 by 2 Pauli spin matrices.

Then the five dimensional Dirac equation may be taken to be

$$i\frac{\partial}{\partial t}\psi(t) = i(\,\overline{\alpha}\cdot\overline{\nabla} - \beta\,)\psi(x) \qquad (6.29)$$

where we have used the four dimensional vector operator. By defining

$$\gamma^0 \equiv \tilde{\beta}\;;\; \gamma^\mu \equiv -\tilde{\beta}\tilde{\alpha}_\mu \;\left(\mu = 1,2,3,4\right),$$

then Equation (6.29) may be written as

$$(i\partial_j\gamma^j + 1)\psi(x) = 0. \qquad (6.30)$$

Taking into consideration Equation (6.30) with the gauge fields of Weyl's Gauge Principle coming from $F_{jk} = \partial_j\phi_k - \partial_k\phi_j$, one arrives at

$$\left[(i\partial_j - \phi_j)(i\partial^k - \phi^k) - 1 - \frac{1}{2}iF_{jk}\sigma^{jk}\right]\psi = 0 \qquad (6.31)$$

where

$$\sigma^{jk} = \begin{vmatrix} 0 & \dot{x}^1 & \dot{x}^2 & \dot{x}^3 & \dot{x}^4 \\ -\dot{x}^1 & 0 & -2is^1 & 2is^2 & u^1 \\ --\dot{x}^2 & 2is^1 & 0 & -2is^3 & u^2 \\ -\dot{x}^3 & -2is^2 & 2is^3 & 0 & u^3 \\ -\dot{x}^4 & -u^1 & -u^2 & -u^3 & 0 \end{vmatrix} \qquad (6.32)$$

and s is the usual intrinsic spin while u is a n ew spin appearing because of the added dimension. By expanding, one finds that Equation (6.31) becomes

$$[(i\partial_j - \phi_j)(i\partial^k - \phi^k) - 1$$
$$+ 2\overline{B}\bullet\overline{s} + 2\overline{V}\bullet\overline{u} + i\overline{E}\bullet\overline{v} - iV_4\dot{x}^4\,]\psi = 0. \qquad (6.33)$$

It may be seen that for the field equations later derived and displayed in Equation (7.10), even a p article without an electric charge (that is an electrically neutral particle) may have a magnetic moment because, for $\rho = J = 0$, one finds

$$\overline{\nabla}\bullet\overline{E} = -a_0\frac{\partial V_4}{\partial\gamma} \quad and \quad \overline{\nabla}\times\overline{B} - \frac{1}{c}\frac{\partial\overline{E}}{\partial t} = -a_0\frac{\partial\overline{V}}{\partial\gamma}. \qquad (6.34)$$

If these new fields are to be interpreted as the gravitational fields then Equation (6.33) may be interpreted as requiring a magnetic moment for spinning, gravitating particles. This gravitational spin component has additional features that play a role in fundamental particles.

Spin Octets

An interesting result occurs when one looks at the allowed fundamental spin states. In the five dimensional quantization of the space-time-mass manifold, three spin vectors appear. One of these is the familiar three component spin vector of relativistic quantum mechanics; the second of the three is a new three component spin vector; the remaining is a four component spin vector defined below.

Using the theorem, if α satisfies $\alpha^2 = a^2$ where a is a number, then the eigenvalues of α are $\pm a$, it is not difficult to show that the component eigenvalues are:

$$s_\alpha = \pm\frac{1}{2}, \; u_\alpha = \pm\frac{1}{2}, \; S_j^{\;2} = \frac{3}{4}; \; \alpha = 1,2,3 \; j = 1,2,3,4. \quad (6.35)$$

If, in analogy with the eigenvalues for the total angular momentum, one writes

$$S_j^{\;2} = \frac{3}{4} = S_j(S_j+1)$$

then the possible eigenvalues become

$$s_\alpha = \pm\frac{1}{2}, \; u_\alpha = \pm\frac{1}{2}, \; S_j = \frac{1}{2}, -\frac{3}{2}. \quad (6.36)$$

However, the following relations, which specify the components of a four dimensional spin vector which, when added to the angular four-momentum, commutes with the specific Hamiltonian, restrict the number of possible combinations of these eigenvalues.

$$S_1 = s_1 - u_2 - u_3, \; S_2 = s_2 + u_1 - u_3$$

$$S_3 = s_3 + u_1 + u_2, \; S_4 = s_1 - s_2 + s_3.$$

The Dynamic Theory

The question to be asked now seems to be, how many combinations of the above eigenvalues are allowed? The answer may be shown to be octets. This predicted result compares with the experimental findings of Gel Mann.

Chapter 7 Fundamental Particle Fields

In the standard model of particle physics particles are classified primarily by electric charge and spin. It is assumed that these properties do not change as the particle moves around in space and time. While the standard model includes Dirac's equations of quantum mechanics that predicts spin, there is no theoretical prediction, within the standard model, of the experimental fact of charge quantization. Weyl's Gauge Principle provides a prediction of particle properties in the absence of external fields, which has no c ounterpart in the standard model, and a separate prediction of the motion of particles in the presence of external fields, including the spin in Dirac's equations. Here we will look at both types of predictions.

The following first looks at the difference in the approaches to determining the fields of a fundamental particle from the classical Maxwell's equations and the five dimensional Weyl Gauge Principle approach. The primary difference in the approaches is that the classical approach starts by taking the Maxwell equations as the starting point and, therefore, by gauge freedom the gauge function is not uniquely determined nor does it enter into consideration. On the other hand, using the Weyl Gauge Principle it is the gauge function that is the starting point and Maxwell's equations, or their five dimensional extensions, follow from the principle.

In the classical approach to determining the electrostatic potential from only Maxwell's equations the problem of a source term appears. It is answered classically by introducing the Dirac delta function. This allows a determination of the potential of a point particle even though no r eal function such as the Dirac delta function exists. It is noted that the radial component of the magnetic field must vary inversely as r, but there is no way, using only Maxwell's equations, that a static solution can relate the magnetic field to the electric field. Nor is it clear why there is a radial dependence to the magnetic field.

Weyl's Derivation of Electromagnetism

The extension of geometry Weyl used to place electromagnetism on a geometrical basis allowed him to use his gauge principle to derive the Maxwell equations of electromagnetism. We shall now present this derivation.

Weyl defined the gauge potentials as

$$\phi_i \equiv \frac{\partial \ln f^{\frac{1}{2}}}{\partial x^i}. \tag{7.1}$$

Now the electromagnetic field tensor is given by

$$F_{ij} \equiv \phi_{i,j} - \phi_{j,i}. \tag{7.2}$$

The field tensor given by Equation (7.2) has 16 components when the indices range over four dimensions. We would like to determine the field equations for these components. The quickest, though not the only, way is to consider the four dimensions to be $x^0=ict$, $x^1=x$, $x^2=y$ and $x^3=z$. The field tensor is then defined to be

$$F_{ij} = \begin{vmatrix} 0 & iE_1 & iE_2 & iE_3 \\ -iE_1 & 0 & B_3 & -B_2 \\ -iE_2 & -B_3 & 0 & B_1 \\ -iE_3 & B_2 & -B_1 & 0 \end{vmatrix}. \tag{7.3}$$

Using Bianchi's identities

$$\frac{\partial F_{ij}}{\partial x^k} + \frac{\partial F_{jk}}{\partial x^i} + \frac{\partial F_{ki}}{\partial x^j} = 0$$

and the various combinations of the indices 0, 1, 2, 3 we obtain the field equations

The Dynamic Theory

$$\overline{\nabla}\bullet\overline{B} = 0$$

$$\overline{\nabla}\times\overline{E}+\frac{1}{c}\frac{\partial\overline{B}}{\partial t} = 0. \qquad (7.4)$$

The definition of the four-vector current density

$$\frac{\partial F_{ij}}{\partial x^i} \equiv \frac{4\pi}{c}J_i \qquad (7.5)$$

yields the equations

$$\overline{\nabla}\bullet\overline{E} = 4\pi\rho$$

$$\overline{\nabla}\times\overline{B}-\frac{1}{c}\frac{\partial\overline{E}}{\partial t} = \frac{4\pi\overline{J}}{c}. \qquad (7.6)$$

In addition to these field equations there is the statement of conservation of charge where

$$\frac{\partial J_i}{\partial x^i} = 0, \quad i=0,1,2,3,$$

so that

$$\frac{\partial\rho}{\partial t}+\overline{\nabla}\bullet\overline{J} = 0. \qquad (7.7)$$

For ease in future reference to these five field equations they may be rewritten as

$$\overline{\nabla}\bullet\overline{B} = 0 \qquad [a]$$

$$\frac{1}{c}\frac{\partial\overline{B}}{\partial t}+\overline{\nabla}\times\overline{E} = \overline{0} \qquad [b]$$

$$\overline{\nabla}\times\overline{B}-\frac{1}{c}\frac{\partial\overline{E}}{\partial t} = \frac{4\pi\overline{J}}{c} \qquad [c] \qquad (7.8)$$

$$\overline{\nabla}\bullet\overline{E} = 4\pi\rho \qquad [d]$$

$$\frac{\partial\rho}{\partial t}+\overline{\nabla}\bullet\overline{J} = 0 \qquad [e]$$

The five field equations in Equations (7.8) are universally known as the Maxwell field equations and the above process shows how they may be derived from Weyl's Gauge Principle. Weyl's derivation of these equations from his gauge function was the origin of the term "gauge field equations" applied to Maxwell's equations.

Electrostatic potential from the field equations

Classically the Maxwell equations were developed from other laws such as Coulomb's Law which stated the functional form of the electric field from which the electrostatic potential could be obtained by integration. However, it is more difficult to arrive at Coulomb's Law from the Maxwell's equations. The difficulty arises in Equation (7.8)[d] when the charge density is taken to zero for a point charge. As it is written Equation (7.8)[d] is a Poisson equation with the charge density as the source for the field. When the charge density is taken to be zero the equation becomes Laplace's homogeneous equation.

The usual approach to the electrostatic potential of a particle is to look for a solution of Laplace's equation that satisfies

$$\Delta u = u_{xx} + u_{yy} + u_{zz} = -\delta(x - x', y - y', z - z'),$$

where the Dirac delta function δ denotes a unit source concentrated at the point (x', y', z'). No function has this property, but it can be thought of as a limit of functions whose integrals over space are unity, and whose support (the region where the function is non-zero) shrinks to a point. The definition of the fundamental solution thus implies that, if the Laplacian of u is integrated over any volume that encloses the source point, then

$$\iiint_V \nabla \cdot \nabla u \, dV = -1.$$

The Laplace equation is unchanged under a rotation of coordinates, and hence we can expect that a fundamental solution may be obtained among solutions that only depend upon the distance r from the source point. If we choose the

174

volume to be a ball of radius a around the source point, then Gauss' divergence theorem implies that

$$-1 = \iiint_V \nabla \cdot \nabla u \, dV = \iint_S u_r dS = 4\pi a^2 u_r(a).$$

It follows that

$$u_r(r) = -\frac{1}{4\pi r^2},$$

on a sphere of radius r that is centered around the source point, and hence

$$u = \frac{1}{4\pi r}.$$

Electrostatic potential from 5D gauge function

If, however, one is operating in a five dimensional manifold of space, time and mass where Weyl's Gauge Principle in five dimensions produces the gauge field tensor that looks like

$$F_{ij} = \begin{vmatrix} 0 & iE_1 & iE_2 & iE_3 & iV_4 \\ -iE_1 & 0 & B_1 & -B_2 & V_1 \\ -iE_2 & -B_1 & 0 & B_3 & V_2 \\ -iE_3 & B_2 & -B_3 & 0 & V_3 \\ -iV_4 & -V_1 & -V_2 & -V_3 & 0 \end{vmatrix}. \qquad (7.9)$$

The same procedure used above yields the field equations

$$\overline{\nabla}\bullet\overline{B} = 0 \qquad [a]$$

$$\frac{1}{c}\frac{\partial\overline{B}}{\partial t} + \overline{\nabla}\times\overline{E} = \overline{0} \qquad [b]$$

$$\overline{\nabla}\times\overline{B} - \frac{1}{c}\frac{\partial\overline{E}}{\partial t} + a_0\frac{\partial\overline{V}}{\partial\gamma} = \frac{4\pi\overline{J}}{c} \qquad [c]$$

$$\overline{\nabla}\bullet\overline{E} + a_0\frac{\partial V_4}{\partial\gamma} = 4\pi\rho \qquad [d]$$

$$\frac{\partial\rho}{\partial t} + \overline{\nabla}\bullet\overline{J} + a_0\frac{\partial J_4}{\partial\gamma} = 0 \qquad [e]$$

$$\overline{\nabla}\times\overline{V} + a_0\frac{\partial\overline{B}}{\partial\gamma} = \overline{0} \qquad [f]$$

$$\overline{\nabla}V_4 + \frac{1}{c}\frac{\partial\overline{V}}{\partial t} = a_0\frac{\partial\overline{E}}{\partial\gamma} \qquad [g]$$

$$\overline{\nabla}\bullet\overline{V} + \frac{1}{c}\frac{\partial V_4}{\partial t} = -\frac{4\pi J_4}{c} \qquad [h] \qquad (7.10)$$

Returning to the definition of the gauge potentials of Equation (7.1) and using the variables $x_0 = ict, x_1 = r, x_2 = r\theta, x_3 = r\sin\theta\phi$, and $x_4 = \gamma/a_o$ plus assuming the gauge function may written as the product of functions each a function of a single variable so that

$$\ln f^{\frac{1}{2}} = f_t(t)f_r(r)f_\theta(\theta)f_\varphi(\varphi)f_\gamma(\gamma). \qquad (7.11)$$

All these functions are unitless.

The gauge potentials then become

$$\phi_0 \equiv \frac{\partial \ln f^{\frac{1}{2}}}{\partial(ict)} = \frac{1}{ic} f_t' f_r f_\theta f_\varphi f_\gamma,$$

$$\phi_r \equiv \frac{\partial \ln f^{\frac{1}{2}}}{\partial r} = f_t f_r' f_\theta f_\varphi f_\gamma,$$

$$\phi_\theta \equiv \frac{\partial \ln f^{\frac{1}{2}}}{\partial(r\theta)} = \frac{1}{r} f_t f_r f_\theta' f_\varphi f_\gamma, \tag{7.12}$$

$$\phi_\varphi \equiv \frac{\partial \ln f^{\frac{1}{2}}}{\partial(r\sin\theta\varphi)} = \frac{1}{r\sin\theta} f_t f_r f_\theta f_\varphi' f_\gamma,$$

$$\phi_\gamma \equiv \frac{\partial \ln f^{\frac{1}{2}}}{\partial\left(\frac{\gamma}{a_o}\right)} = a_o f_t f_r f_\theta f_\varphi f_\gamma',$$

where the prime means ordinary differentiation with respect to time and all the gauge potentials have units of 1/meter.

Now consider the definition of the field quantities as given by Equation (7.2) which is repeated here

$$F_{ij} \equiv \phi_{i,j} - \phi_{j,i}.$$

The units of all the field components are $1/m^2$.

Weyl's scale factor which represents the basis of his gauge field development is

$$l_f = l_o e^{\int(\phi_i dx_i)}. \tag{7.13}$$

The quantum condition that London showed led to Schrodinger's quantum mechanics was the condition that the final length of the vector, l_f, equal the initial vector length, l_o. This condition requires quantization of the exponent, or

$$2\pi i n = \int \phi_i dx_i. \tag{7.14}$$

London showed that this condition may be met for an electrostatic potential, ϕ_o, only by paths given by Schrodinger's quantum mechanics.

Here the aim is to determine what gauge potentials may be allowed for a particle by the unity scale factor when these gauge potentials cannot change when particle is moved around in space and time. This requires that each gauge potential must itself be quantized since to be independent of the path the gauge potentials must be quantized for any arbitrary path. Then

$$\int N_j \phi_j dx^j = 2\pi i N \tag{7.15}$$

for all j without summation on j.

Because Equation (7.15) must be satisfied for all the particles that satisfy Weyl's unity scale factor, or we might call them Weyl's Unity Particles (WUPs), then we might just as well multiply Equation (7.15) by a constant such as

$$\frac{e}{4\pi\varepsilon_o} \int N_j \phi_j dx^j = \frac{e2\pi i N}{4\pi\varepsilon_o} \tag{7.16}$$

This requires that the gauge potentials of Equation (7.12) be written as

$$\Phi_0 = \frac{e}{4\pi\varepsilon_o}\phi_0 \equiv \frac{e}{4\pi\varepsilon_o}\frac{\partial \ln f^{\frac{1}{2}}}{\partial(ict)} = \frac{N_0}{ic}\frac{e}{4\pi\varepsilon_o}f_t'f_r f_\theta f_\varphi f_\gamma$$

$$\Phi_r = \frac{e}{4\pi\varepsilon_o}\phi_r \equiv \frac{e}{4\pi\varepsilon_o}\frac{\partial \ln f^{\frac{1}{2}}}{\partial r} = N_1\frac{e}{4\pi\varepsilon_o}f_t f_r' f_\theta f_\varphi f_\gamma$$

$$\Phi_\theta = \frac{e}{4\pi\varepsilon_o}\phi_\theta \equiv \frac{e}{4\pi\varepsilon_o}\frac{\partial \ln f^{\frac{1}{2}}}{\partial(r\theta)} = \frac{N_2}{r}\frac{e}{4\pi\varepsilon_o}f_t f_r f_\theta' f_\varphi f_\gamma \qquad .(7.17)$$

$$\Phi_\varphi = \frac{e}{4\pi\varepsilon_o}\phi_\varphi \equiv \frac{e}{4\pi\varepsilon_o}\frac{\partial \ln f^{\frac{1}{2}}}{\partial(r\sin\theta\varphi)} = \frac{N_3}{r\sin\theta}\frac{e}{4\pi\varepsilon_o}f_t f_r f_\theta f_\varphi' f_\gamma$$

$$\Phi_\gamma = \frac{e}{4\pi\varepsilon_o}\phi_\gamma \equiv \frac{e}{4\pi\varepsilon_o}\frac{\partial \ln f^{\frac{1}{2}}}{\partial(\gamma/a_o)} = N_4 a_o\frac{e}{4\pi\varepsilon_o}f_t f_r f_\theta f_\varphi f_\gamma'$$

Forming the fields from the potentials given in Equation (7.17) produces

$$E_r = \frac{e}{4\pi\varepsilon_o}F_{01} = \frac{e}{4\pi\varepsilon_o}\left(\phi_{0,1} - \phi_{1,0}\right) = \Phi_{0,r} - \Phi_{r,0}$$

$$= \frac{d}{dr}\left(\frac{N_0}{ic}\frac{e}{4\pi\varepsilon_o}f_t'f_r f_\theta f_\varphi f_\gamma\right) - \frac{d}{d(ict)}\left(N_1 f_t f_r' f_\theta f_\varphi f_\gamma\right) \qquad (7.18)$$

$$= \frac{\left(N_0 - N_1\right)}{ic}\frac{e}{4\pi\varepsilon_o}f_t' f_r' f_\theta f_\varphi f_\gamma$$

The quantization of the gauge potentials allows for different quantum numbers that produces non-zero fields.

In this fashion all ten field components may be determined.

The remaining electric field components are

$$E_\theta = \frac{e}{4\pi\varepsilon_o} F_{02} = \frac{e}{4\pi\varepsilon_o}\left(\phi_{0,2} - \phi_{2,0}\right) = \Phi_{0,\theta} - \Phi_{\theta,0}$$

$$= \frac{d}{d(r\theta)}\left(\frac{N_0}{ic}\frac{e}{4\pi\varepsilon_o} f_t' f_r f_\theta f_\varphi f_\gamma\right) - \frac{d}{d(ict)}\left(\frac{N_2}{r}\frac{e}{4\pi\varepsilon_o} f_t f_r f_\theta' f_\varphi f_\gamma\right) \quad (7.19)$$

$$= \frac{(N_0 - N_2)}{icr}\frac{e}{4\pi\varepsilon_o} f_t' f_r f_\theta' f_\varphi f_\gamma$$

$$E_\varphi = \frac{e}{4\pi\varepsilon_o} F_{03} = \frac{e}{4\pi\varepsilon_o}\left(\phi_{0,3} - \phi_{3,0}\right) = \Phi_{0,\varphi} - \Phi_{\varphi,0}$$

$$= \frac{d}{d(r\sin\theta\varphi)}\left(\frac{N_0}{ic}\frac{e}{4\pi\varepsilon_o} f_t' f_r f_\theta f_\varphi f_\gamma\right) - \frac{d}{d(ict)}\left(\frac{N_3}{r\sin\theta}\frac{e}{4\pi\varepsilon_o} f_t f_r f_\theta f_\varphi' f_\gamma\right) \quad (7.20)$$

$$= \frac{(N_0 - N_3)}{icr\sin\theta}\frac{e}{4\pi\varepsilon_o} f_t' f_r f_\theta f_\varphi' f_\gamma$$

The magnetic fields may be shown to be given by

$$B_r = \frac{e}{4\pi\varepsilon_o} F_{23} = \frac{e}{4\pi\varepsilon_o}\left(\phi_{2,3} - \phi_{3,2}\right) = \Phi_{\theta,\varphi} - \Phi_{\varphi,\theta}$$

$$= \frac{d}{d(r\sin\theta\varphi)}\left(\frac{N_2}{r}\frac{e}{4\pi\varepsilon_o} f_t f_r f_\theta' f_\varphi f_\gamma\right) - \frac{d}{d(r\theta)}\left(\frac{N_3}{r\sin\theta}\frac{e}{4\pi\varepsilon_o} f_t f_r f_\theta f_\varphi' f_\gamma\right)$$

$$= \frac{1}{r\sin\theta}\left(\frac{N_2}{r}\frac{e}{4\pi\varepsilon_o} f_t f_r f_\theta' f_\varphi' f_\gamma\right) - \frac{1}{r}\left(-\frac{N_3\cos\theta}{r\sin^2\theta}\frac{e}{4\pi\varepsilon_o} f_t f_r f_\theta f_\varphi' f_\gamma + \frac{N_3}{r\sin\theta}\frac{e}{4\pi\varepsilon_o} f_t f_r f_\theta' f_\varphi' f_\gamma\right) \quad (7.21)$$

$$= \frac{1}{r\sin\theta}\left(\frac{N_2}{r}\frac{e}{4\pi\varepsilon_o} f_t f_r f_\theta' f_\varphi' f_\gamma\right) - \frac{N_3}{r^2\sin\theta}\frac{e}{4\pi\varepsilon_o}\left(f_t f_r f_\theta' f_\varphi' f_\gamma - \cot\theta f_t f_r f_\theta f_\varphi' f_\gamma\right)$$

$$= \frac{(N_2 - N_3)}{r^2\sin\theta}\frac{e}{4\pi\varepsilon_o}\left(f_t f_r f_\theta' f_\varphi' f_\gamma\right) + \frac{N_3}{r^2\sin\theta}\frac{e}{4\pi\varepsilon_o}\left(\cot\theta f_t f_r f_\theta f_\varphi' f_\gamma\right)$$

$$B_\theta = \frac{(N_3 - N_1)}{r\sin\theta}\frac{e}{4\pi\varepsilon_o}\left[f_t f_r' f_\theta f_\varphi' f_\gamma - \frac{N_3 f_t f_r f_\theta f_\varphi' f_\gamma}{r}\right]$$

$$B_\varphi = \frac{(N_1 - N_2)}{r}\frac{e}{4\pi\varepsilon_o}\left[f_t f_r' f_\theta' f_\varphi f_\gamma - \frac{N_2 f_t f_r f_\theta' f_\varphi f_\gamma}{r}\right] \quad (7.22)$$

Now let us turn to the fields that arise in the fifth dimension. We have not yet discussed these fields and therefore do not have a physical interpretation for them. However, we do s ee that whatever they are they will depend upon the radial function f_r as does the field we have called the electric field. The only other field that has the long range $1/r^2$ radial dependence as the electric field is the gravitational field. In a later chapter we shall show that the interpretation of the fields of the fifth dimension must be the gravitational fields. These fifth dimension, or gravitational, fields are

$$V_4 = \frac{e}{4\pi\varepsilon_o} F_{04} = \frac{e}{4\pi\varepsilon_o}\left(\phi_{0,4} - \phi_{4,0}\right) = \Phi_{0,\gamma} - \Phi_{\gamma,0}$$

$$= \frac{d}{dr}\left(\frac{N_0}{ic}\frac{e}{4\pi\varepsilon_o} f_i' f_r f_\theta f_\varphi f_\gamma\right) - \frac{d}{d(ict)}\left(a_o N_4 \frac{e}{4\pi\varepsilon_o} f_i f_r f_\theta f_\varphi f_\gamma'\right) \quad (7.23)$$

$$= \frac{a_o\left(N_0 - N_4\right)}{ic}\frac{e}{4\pi\varepsilon_o} f_i' f_r f_\theta f_\varphi f_\gamma'$$

$$V_1 = \frac{e}{4\pi\varepsilon_o} F_{14} = \frac{e}{4\pi\varepsilon_o}\left(\phi_{1,4} - \phi_{4,1}\right) = \Phi_{r,\gamma} - \Phi_{\gamma,r}$$

$$= \frac{a_o d}{d\gamma}\left(N_1 \frac{e}{4\pi\varepsilon_o} f_i f_r' f_\theta f_\varphi f_\gamma\right) - \frac{d}{dr}\left(a_o N_4 \frac{e}{4\pi\varepsilon_o} f_i f_r f_\theta f_\varphi f_\gamma'\right) \quad (7.24)$$

$$= a_o\left(N_1 - N_4\right)\frac{e}{4\pi\varepsilon_o} f_i f_r' f_\theta f_\varphi f_\gamma'$$

$$V_2 = \frac{e}{4\pi\varepsilon_o} F_{24} = \frac{e}{4\pi\varepsilon_o}\left(\phi_{2,4} - \phi_{4,2}\right) = \Phi_{\theta,\gamma} - \Phi_{\gamma,\theta}$$

$$= \frac{a_o d}{d\gamma}\left(\frac{N_2}{r}\frac{e}{4\pi\varepsilon_o} f_i f_r f_\theta' f_\varphi f_\gamma\right) - \frac{d}{d(r\theta)}\left(a_o N_4 \frac{e}{4\pi\varepsilon_o} f_i f_r f_\theta f_\varphi f_\gamma'\right) \quad (7.25)$$

$$= \frac{a_o\left(N_2 - N_4\right)}{r}\frac{e}{4\pi\varepsilon_o} f_i f_r f_\theta' f_\varphi f_\gamma'$$

$$V_3 = \frac{e}{4\pi\varepsilon_o} F_{34} = \frac{e}{4\pi\varepsilon_o}\left(\phi_{3,4} - \phi_{4,3}\right) = \Phi_{\varphi,\gamma} - \Phi_{\gamma,\varphi}$$

$$= \frac{a_o d}{d\gamma}\left(\frac{N_3}{r\sin\theta}\frac{e}{4\pi\varepsilon_o} f_i f_r f_\theta f_\varphi' f_\gamma\right) - \frac{d}{d(r\sin\theta\varphi)}\left(a_o N_4 \frac{e}{4\pi\varepsilon_o} f_i f_r f_\theta f_\varphi f_\gamma'\right) \quad (7.26)$$

$$= \frac{a_o\left(N_3 - N_4\right)}{r\sin\theta}\frac{e}{4\pi\varepsilon_o} f_i f_r f_\theta f_\varphi' f_\gamma'$$

To summarize

$$\frac{F_{ij}}{\left(\frac{e}{4\pi\varepsilon_o}\right)} = \begin{vmatrix} 0 & \frac{(N_0-N_1)}{ic}F_{tr} & \frac{(N_0-N_2)}{icr}F_{t\theta} & \frac{(N_0-N_3)}{icr\sin\theta}F_{t\varphi} & \frac{a_o(N_0-N_4)}{ic}F_{t\gamma} \\ -\frac{(N_0-N_1)}{ic}F_{tr} & 0 & \frac{(N_2-N_3)}{r\sin\theta}\left[\frac{F_{\theta\varphi}}{r}-N_3\cot\theta\,F_\varphi\right] & \frac{(N_3-N_1)}{r\sin\theta}\left[F_{r\varphi}-\frac{N_3F_\varphi}{r}\right] & a_o(N_0-N_4)F_{r\gamma} \\ -\frac{(N_0-N_2)}{icr}F_{t\theta} & -\frac{(N_2-N_3)}{r\sin\theta}\left[\frac{F_{\theta\varphi}}{r}-N_3\cot\theta\,F_\varphi\right] & 0 & \frac{(N_1-N_2)}{r}\left[F_{r\theta}-\frac{N_2F_\theta}{r}\right] & \frac{a_o(N_2-N_4)}{r}F_{\theta\gamma} \\ -\frac{(N_0-N_3)}{icr\sin\theta}F_{t\varphi} & -\frac{(N_3-N_1)}{r\sin\theta}\left[F_{r\varphi}-\frac{N_3F_\varphi}{r}\right] & -\frac{(N_1-N_2)}{r}\left[F_{r\theta}-\frac{N_2F_\theta}{r}\right] & 0 & \frac{a_o(N_3-N_4)}{r\sin\theta}F_{\varphi\gamma} \\ -\frac{a_o(N_0-N_4)}{ic}F_{t\gamma} & -a_o(N_0-N_4)F_{r\gamma} & -\frac{a_o(N_2-N_4)}{r}F_{\theta\gamma} & -\frac{a_o(N_3-N_4)}{r\sin\theta}F_{\varphi\gamma} & 0 \end{vmatrix} \quad (7.27)$$

Now the fields of Equation (7.27) must satisfy all of Equations (7.10). First, note that in spherical coordinates Equation (7.10)[a] is

$$\left(\frac{1}{r^2}\right)\frac{\partial\left(r^2B_r\right)}{\partial r}+\left(\frac{1}{r\sin\theta}\right)\frac{\partial\left(\sin\theta B_\theta\right)}{\partial\theta}+\left(\frac{1}{r\sin\theta}\right)\frac{\partial B_\phi}{\partial\phi}=0. \quad (7.28)$$

Substituting Equation (7.22) into (7.28) and simplifying produces

$$\left(\frac{1}{r^2}\right)\frac{\partial\left(r^2B_r\right)}{\partial r}+\left(\frac{1}{r\sin\theta}\right)\frac{\partial\left(\sin\theta B_\theta\right)}{\partial\theta}+\left(\frac{1}{r\sin\theta}\right)\frac{\partial B_\phi}{\partial\phi}=0$$

$$\left(\frac{1}{r^2}\right)\frac{e}{4\pi\varepsilon_o}\left\{\frac{(N_2-N_3)}{\sin\theta}\left[f_tf_r'f_\theta f_\varphi'f_\gamma-N_3\cot\theta\,f_tf_rf_\theta f_\varphi'f_\gamma-rN_3\cot\theta\,f_tf_r'f_\theta f_\varphi'f_\gamma\right]\right\}$$

$$+\left(\frac{1}{r\sin\theta}\right)\left\{\frac{(N_3-N_1)}{r}\left[f_tf_r'f_\theta f_\varphi'f_\gamma-\frac{N_3f_tf_rf_\theta f_\varphi'f_\gamma}{r}\right]\right\}$$

$$+\left(\frac{1}{r\sin\theta}\right)\left\{\frac{(N_1-N_2)}{r}\left[f_tf_r'f_\theta f_\varphi'f_\gamma-\frac{N_2f_tf_rf_\theta f_\varphi'f_\gamma}{r}\right]\right\}=0 \qquad .(7.29)$$

$$\left(\frac{1}{r^3\sin\theta}\right)\frac{e}{4\pi\varepsilon_o}\left\{\begin{array}{l}(N_2-N_3)\left[-rN_3\cot\theta\,f_tf_rf_\theta f_\varphi'f_\gamma-r^2N_3\cot\theta\,f_tf_r'f_\theta f_\varphi'f_\gamma\right] \\ +N_3\left[(N_1-N_3)f_tf_rf_\theta f_\varphi'f_\gamma\right] \\ (N_2-N_1)\left[N_2f_tf_rf_\theta f_\varphi'f_\gamma\right]\end{array}\right\}=0$$

$$\left(\frac{1}{r^3\sin\theta}\right)\left\{\begin{array}{l}(N_3-N_2)N_3r\cot\theta\left(f_tf_rf_\theta f_\varphi'f_\gamma+rf_tf_r'f_\theta f_\varphi'f_\gamma\right) \\ +f_tf_rf_\theta'f_\varphi'f_\gamma\left[(N_2-N_1)N_2+(N_1-N_3)N_3\right]\end{array}\right\}=0$$

when $r^3\sin\theta\neq0$ Equation (7.29) requires that

$$\left(N_3-N_2\right)N_3\cot\theta\left(rf_tf_rf_\theta f_\varphi'f_\gamma+r^2f_tf_r'f_\theta f_\varphi'f_\gamma\right)$$
$$=f_tf_rf_\theta'f_\varphi'f_\gamma\left[\left(N_2-N_1\right)N_2+\left(N_1-N_3\right)N_3\right] \qquad (7.30)$$

If $f_tf_\varphi'f_\gamma\neq0$ Equation (7.30) becomes

$$(N_3 - N_2)N_3 \cot\theta\left(rf_r f_\theta + r^2 f_r' f_\theta\right)$$
$$= f_r f_\theta'\left[(N_2 - N_1)N_2 + (N_1 - N_3)N_3\right] \tag{7.31}$$

The variables may be separated by dividing Equation (7.31) by $f_r f_\theta' \cot\theta$ leaving

$$\left(r + r^2 \frac{f_r'}{f}\right) = \frac{f_\theta'}{f_\theta \cot\theta} \frac{\left[(N_2 - N_1)N_2 + (N_1 - N_3)N_3\right]}{(N_3 - N_2)N_3}$$
$$\equiv \frac{Kf_\theta'}{f_\theta \cot\theta} \tag{7.32}$$

The left hand side of Equation (7.32) is a function of r only, while the right hand side is only a function of θ. Therefore, Equation (7.32) must be a constant. Then

$$\left(r + r^2 \frac{f_r'}{f}\right) = \frac{\left[(N_2 - N_1)N_2 + (N_1 - N_3)N_3\right]}{(N_3 - N_2)N_3}\lambda_o \tag{7.33}$$
$$\equiv K\lambda_o \equiv \lambda_N$$

The constant, λ_N, depends upon t he set of quantum numbers, N_1, N_2, and N_3, for the particle. Thus, λ_N depends upon the particle under consideration.

The radial equation in Equation (7.33) may be integrated immediately with the result

$$f_r = \left(\frac{r_o}{r}\right)e^{-\left(\frac{\lambda_N}{r}\right)}. \tag{7.34}$$

In Equation (7.34) r_o is a co nstant of integration to be determined with units of m.

Then we may write the relation

$$\ln f^{\frac{1}{2}} = f_t(t)\left(\frac{r_o}{r}\right)e^{-\left(\frac{\lambda_N}{r}\right)} f_\theta(\theta) f_\varphi(\varphi) f_\gamma(\gamma). \tag{7.35}$$

The electrostatic potential thus becomes

$$\Phi_0 \equiv \frac{e}{4\pi\varepsilon_o}\frac{\partial \ln f^{\frac{1}{2}}}{\partial(ict)} = \frac{N_0}{ic}\frac{e}{4\pi\varepsilon_o}f_t'\left(\frac{r_o}{r}\right)e^{-\left(\frac{\lambda_N}{r}\right)}f_\theta.f_\varphi.f_\gamma \ .(7.36)$$

Now return to Equation (7.32) to find the solution to f_θ where we have

$$\frac{f_\theta'}{f_\theta \cot\theta} = \lambda_0$$

$$\Rightarrow \frac{f_\theta'}{f_\theta} = \lambda_0 \cot\theta = \lambda_0 \frac{\cos\theta}{\sin\theta}$$

$$\Rightarrow \frac{df_\theta}{f_\theta} = \lambda_0 \frac{\cos\theta}{\sin\theta}d\theta = \lambda_0 \frac{d\sin\theta}{\sin\theta} \ . \qquad (7.37)$$

$$\Rightarrow \ln f_\theta = \lambda_0 \ln\sin\theta + \ln D$$

$$\Rightarrow f_\theta = D\sin^{\lambda_0}\theta$$

D must be a unitless constant.

But will this solution satisfy all of Equations (7.10)? If these relations for the functions f_r and f_θ are to be valid candidates for our WUP solution then they must also satisfy all of the field equations. Indeed that is how we may determine the solutions for the rest of the functions. Start with [b]

$$\frac{1}{c}\frac{\partial \overline{B}}{\partial t} + \overline{\nabla}\times\overline{E} = \overline{0} \qquad (7.38)$$

Which is a statement of three equations since it is a vector equation. First

The Dynamic Theory

$$\frac{1}{c}\frac{\partial B_r}{\partial t}+\frac{1}{r\sin\theta}\left[\frac{\partial}{\partial\theta}\left(\sin\theta E_\varphi\right)-\frac{\partial E_\theta}{\partial\varphi}\right]=0$$

$$\frac{\left(N_2-N_3\right)}{icr^2\sin\theta}\frac{e}{4\pi\varepsilon_o}\left[F_{\theta\varphi t}-N_3\cot\theta\,F_{\varphi t}\right]+\frac{1}{r\sin\theta}\frac{e}{4\pi\varepsilon_o}\left[\frac{\left(N_0-N_3\right)}{icr}F_{t\varphi\theta}-\frac{\left(N_0-N_2\right)}{icr}F_{t\theta\varphi}\right]=0$$

$$\left(N_2-N_3\right)\left[1-\frac{N_3\cot\theta\,F_{\varphi t}}{F_{t\varphi\theta}}\right]+\left(N_2-N_3\right)=0 \qquad (7.39)$$

$$\left(N_2-N_3\right)\left[2+\frac{N_3\cot\theta\,f_t'f_r f_\theta f_\varphi' f_\gamma}{f_t'f_r f_\theta' f_\varphi' f_\gamma}\right]=0$$

This is true if $N_2=N_3$,

The next component of the vector obtains

$$\frac{1}{c}\frac{\partial B_\theta}{\partial t}+\frac{1}{r}\left[\frac{1}{\sin\theta}\frac{\partial E_r}{\partial\varphi}-\frac{\partial\left(rE_\varphi\right)}{\partial r}\right]=0 \qquad (7.40)$$

$$\frac{1}{c}\frac{e}{4\pi\varepsilon_o}\frac{\partial\left[\frac{\left(N_3-N_1\right)}{r\sin\theta}\left[F_{r\varphi}-\frac{N_3 F_\varphi}{r}\right]\right]}{\partial t}+\frac{1}{r}\frac{e}{4\pi\varepsilon_o}\left[\frac{1}{\sin\theta}\frac{\partial\left[\frac{\left(N_0-N_1\right)}{ic}F_{tr}\right]}{\partial\varphi}-\frac{\partial\left(r\left[\frac{\left(N_0-N_3\right)}{icr\sin\theta}F_{t\varphi}\right]\right)}{\partial r}\right]=0$$

$$\left(N_3-N_1\right)\left[F_{r\varphi t}-\frac{N_3 F_{\varphi t}}{r}\right]+\left(N_0-N_1\right)F_{tr\varphi}-\frac{\left(N_0-N_3\right)}{r}F_{t\varphi}+\left(N_0-N_3\right)F_{t\varphi r}=0$$

$$2\left(N_0-N_1\right)f_t'f_r f_\theta f_\varphi f_\gamma-\frac{\left[\left(N_3-N_1\right)N_3-\left(N_0-N_3\right)\right]f_t'f_r f_\theta f_\varphi f_\gamma}{r}=0$$

This is true if $f_\varphi'=0$. This implies that f_φ is a constant.

The last component is

$$\frac{1}{c}\frac{\partial B_\varphi}{\partial t}+\frac{1}{r}\left\{\frac{\partial\left[rE_\theta\right]}{\partial r}-\frac{\partial E_r}{\partial\theta}\right\}=0$$

$$\frac{1}{ic}\frac{e}{4\pi\varepsilon_o}\frac{\partial\left[\frac{\left(N_1-N_2\right)}{r}\left(F_{r\theta}-\frac{N_2 F_\theta}{r}\right)\right]}{\partial t}+\frac{1}{r}\frac{e}{4\pi\varepsilon_o}\left\{\frac{\partial\left[r\frac{\left(N_0-N_2\right)}{icr}F_{t\theta}\right]}{\partial r}-\frac{\partial\frac{\left(N_0-N_1\right)}{ic}F_{tr}}{\partial\theta}\right\}=0 \qquad (7.41)$$

$$\left[\frac{\left(N_1-N_2\right)}{r}\left(F_{r\theta t}-\frac{N_2 F_{\theta t}}{r}\right)\right]+\frac{1}{r}\left\{\left(N_0-N_2\right)F_{t\theta r}-\left(N_0-N_1\right)F_{tr\theta}\right\}=0$$

$$\left(N_1-N_2\right)\left(2F_{r\theta t}-\frac{N_2 F_{\theta t}}{r}\right)=0$$

$$\left(N_1-N_2\right)\left(2f_t'f_r'f_\theta'f_\varphi f_\gamma-\frac{N_2 f_t'f_r f_\theta f_\varphi f_\gamma}{r}\right)=0$$

This is true if $f_\theta'=0$ or if $N_1=N_2$. Since $f_\theta'\neq 0$ by Equation (7.37) we should choose $N_1=N_2$.

Therefore Equation (7.10)[b] requires $N_2=N_3$, $f'_\varphi = 0$ and $N_1=N_2$.

After two of the eight equations we must have

$$\ln f^{\frac{1}{2}} = f_t(t)\left(\frac{r_o}{r}\right)e^{-\left(\frac{\lambda_N}{r}\right)}D\sin^{\lambda_0}\theta f_\gamma(\gamma).$$

$$(7.42)$$

and $N_1 = N_2 = N_3$

Since the unitless constant D is unknown we may combine it with the integration constant for the radial equation and write

$$\ln f^{\frac{1}{2}} = f_t(t)\left(\frac{r_o}{r}\right)e^{-\left(\frac{\lambda_N}{r}\right)}\sin^{\lambda_0}\theta f_\gamma(\gamma)$$

$$(7.43)$$

and $N_1 = N_2 = N_3$

Further, we may summarize by looking at the field matrix

$$\frac{F_{ij}}{\left(\dfrac{e}{4\pi\varepsilon_o}\right)} = \begin{vmatrix} 0 & \dfrac{(N_0-N_1)}{ic}F_{tr} & \dfrac{(N_0-N_2)}{icr}F_{t\theta} & 0 & \dfrac{a_o(N_0-N_4)}{ic}F_{t\gamma} \\[2ex] -\dfrac{(N_0-N_1)}{ic}F_{tr} & 0 & 0 & 0 & a_o(N_1-N_4)F_{r\gamma} \\[2ex] -\dfrac{(N_0-N_2)}{icr}F_{t\theta} & 0 & 0 & 0 & \dfrac{a_o(N_2-N_4)}{r}F_{\theta\gamma} \\[2ex] 0 & 0 & 0 & 0 & 0 \\[2ex] -\dfrac{a_o(N_0-N_4)}{ic}F_{t\gamma} & -a_o(N_1-N_4)F_{r\gamma} & -\dfrac{a_o(N_2-N_4)}{r}F_{\theta\gamma} & 0 & 0 \end{vmatrix} \quad (7.44)$$

Therefore, our field tensor has a reduced number of non-zero components.

Next consider Equation (7.10)[c]

$$\overline{\nabla}\times\overline{B} - \frac{1}{c}\frac{\partial\overline{E}}{\partial t} + a_0\frac{\partial\overline{V}}{\partial\gamma} = \frac{4\pi\overline{J}}{c}$$

$$(7.45)$$

where we have not specified anything about the current that may be flowing. If we now suppose the there is no current flowing the $\hat{r}$ component of Equation (7.45) gives us

$$\frac{1}{r\sin\theta}\left[\frac{\partial}{\partial\theta}\left(\sin\theta B_{\varphi}\right)-\frac{\partial B_{\theta}}{\partial\varphi}\right]-\frac{1}{c}\frac{\partial E_{r}}{\partial t}+a_{0}\frac{\partial V_{r}}{\partial\gamma}=0$$

$$-\frac{1}{ic}\frac{e}{4\pi\varepsilon_{o}}\frac{\partial\left\{\dfrac{(N_{0}-N_{1})}{ic}F_{ir}\right\}}{\partial t}+a_{0}\frac{e}{4\pi\varepsilon_{o}}\frac{\partial\left\{a_{o}(N_{1}-N_{4})F_{r\gamma}\right\}}{\partial\gamma}=0 \qquad (7.46)$$

$$\frac{(N_{0}-N_{1})}{c^{2}}F_{itr}+a_{o}^{2}(N_{1}-N_{4})F_{r\gamma\gamma}=0$$

$$\frac{(N_{0}-N_{1})}{c^{2}}\frac{f_{t}''}{f_{t}}+a_{o}^{2}(N_{1}-N_{4})\frac{f_{\gamma}''}{f_{\gamma}}=0$$

$$\frac{f_{t}''}{f_{t}}=\frac{-a_{o}^{2}c^{2}(N_{1}-N_{4})}{(N_{0}-N_{1})}\frac{f_{\gamma}''}{f_{\gamma}}=\text{constant}\equiv H_{o}^{2}$$

due to the vanishing of the magnetic components.

The $\hat{\theta}$ component implies that

$$\frac{1}{r}\left[\frac{1}{\sin\theta}\frac{\partial B_{r}}{\partial\varphi}-\frac{\partial\left(rB_{\varphi}\right)}{\partial r}\right]-\frac{1}{c}\frac{\partial E_{\theta}}{\partial t}+a_{o}\frac{\partial V_{\theta}}{\partial\gamma}=0$$

$$\Rightarrow \frac{1}{c}\frac{\partial E_{\theta}}{\partial t}=a_{o}\frac{\partial V_{\theta}}{\partial\gamma}$$

$$\frac{\partial\left\{-\dfrac{(N_{0}-N_{2})}{cr}\dfrac{e}{4\pi\varepsilon_{o}}F_{t\theta}\right\}}{\partial t}=a_{o}c\frac{\partial\left\{\dfrac{a_{o}(N_{2}-N_{4})}{r}\dfrac{e}{4\pi\varepsilon_{o}}F_{\theta\gamma}\right\}}{\partial\gamma}$$

$$-\frac{(N_{0}-N_{2})}{cr}\frac{e}{4\pi\varepsilon_{o}}\frac{\partial\left\{f_{t}'\left(\dfrac{r_{0}}{r}\right)e^{-\left(\frac{\lambda_{N}}{r}\right)}\lambda_{0}\sin^{(\lambda_{0}-1)}\theta\cos\theta f_{\gamma}\right\}}{\partial t}$$

$$=a_{o}^{2}c\frac{(N_{2}-N_{4})}{r}\frac{e}{4\pi\varepsilon_{o}}\frac{\partial\left\{f_{t}\left(\dfrac{r_{0}}{r}\right)e^{-\left(\frac{\lambda_{N}}{r}\right)}\lambda_{0}\sin^{(\lambda_{0}-1)}\theta\cos\theta f_{\gamma}'\right\}}{\partial\gamma}$$

$$\frac{f_{t}''}{f_{t}}=-a_{o}^{2}c^{2}\frac{(N_{2}-N_{4})}{(N_{0}-N_{2})}\frac{f_{\gamma}''}{f_{\gamma}}=\text{constant}\equiv H_{o}^{2} \qquad (7.47)$$

Since $N_{1}=N_{2}$ the equation that the time function must satisfy is

$$\frac{f_t''}{f_t} = H_o^2$$

$$\Rightarrow f_t'' = H_o^2 f_t \qquad\qquad (7.48)$$

$$\Rightarrow f_t = e^{\pm H_o t} = e^{-H_o t}$$

The units of H_o are 1/sec. It may be noted that if $H_o t$ is small compared to one then

$$f_t = e^{-H_o t} = 1 - H_o t + \frac{H_o^2 t^2}{2!} \pm \ldots \approx (1 - H_o t) \qquad (7.49)$$

This argues that the gravitational field, which is the only field component left that is not a time derivative, must be a function of time. This is borne out in later studies and is the reason for the assignment of the constant in Equation (7.49) as the Hubble constant, H_o.

Recalling Equation (7.47) we find we can replace the second order derivative with respect to y with

$$f_\gamma'' = \frac{H_o^2 \left(N_0 - N_2 \right)}{a_o^2 c^2 \left(N_4 - N_2 \right)} f_\gamma \equiv K_\gamma^2 f_\gamma$$

$$\Rightarrow f_\gamma = e^{-K_\gamma y} \approx 1 - \frac{H_o}{a_o c} \sqrt{\frac{\left(N_0 - N_2 \right)}{\left(N_4 - N_2 \right)}} y \qquad (7.50)$$

K_γ has units of m³/kg.

The $\hat{\varphi}$ component is

$$\frac{1}{r}\left\{ \frac{\partial [rB_\theta]}{\partial r} - \frac{\partial B_r}{\partial \theta} \right\} - \frac{1}{c}\frac{\partial E_\varphi}{\partial t} + a_0 \frac{\partial V_\varphi}{\partial \gamma} = 0$$

$$\frac{1}{r}\left\{ -\frac{\partial B_r}{\partial \theta} \right\} + a_0 \frac{\partial V_\varphi}{\partial \gamma} = 0$$

$$\frac{1}{r}\left\{ -\frac{\partial 0}{\partial \theta} \right\} + a_0 \frac{\partial V_\varphi}{\partial \gamma} = 0 \qquad (7.51)$$

$$a_0 \frac{\partial V_\varphi}{\partial \gamma} = 0$$

This is true because $f'_\varphi = 0$ which means that $V_\varphi = 0$.

Now a summary of our solution is

$$\ln f^{\frac{1}{2}} = \left(\frac{r_o}{r}\right) e^{-\left(\frac{\lambda_N}{r}\right)} \sin^{\lambda_0}\theta\, e^{-H_o t}\, e^{-K_\gamma \gamma}$$

$$K_\gamma^2 = \frac{H_o^2\left(N_0 - N_2\right)}{a_o^2 c^2\left(N_4 - N_2\right)} \qquad (7.52)$$

and $N_1 = N_2 = N_3$

Now consider Equation (7.10)[d] which is

$$\overline{\nabla} \bullet \overline{E} + a_0 \frac{\partial V_4}{\partial \gamma} = 4\pi\rho . \qquad (7.53)$$

Thus,

$$\left(\frac{1}{r^2}\right)\frac{\partial\left(r^2 E_r\right)}{\partial r} + \left(\frac{1}{r\sin\theta}\right)\frac{\partial\left(\sin\theta E_\theta\right)}{\partial\theta} + \left(\frac{1}{r\sin\theta}\right)\frac{\partial E_\phi}{\partial\phi} + a_0\frac{\partial V_4}{\partial \gamma} = 4\pi\varepsilon_o\rho$$

$$. \ (7.54)$$

$$\left(\frac{1}{r^2}\right)\frac{\partial\left(r^2 E_r\right)}{\partial r} + a_0\frac{\partial V_4}{\partial \gamma} = 4\pi\varepsilon_o\rho$$

But suppose $N_0 = N_4$ so that $V_4 = 0$, then Equation (7.54) reads

$$\rho = \left(\frac{1}{4\pi r^2}\right)\frac{\partial\left(r^2 E_r\right)}{\partial r} . \qquad (7.55)$$

Now we can integrate this over all space to get the total charge as

$$q = \iiint_{vol} \rho r^2 dr d\theta \sin\theta d\varphi = \iiint_{vol} \left\{ \left(\frac{1}{4\pi r^2}\right) \frac{\partial\left(r^2 E_r\right)}{\partial r} \right\} r^2 dr d\theta \sin\theta d\varphi$$

$$= \iiint_{vol} \left\{ \frac{\left(N_0 - N_1\right) r_o H_o}{4\pi c} ee^{-H_o t} e^{-K_\gamma \gamma} \left\{ \left(\frac{1}{r^2}\right) \frac{\partial\left[\left(1-\frac{\lambda_N}{r}\right)e^{-\frac{\lambda_N}{r}}\right]}{\partial r} \right\} \right\} r^2 dr d\theta \sin\theta d\varphi$$

$$= \frac{\left(N_0 - N_1\right) r_o H_o}{c} ee^{-H_o t} e^{-K_\gamma \gamma} \int_{r=0}^{r=\infty} \left\{ \frac{d\left[\left(1-\frac{\lambda_N}{r}\right)e^{-\frac{\lambda_N}{r}}\right]}{dr} \right\} dr$$

$$= \frac{\left(N_0 - N_1\right) r_o H_o}{c} ee^{-H_o t} e^{-K_\gamma \gamma} \left\{ \int_{r=0}^{r=\infty} d\left[\left(1-\frac{\lambda_N}{r}\right)e^{-\frac{\lambda_N}{r}}\right] \right\}$$

$$= \frac{\left(N_0 - N_1\right) r_o H_o}{c} ee^{-H_o t} e^{-K_\gamma \gamma} \left\{ \left[\left(1-\frac{\lambda_N}{r}\right)e^{-\frac{\lambda_N}{r}} - \left(1-\frac{\lambda_N}{r}\right)e^{-\frac{\lambda_N}{r}}\right]_{r=0}^{r=\infty} \right\}$$

$$= \frac{\left(N_0 - N_1\right) r_o H_o}{c} ee^{-H_o t} e^{-K_\gamma \gamma} \left\{ \left[\left(1-\frac{\lambda_N}{\infty}\right)e^{-\frac{\lambda_N}{\infty}} - \left(1-\frac{\lambda_N}{0}\right)e^{-\frac{\lambda_N}{0}}\right]_{r=0}^{r=\infty} \right\} \qquad (7.56)$$

$$= \frac{\left(N_0 - N_1\right) r_o H_o}{c} ee^{-H_o t} e^{-K_\gamma \gamma} \left\{ (1-0)e^0 - 0 \right\}$$

$$= \frac{\left(N_0 - N_1\right) r_o H_o}{c} ee^{-H_o t} e^{-K_\gamma \gamma}$$

But Equation (7.10)[g] also involves V_4, or

The Dynamic Theory

$$\overline{\nabla}V_4 + \frac{1}{c}\frac{\partial \overline{V}}{\partial t} = a_0 \frac{\partial \overline{E}}{\partial \gamma}$$

$$\left(\frac{\partial 0}{\partial r} + \frac{1}{c}\frac{\partial V_r}{\partial t} - a_0 \frac{\partial E_r}{\partial \gamma} \right)\hat{r}$$

$$+\left(\frac{1}{c}\frac{\partial V_\theta}{\partial t} - a_0 \frac{\partial E_\theta}{\partial \gamma} \right)\hat{\theta} \qquad\qquad . \qquad (7.57)$$

$$+\left(\frac{1}{c}\frac{\partial V_\varphi}{\partial t} - a_0 \frac{\partial E_\varphi}{\partial \gamma} \right)\hat{\varphi} \equiv \overline{0}$$

$$\left(\frac{1}{c}\frac{\partial V_r}{\partial t} - a_0 \frac{\partial E_r}{\partial \gamma} \right)\hat{r} + (0)\hat{\theta} + (0\ )\hat{\varphi} = \overline{0}$$

Thus, Equation (7.57) reduces to the single equation

$$\frac{1}{c}\frac{\partial V_r}{\partial t} = a_0 \frac{\partial E_r}{\partial \gamma}$$

$$\frac{1}{ic}\frac{\partial \left\{ a_o \dfrac{e}{4\pi\varepsilon_o}(N_1 - N_4)F_{r\gamma} \right\}}{\partial t} = a_0 \frac{\partial \left\{ \dfrac{(N_0 - N_1)}{ic}\dfrac{e}{4\pi\varepsilon_o}F_{tr} \right\}}{\partial \gamma} . \quad (7.58)$$

$$\left(N_1 - N_4 \right)F_{r\gamma t} = -\left(N_0 - N_1 \right)F_{tr\gamma}$$

which is satisfied because $N_0 = N_4$.

The continuity equation, Equation (7.10)[e] also provides a tie between these two, as

$$0 = \frac{\partial \rho}{\partial t} + \overline{\nabla} \bullet \overline{J} + a_0 \frac{\partial J_4}{\partial \gamma}$$

$$\Rightarrow \frac{\partial \left\{ \left(\dfrac{1}{4\pi r^2} \right) \dfrac{\partial \left(r^2 E_r \right)}{\partial r} \right\}}{\partial t} + a_0 \frac{\partial J_4}{\partial \gamma} = 0$$

$$\frac{\partial J_4}{\partial \gamma} = \frac{H_o}{a_o} \left(\frac{1}{4\pi r^2} \right) \frac{\partial \left(r^2 E_r \right)}{\partial r} \qquad . \text{(7.59)}$$

$$J_4 = \int \frac{H_o}{a_o} \left(\frac{1}{4\pi r^2} \right) \frac{\partial \left(r^2 E_r \right)}{\partial r} d\gamma = \frac{-H_o}{a_o K_\gamma} \left(\frac{1}{4\pi r^2} \right) \frac{\partial \left(r^2 E_r \right)}{\partial r}$$

$$= \frac{-H_o}{a_o K_\gamma} \rho$$

Equation (7.10)[f] may be considered next. Thus,

$$\overline{\nabla} \times \overline{V} + a_0 \frac{\partial \overline{B}}{\partial \gamma} = \overline{0}. \qquad \text{(7.60)}$$

This is a vector equation and has three components, The $\hat{r}$ component is

$$\frac{1}{r \sin \theta} \left[\frac{\partial}{\partial \theta} \left(\sin \theta V_\varphi \right) - \frac{\partial B_\theta}{\partial \varphi} \right] + a_0 \frac{\partial B_r}{\partial \gamma} = 0 \qquad \text{(7.61)}$$

which is satisfied identically due to the vanishing of the magnetic field components and the V_φ.

The $\hat{\theta}$ component is

$$\frac{1}{r} \left[\frac{1}{\sin \theta} \frac{\partial V_r}{\partial \varphi} - \frac{\partial \left(r V_\varphi \right)}{\partial r} \right] + a_0 \frac{\partial B_\theta}{\partial \gamma} = \overline{0} \qquad \text{(7.62)}$$

which is also satisfied identically.

The $\hat{\varphi}$ component is also an identity since

$$\frac{1}{r} \left\{ \frac{\partial \left[r V_\theta \right]}{\partial r} - \frac{\partial V_r}{\partial \theta} \right\} + a_0 \frac{\partial B_\varphi}{\partial \gamma} = 0. \qquad \text{(7.63)}$$

192

The remaining equation to be checked is Equation (7.10)[h], which is

$$\bar{\nabla}\bullet\bar{V}+\frac{1}{c}\frac{\partial V_4}{\partial t}=-\frac{4\pi\mu J_4}{c}$$

$$\left(\frac{1}{r^2}\right)\frac{\partial\left(r^2 V_r\right)}{\partial r}+\left(\frac{1}{r\sin\theta}\right)\frac{\partial\left(\sin\theta V_\theta\right)}{\partial\theta}+\left(\frac{1}{r\sin\theta}\right)\frac{\partial V_\phi}{\partial\phi}+\frac{1}{c}\frac{\partial V_4}{\partial t}=-\frac{4\pi\mu J_4}{c} \tag{7.64}$$

$$\left(\frac{1}{r^2}\right)\frac{\partial\left(r^2\left\{a_o\left(N_1-N_4\right)\dfrac{e}{4\pi\varepsilon_o}r_o K_\gamma\left(\dfrac{1}{r^2}\right)\left(1-\dfrac{\lambda_N}{r}\right)e^{-\frac{\lambda_N}{r}}D\sin^{\lambda_o}\theta e^{-K_\gamma\gamma}e^{-H_o t}\right\}\right)}{\partial r}=-\frac{4\pi\mu J_4}{c}$$

$$J_4=-\frac{a_o c^2\left(N_1-N_4\right)K_\gamma}{4\pi\left(N_0-N_1\right)H_o}\left\{\left(\frac{e}{4\pi\varepsilon_o}\right)\left(\frac{r_o}{r^2}\right)\frac{\partial\left(r^2\left\{-\left(N_0-N_1\right)\dfrac{H_o}{c}\left(\dfrac{1}{r^2}\right)\left(1-\dfrac{\lambda_N}{r}\right)e^{-\frac{\lambda_N}{r}}D\sin^{\lambda_o}\theta e^{-K_\gamma\gamma}e^{-H_o t}\right\}\right)}{\partial r}\right\}$$

$$J_4=-\frac{a_o c^2\left(N_1-N_4\right)K_\gamma}{4\pi\left(N_0-N_1\right)H_o}\rho=\frac{a_o c^2 K_\gamma}{H_o}\rho$$

Summarizing our solution is now

$$\ln f^{\frac{1}{2}}=\left(\frac{r_o}{r}\right)e^{-\left(\frac{\lambda_N}{r}\right)}\sin^{\lambda_o}\theta e^{-H_o t}e^{-K_\gamma\gamma}$$

$$\lambda_o\ll 1\Rightarrow\sin^{\lambda_o}\theta\cong 1$$

$$N_0=N_4$$

$$K_\gamma=\frac{H_o}{a_o c} \tag{7.65}$$

$$Z=\left(N_0-N_1\right)$$

$$J_4=\frac{a_o c^2 K_\gamma}{H_o}\rho=c\rho$$

and $N_1=N_2=N_3$

In order for the charge of the particle to be equal to the experimentally determined value of Ze then we must have

$$q = Ze = \frac{(N_0 - N_1)r_o H_o}{c} e e^{-H_o t} e^{-K_\gamma \gamma}$$

$$\Rightarrow r_o = \frac{c}{H_o} \tag{7.66}$$

Then our summary may be written as

$$\ln f^{\frac{1}{2}} = \left(\frac{c}{H_o}\right)\left(\frac{1}{r}\right) e^{-\left(\frac{\lambda_N}{r}\right)} \sin^{\lambda_0}\theta\, e^{-H_o t} e^{-\left(\frac{H_o}{a_o c}\right)\gamma}$$

$$\lambda_o \ll 1 \Rightarrow \sin^{\lambda_0}\theta \cong 1$$

$$N_0 = N_4$$

$$Z = (N_0 - N_1) \tag{7.67}$$

$$q = Zee^{-H_o t} e^{-K_\gamma \gamma}$$

$$J_4 = c\rho$$

and $N_1 = N_2 = N_3$

Our field tensor is now given by

$$\frac{F_{ij}}{\left(\frac{e}{4\pi\varepsilon_o}\right)} \approx \begin{vmatrix} 0 & Z\left(\frac{1}{r^2}\right)\left(1-\frac{\lambda_N}{r}\right)e^{-\left(\frac{\lambda_N}{r}\right)} & Zr_o\lambda_0\left(\frac{1}{cr^2}\right)e^{-\left(\frac{\lambda_N}{r}\right)}\cot\theta & 0 & 0 \\ -Z\left(\frac{1}{r^2}\right)\left(1-\frac{\lambda_N}{r}\right)e^{-\left(\frac{\lambda_N}{r}\right)} & 0 & 0 & 0 & -Z\left(\frac{1}{r^2}\right)\left(1-\frac{\lambda_N}{r}\right)e^{-\frac{\lambda_N}{r}} \\ -Zr_o\lambda_0\left(\frac{1}{cr^2}\right)e^{-\left(\frac{\lambda_N}{r}\right)}\cot\theta & 0 & 0 & 0 & -Z\left(\frac{1}{r^2}\right)e^{-\left(\frac{\lambda_N}{r}\right)}\cot\theta \\ 0 & 0 & 0 & 0 & 0 \\ 0 & Z\left(\frac{1}{r^2}\right)\left(1-\frac{\lambda_N}{r}\right)e^{-\frac{\lambda_N}{r}} & Z\left(\frac{1}{r^2}\right)e^{-\left(\frac{\lambda_N}{r}\right)}\cot\theta & 0 & 0 \end{vmatrix} \tag{7.68}$$

This forms a complete solution set.

Electric to Gravitational Force Ratio

The electric field is

$$E_r = Z\left(\frac{e}{4\pi\varepsilon_o r^2}\right)\left(1-\frac{\lambda_N}{r}\right)e^{-\left(\frac{\lambda_N}{r}\right)} e^{-H_o t} e^{-K_\gamma \gamma} \tag{7.69}$$

The electrical force between identical particles would then be given by

194

$$F_e = eE_r \approx Z\left(\frac{e^2}{4\pi\varepsilon_o r^2}\right)\left(1-\frac{\lambda_N}{r}\right)e^{-\left(\frac{\lambda_N}{r}\right)}e^{-H_o t}e^{-K_\gamma \gamma}. \quad (7.70)$$

The gravitational field is

$$V_r = -Z\left(\frac{e}{4\pi\varepsilon_o r^2}\right)\left(1-\frac{\lambda_N}{r}\right)e^{-\left(\frac{\lambda_N}{r}\right)}e^{-H_o t}e^{-K_\gamma \gamma}. \quad (7.71)$$

However the units of the gravitational field are volt/m here. We need to look at the units conversions. To convert the units of the gravitational field from volts/m to N/kg we need to multiply by the charge to mass ratio, $\beta \equiv \sqrt{4\pi\varepsilon_o G}$, to get

$$V_g = V_r\beta = -Z\beta\left(\frac{e}{4\pi\varepsilon_o r^2}\right)\left(1-\frac{\lambda_N}{r}\right)e^{-\left(\frac{\lambda_N}{r}\right)}e^{-H_o t}e^{-K_\gamma \gamma}. \quad (7.72)$$

To get the gravitational mass we need to divide the gravitational charge by the charge to mass ratio, β, to get

$$m = \frac{M}{c\beta} \approx -\frac{e}{\beta}. \quad (7.73)$$

The gravitational field then is given by

$$V_g = V_r\beta = -Z\beta\left(\frac{m\beta}{4\pi\varepsilon_o r^2}\right)\left(1-\frac{\lambda_N}{r}\right)e^{-\left(\frac{\lambda_N}{r}\right)}e^{-H_o t}e^{-K_\gamma \gamma}. \quad (7.74)$$

Now we can form the gravitational force by multiplying Equation (7.74) by the gravitational mass of our second particle which is also m and obtain

$$F_g = mV_g = -Zm^2\beta^2\left(\frac{1}{4\pi\varepsilon_o r^2}\right)\left(1-\frac{\lambda_N}{r}\right)e^{-\left(\frac{\lambda_N}{r}\right)}e^{-H_o t}e^{-K_\gamma \gamma}. \quad (7.75)$$

Forming the absolute value of the ratio of the electrical force and the gravitational force produces

$$|F_{ratio}| \approx \left| \frac{Z\left(\dfrac{e^2}{4\pi\varepsilon_o r^2}\right)\left(1-\dfrac{\lambda_N}{r}\right)e^{-\left(\frac{\lambda_N}{r}\right)}e^{-H_o t}e^{-K_\gamma \gamma}}{-Zm^2\beta^2\left(\dfrac{1}{4\pi\varepsilon_o r^2}\right)\left(1-\dfrac{\lambda_{N'}}{r}\right)e^{-\left(\frac{\lambda_N}{r}\right)}e^{-H_o t}e^{-K_\gamma \gamma}} \right|. \quad (7.76)$$

$$= \frac{e^2}{m^2\beta^2} = \frac{e^2}{m^2\left(4\pi\varepsilon_o G\right)} = \frac{\dfrac{e^2}{4\pi\varepsilon_o}}{Gm^2}$$

This is the classical ratio of the electric and gravitational forces.

Our expression for the mass, m, is only good for Weyl Unity Particles (WUP) and not for a composite particle. Therefore, we should only use a particle such as the electron in Equation (7.76).

A few things may be seen from this exercise. These include:

1. The five dimensional gauge fields and the Weyl Quantum Principle give us both the electrostatic and the gravitational fields that have been measured on fundamental particles such as the electron. These descriptions are complete to the point of specifying the value of the ratio of the electric force to the gravitational force between two identical particles.

2. The gravitational and electric fields depend on both time and mass density. It is the time dependence that keeps the electrostatic potential from vanishing and it is the mass density dependence that keeps the gravitational field non-zero.

3. The exponential time dependence leads to decreasing electric and gravitational field with the passing of time. This in turn leads to an expanding universe due to the decreasing gravitational field with the exponential coefficient, Ho, acting as the first order expansion coefficient which is Hubble's

constant. Indeed if we watched the gravitational field as time proceeded to infinity we would find the gravitational field would become zero, or vanish.

4. Hubble's constant is a small number. Yet the dependence of the field upon the mass density is an exponential dependence that is smaller still. The exponential coefficient of the mass density is the small Hubble's constant divided by two large positive numbers. One of these numbers is the speed of light which determines the limiting rate of change of position with respect to time. The other is the limiting mass density gradient that may be found in strong shocks. The product of this limiting mass density gradient and the speed of light is the limiting rate at which mass may be converted into energy.

5. We see that the gravitational field depends in two ways upon the mass density. The first dependence is as a multiplicative product of the radial dependence and when the gravitational field is integrated over a volume this dependence is the primary role of mass density in determining the mass contained in the volume. However, we see that the exponential dependence of the gravitational field upon m ass density is such that as the mass density increases to infinity the gravitational field vanishes.

6. The above conclusions are made using the fields of a fundamental particle we have defined and called WUPs as they were particles required to have a unitary scale factor. However, a macroscopic mass made up of these fundamental particles should be expected to exhibit these same variations with respect to time and mass. Indeed in other parts of this book has been, or will be, shown that the time

dependence of the gravitational field shows up in stellar red shifts, dark matter and dark energy data.

Proton-Proton Scattering

There is an astonishing effect of the non-singular potential upon how one might describe nuclear phenomena. One of the first features noted about the potential was its return to a zero value as the radial value approaches zero. This has the effect of producing a force given by,

$$F_r = \frac{eZk}{r^2}\left(1-\frac{\lambda}{r}\right)e^{-\frac{\lambda}{r}}. \qquad (7.77)$$

If this force is repulsive when r is infinite for like particles, it becomes zero when the separation is at the distance lambda and becomes a strongly attractive force when the separation becomes less than lambda. This is just the sort of behavior found when proton-proton scattering was first done at high enough energies to see a deviation from Coulombic scattering. The expression for the non-singular scattering cross-section was found to be

$$d\sigma = \left(\frac{q_1 q_2}{2mV_0^2}\right)\left[\frac{2\pi\sin\theta\, d\theta}{\sin^4(\frac{\theta}{2})}\right]\delta,$$

where

$$\delta = \frac{1+6\left(\frac{4\lambda E}{k}\right)^2 \sin^4(\frac{\theta}{2})\left[1+\frac{1}{2}(\pi-\theta)\tan(\frac{\theta}{2})\right]}{\left[1+\frac{3}{2}\left(\frac{4\lambda E}{k}\right)^2 \sin^2(\frac{\theta}{2})\sin\theta(\pi-\theta)\right]^4}.$$

This scattering cross section for like-particle interaction appears to have the right dependencies to explain the scattering data. It remains to compare this prediction with existing experimental data to determine the validity of the prediction and the ability of the non-singular potential to explain the proto-proton scattering with that portion of its radial dependence that causes the value of the potential to return to zero.

Fundamental Particle Fields

Chapter 8 Electromagnetic Fields and Wave-Particle Duality

In the previous chapter we discussed Einstein's use of a curved geometry instead of forces to describe the motion of a particle under the influence of gravity. Then we described how Weyl extended the geometrical theme by allowing the length of a vector to vary from point to point in the space. We discussed how Schrödinger and London noticed that quantum mechanics appeared as a subset of Weyl's Gauge Principle when one restricts their attention to a unity scale factor. We used this restriction to state Weyl's Quantum Principle (WQP) and showed how the WQP led to the functional form for the electrostatic and gravitational fields for a Weyl Unitary Particle.

The objective of this chapter is to present a couple of immediate results of the WQP that offer a different basis for interpretation of quantum phenomena and the prediction, with a means for its experimental verification, of very different energy levels of light quanta. Not only does this new light quanta energy level prediction expand Einstein's $\varepsilon = h\nu$ prediction, but it also presents an entirely new view of the almost century old wave-particle duality of light.

In Chapter 6 the extension of geometry Weyl used to place electromagnetism on a geometrical basis was presented. It was also displayed how he used his gauge principle to derive the Maxwell equations of electromagnetism. As we proceed this derivation may be referred to from time to time.

Uncertainty for Gauge Fields

In Chapter 6 we saw that Weyl's Quantum Principle required both Maxwell's electromagnetism and

Schrödinger's quantum mechanics. We have showed Weyl's derivation of Maxwell's electromagnetism and discussed London's derivation of Schrödinger's quantum mechanics. However, we shall skip London's derivation due to its length and complexity.

In Chapter 7 we presented the derivation of the gauge potentials and fields for a fundamental particle that satisfied Weyl's Quantum Principle and had a unity scale factor. These potentials were gauge potentials that came from a gauge function and led to gauge fields. There is another concept that generally is conceptually tied to quantum mechanics that we should present at this time because it depends upon the gauge function. This is the concept termed Heisenberg's Uncertainty Principle.

Heisenberg's Uncertainty Principle has been employed in a myriad of ways in relation to numerous situations. Interpretations resulting from the use of Heisenberg's Uncertainty Principle may change after noting the following results of the influence that a gauge function has upon the quantum Poisson brackets.

A gauge function, such as the electromagnetic gauge, has a g eometrical effect that could be thought of as effectively changing the unit of action in quantum mechanics. To show this we shall begin with the classical Poisson bracket.

The classical Poisson bracket is defined by

$$\{F,G\} = \sum_{j} \left(\frac{\partial F}{\partial q_j} \frac{\partial G}{\partial p_j} - \frac{\partial F}{\partial p_j} \frac{\partial G}{\partial q_j} \right) \, ,$$

where F and G are any two functions of the canonically conjugate variables q_j and p_j. The special relations that occur when F and G are q_j and p_j, respectively, are especially important in quantum mechanics; these are, classically:

The Dynamic Theory

$$\left\{q_j, q_k\right\} = 0$$

$$\left\{p_j, p_k\right\} = 0, \text{ and} \qquad (8.1)$$

$$\left\{q_j, p_k\right\} = \delta_{jk},$$

where δ_{jk} is the Kronecker delta. The classical Poisson brackets of Equation (8.1) are obtained when Euclidean spaces are assumed. However, the definition of Poisson brackets remains valid for general metric spaces, where the notion of covariant differentiation is used. If we now consider the components of the momentum, expressed in a general coordinate system, the covariant and contravariant components are,

$$p_j = mg_{jk}\dot{x}^k, \text{ and}$$

$$p^j = g^{jk}p_k. \qquad (8.2)$$

Covariant differentiation must be carried out with respect to contravariant vector components. In a general space the canonically conjugate variables to be considered are x^j and p^k, and the Poisson bracket of the position and the components of the momentum become

$$\left\{x^j, p^k\right\} = \left[\frac{\partial x^j}{\partial x^l} + \left\{\begin{matrix}j\\s\,l\end{matrix}\right\}x^2\right]\frac{\partial p^k}{\partial p^l}$$

$$-\frac{\partial x^j}{\partial p^l}\left[\frac{\partial p^k}{\partial x^l} + \left\{\begin{matrix}k\\n\,l\end{matrix}\right\}p^n\right] \qquad (8.3)$$

$$= \left[\delta_{jl} + \left\{\begin{matrix}j\\s\,l\end{matrix}\right\}x^s\right]\delta_{lk}.$$

or

$$\left\{x^j, p^k\right\} = \delta_{jk} + \left\{\begin{matrix}j\\s\,k\end{matrix}\right\}x^s.$$

Quantum mechanics adopts the operator,

Electromagnetic Fields and Wave Particle Duality

$$\left(\frac{\hbar}{i}\right)\frac{\partial}{\partial x^j} \to p_j \ ,$$

for the momentum. This, in the general space, becomes the covariant operator

$$\left(\frac{\hbar}{i}\right)(\quad)_{,j} \to p_j \ . \tag{8.4}$$

The operator for the contravariant momentum components is then

$$\left(\frac{\hbar}{i}\right)g^{jl}(\quad)_{,l} \to p^j \ . \tag{8.5}$$

Now if we look at the quantum Poisson bracket, where the operators are operating on a scalar ψ, then

$$[x^j, p^k]\psi = \left[x^j\left(\frac{\hbar}{i}\right)g^{kl}\frac{\partial\psi}{\partial x^l} - \left(\frac{\hbar}{i}\right)g^{kl}(x^j\psi)_{,l}\right]$$

$$= x^j\left(\frac{\hbar}{i}\right)g^{kl}\frac{\partial\psi}{\partial x^l} - g^{kl}\left(\frac{\hbar}{i}\right)\left[\frac{\partial x^j}{\partial x^l} + \left\{\begin{matrix} j \\ s\,l \end{matrix}\right\}x^s\right]\psi$$

$$- \left(\frac{\hbar}{i}\right)g^{jl}x^j\frac{\partial\psi}{\partial x^l} \tag{8.6}$$

$$= i\hbar\, g^{kl}\left[\delta_{jl} + \left\{\begin{matrix} j \\ s\,l \end{matrix}\right\}x^s\right]\psi \ .$$

This may be written in terms of the classical Poisson bracket, Equation (2.10), as

$$\left[x^j, p^k\right]\psi = i\,\hbar\, g^{kl}\left\{x^j, p^l\right\}\psi \ .$$

If the space is Euclidean the g^{kl} become the Kronecker delta and the Christoffel symbols vanish and the quantum Poisson bracket of Equation (8.6) becomes

$$\left[x^j, p_k\right]\psi = i\hbar\,\delta_{jk}\,\psi \ ,$$

because $p^k = g^{kl}p_l = \delta^{kl}p_l = p_k$. However, from Equation (8.6) we see that the metric does play a role in the quantum operators. This should also be seen in the use of the operators in the Schrödinger Hamiltonian operator, because

$$p_j p^j = \frac{\hbar}{i} \frac{1}{\sqrt{g}} \frac{\partial}{\partial x^j} \left(\sqrt{g}\, p^j \right)$$

$$= \frac{\hbar}{i} \frac{1}{\sqrt{g}} \frac{\partial}{\partial x^j} \left(\sqrt{G} \left(\frac{\hbar}{i} \right) g^{jl} \frac{\partial}{\partial x^l} \right) \qquad (8.7)$$

$$= \frac{\hbar^2}{\sqrt{g}} \frac{\partial}{\partial x^j} \left(\sqrt{g}\, g^{jl} \frac{\partial}{\partial x^l} \right)$$

becomes the operator to be used in a general space and, of course, is the operator used in applying Schrödinger's equation to the hydrogen atom.

The geometrical effect may be seen also in Dirac's equation by considering that the restrictions,

$$(\overline{\alpha} \bullet \overline{p})^2 = p_j p^j \quad ,$$

$$\beta^2 = 1 \quad , \qquad (8.8)$$

$$and$$

$$\overline{\alpha}\beta + \beta\overline{\alpha} = 0 \quad ,$$

must be met in order for solutions of

$$p^o |> = H |> \quad ,$$

where

$$H = -\left(\overline{\alpha} \bullet \overline{p} + \beta m \right) \quad ,$$

is also a solution of $p_j p^j = m^2$ in natural units.

The first restriction may be rewritten as

$$(\alpha p^j)^2 = p_i p^j = m^2 \, g_{jk} \dot{x}^j \dot{x}^k = g^{jk} p^j p^k$$

by the definition of the momentum components. Then, if we expand the left-hand side and equate coefficients of the $p^j p^k$, we find that

Electromagnetic Fields and Wave Particle Duality

$$(\alpha)^2 = -g_{11},$$

$$(\alpha)^2 = -g_{22},$$

$$(\alpha)^2 = -g_{33}, \tag{8.9}$$

$$\alpha_1\alpha_2 + \alpha_2\alpha_3 = \{\alpha_1\alpha_2\} = -2g_{12}\ ,$$

$$\alpha_1\alpha_3 + \alpha_3\alpha_1 = \{\alpha_1\alpha_3\} = -2g_{13}\ ,\ \text{and}$$

$$\alpha_2\alpha_3 + \alpha_3\alpha_2 + \{\alpha_2\alpha_3\} = -2g_{23}\ .$$

From Equations (8.9), for a Euclidian metric where $g_{jk} = \delta_{jk}$, these restrictions reduce to the usual restrictions. Any metric property will affect these restrictions and will therefore feed into the solutions.

Now, of what benefit is this discussion of geometrical effect upon quantum mechanics? A gauge function has a geometrical effect that could be thought of as effectively changing the unit of action in quantum mechanics. To see the basis for this statement, let us restate the quantum Poisson bracket operations on a scalar,

$$\left[x^j, p^k\right]\psi = i\,\hbar g^{kl}\left[\delta_{jl} + \left\{\begin{matrix} j \\ s\ l \end{matrix}\right\}x^s\right]\psi\ , \tag{8.10}$$

and let us define

$$\hbar'\,\delta_{jk} = \hbar\,g^{kl}\left[\delta_{jl} + \left\{\begin{matrix} j \\ s\ l \end{matrix}\right\}x^s\right]\ , \tag{8.11}$$

then we can write

$$\left[x^j, p^k\right]\psi = i\,\hbar'\,\delta_{jk}\ , \tag{8.12}$$

which has the same form as the classical quantum Poisson bracket, but the effective unit of action $\hbar'$ depends on the geometry as seen by Equation (8.11).

Equation (8.12) becomes the statement of Heisenberg's Uncertainty Principle in a general space. It is easy to use Equation (8.11) to see that any uncertainty predicted by this uncertainty principle will depend upon the gauge function at the point in space the principle is to be applied.

How is this conclusion important to us? In all of the interpretations of physical phenomena subsequent to Heisenberg announcing his uncertainty principle a universal value of the uncertainty has been used. Equation (8.12) tells us that a universal uncertainty may not apply to phenomena involving a gauge function. This means that when considering the uncertainty of phenomena of electromagnetism, we must use the effective unit of action, at the point in question, as the value of the uncertainty. This is so because we have just shown that electromagnetism comes from a gauge function. Later we will see just how important a difference this will make when considering nuclear phenomena.

Light Quanta

Phat Photons

The two theories, Maxwell's electromagnetism and Schrödinger's quantum mechanics, that we have shown as being required by Weyl's Gauge and Quantum Principles lead to the resolution of a long standing theoretical problem. The theoretical problem is the wave-particle duality of light.

The concept of a quantum of energy for light started with Einstein's light quanta. The concept has been the subject of many articles since 1905. The name "photon" was first introduced by Lewis 21 years after Einstein's introduction of the concept. However, both Planck's and Einstein's derivations of the famous relation between energy and frequency, $\varepsilon = h\nu$, came from studies of radiation in thermal equilibrium with a system described by statistical thermodynamics. Planck quantized the equilibrium energy, U, of an oscillator while Einstein quantized the entropy density per unit volume. In 1917 Einstein wrote, "The properties of elementary processes required by [his momentum fluctuation relation] make it seem almost inevitable to formulate a truly quantized theory of radiation." Einstein was not, and never would be, satisfied with his, and others', inability to obtain such a theory. In 1924, a fter the experimental evidence of the Compton Effect provided proof of the quantization of light, he wrote, "There are therefore now two theories of light, both indispensable, and-as one must admit today despite twenty years of tremendous effort on the part of theoretical physicists-without any logical connection".

Weyl used his gauge principle to derive Maxwell's electromagnetism. Therefore, the WGP requires a w ave description of all light. However, when the unity scale factor is also required a quantization is introduced. London's work was done in order to look at the possible paths the WQP specified when a charged particle moved in the presence of an electrostatic potential. But we might ask "What about the vector potential components of the gauge potential?" We already know the WGP requires the vector potential components to satisfy the electromagnetic wave equations. We might easily suspect the WQP to also impose a quantization. What might be the resulting description of light waves when the WQP is imposed?

Weyl's Gauge Principle has been used to derive Maxwell's electromagnetic field equations. These, in turn, may be used to derive the electromagnetic wave equations. At the same time Weyl's Quantum Principle requires that for any isentropic propagation of those electromagnetic waves the gauge potentials be quantized since

$$\int \tilde{\phi}_j dx^j = 2\pi i N \tag{8.13}$$

where $i = \sqrt{-1}$, there is no summation over the j's and this

must be true for any isentropic path. If the quantum principle must hold for any path, then by choosing all dx^j to be zero except one, it may be seen that each gauge potential must be quantized and, therefore, the principle may be written as

$$\int N_j \phi_j dx^j = 2\pi i N .\qquad(8.14)$$

The radial electrostatic dependence may be investigated by considering N_0 to be nonzero and $N_x=N_y=N_z=0$. The concept of a photon, as a particle, is one that is electrically neutral. The wave description allows the discussion of polarization such that an electromagnetic wave traveling along the x-axis may have its electric field directed along the y-axis. Therefore, consider $N_0=N_x=N_z=0$. The quantum principle requires that the y component of the gauge (vector) potential to be quantized, as

$$\phi_y = NB \cos 2\pi \left(\frac{x}{\lambda} - vt \right)\qquad(8.15)$$

where the dependence upon x and t was chosen to ensure that the electric field, given by

$$E_y(x,t) = \frac{\partial \phi_y(x,t)}{\partial(ct)} = \frac{NBv}{c} \sin 2\pi \left(\frac{x}{\lambda} - vt \right),\quad(8.16)$$

satisfies the wave equations. This expression may be used to find the average value of the Poynting vector, or

$$I = \left(\frac{1}{\mu_o} \right) \left\langle E^2 \right\rangle .\qquad(8.17)$$

The average value of the square of electric field is given by

$$\left\langle E^2(x,t) \right\rangle = \frac{N^2 B^2 v^2}{c^3} \int_{t=0}^{t=\frac{1}{v}} \sin^2 \left(\frac{x}{\lambda} - vt \right) dt = \frac{N^2 B^2 v}{2c^3} .\qquad(8.18)$$

Therefore,

$$I = \left(\frac{N^2 B^2}{2\mu_o hc^3} \right) v .\qquad(8.19)$$

Electromagnetic Fields and Wave Particle Duality

Now a quantum of light for which $N=1$ have an would energy flow of $I=h\nu$ when

$$B = \sqrt{2\mu_o hc^3} \; . \qquad (8.20)$$

Einstein's energy relation, for a single light quantum passing through a unit area and with $N=1$, is then $\varepsilon=h\nu$. Photons for which $N>1$ have greater energy for their frequency than photons for which $N=1$ and may be called Phat Photons to indicate their higher energy and distinguish them from the $N=1$ photons that are commonly known. Phat photons have an energy given by

$$\varepsilon = N^2 h\nu \qquad (8.21)$$

with $N>1$.

It is a remarkable discovery to learn that the WQP requires that the energy of a photon be quantized. First, it is remarkable that such an approach to the resolution of the wave-particle duality had not been addressed from this point of view. Both Planck and Einstein started their approaches to the quantization of energy using thermal equilibrium considerations. Here the only fundamental principle needed was the WQP which has the WGP embedded within it. The WGP provided electromagnetism and the WQP provided the quantization. In this manner the Weyl Gauge Principle and its restricted subset of unity scale factor provides both the wave and the particle aspects of light.

Should the scale factor be allowed to vary from point to point, electromagnetic waves are allowed with no quantization, or particle-like behavior of photons being required. Indeed this may make up t he majority of electromagnetic waves in nature.

Second, it is remarkable that such photon energy has not been measured. However, on second thought, this is not so remarkable. No one would have thought to look for such energy of a photon since theory did not predict it.

210

The Dynamic Theory

One of the first things that might come to mind after hearing of the concept of phat photons is, "Would a laser emitting phat photons be possible?" We shall now develop the equations related to a phat photon laser.

Phat Photon Energy Density

Planck quantized the allowed energy levels in order to achieve a correct description of the frequencies of light emitted by 'black bodies.' Though the term black body seems to imply no light being emitted, the term is also used to describe the energy density of light emitted by heated, glowing objects.

The quantization of light eventually led to the very useful concept of a laser. The possibility of a laser emitting the more energetic phat photon will now be considered.

Before looking into the emission and absorption ratios needed for the phat photon laser concept, we must first develop the phat photon energy density that will be needed in this investigation. This may be done using the same procedure as used in deriving Planck's energy density. Boltzmann's distribution is used unaltered. The energy of the electromagnetic radiation is discrete with the phat photon requirement that $E = N^2 h\nu$.

The quantity $\overline{E}$ is evaluated from the ratio of sums

$$\overline{E} = \frac{\sum_{n=0}^{\infty} EP(E)}{\sum_{n=0}^{\infty} P(E)}.$$

Sums must be used because with phat photons the energy becomes a discrete variable that takes on only the values $E = 0, h\nu, 4h\nu, 9h\nu,...$ That is, $E = N^2 h\nu$ where $N = 0, 1, 2, 3,...$ Evaluating the Boltzmann distribution $P(E) = e^{-E/kT} / kT$, we have

Electromagnetic Fields and Wave Particle Duality

$$\overline{E} = \frac{\displaystyle\sum_{n=0}^{\infty} \frac{n^2 hv}{kT} e^{-n^2 hv/kT}}{\displaystyle\sum_{n=0}^{\infty} \frac{1}{kT} e^{-n^2 hv/kT}} = kT \frac{\displaystyle\sum_{n=0}^{\infty} n^2 \alpha e^{-n^2 \alpha}}{\displaystyle\sum_{n=0}^{\infty} e^{-n^2 \alpha}},$$

where $\alpha = hv/kT$. This, in turn, can be evaluated most easily by noting that

$$-\alpha \frac{d}{d\alpha} \ln \sum_{n=0}^{\infty} e^{-n^2 \alpha} = \frac{-\alpha \dfrac{d}{d\alpha} \displaystyle\sum_{n=0}^{\infty} e^{-n^2 \alpha}}{\displaystyle\sum_{n=0}^{\infty} e^{-n^2 \alpha}} = \frac{-\displaystyle\sum_{n=0}^{\infty} \alpha \dfrac{d}{d\alpha} e^{-n^2 \alpha}}{\displaystyle\sum_{n=0}^{\infty} e^{-n^2 \alpha}}$$

$$= \frac{\displaystyle\sum_{n=0}^{\infty} n^2 \alpha e^{-n^2 \alpha}}{\displaystyle\sum_{n=0}^{\infty} e^{-n^2 \alpha}}$$

so that

$$\overline{E} = kT \left(-\alpha \frac{d}{d\alpha} \ln \sum_{n=0}^{\infty} e^{-n^2 \alpha} \right) = -hv \frac{d}{d\alpha} \ln \sum_{n=0}^{\infty} e^{-n^2 \alpha}$$

Now

$$\sum_{n=0}^{\infty} e^{-n^2 \alpha} = 1 + e^{-\alpha} + e^{-4\alpha} + e^{-9\alpha} + \ldots = 1 + X + X^4 + X^9 + \ldots$$

where $X \equiv e^{-\alpha}$.

We now need a closed form for this infinite sum. In Planck's case where the energy was assumed to be in integer steps the sum was $(1-X)^{-1} = 1 + X + X^2 + X^3 + \ldots$ In the absence of the knowing closed form we shall adopt the form of $(1-X)^{-(1+\delta)} = 1 + X + X^4 + X^9 + \ldots$ where $0<\delta<1$. A few trials using various values of X from zero to 1 will display the utility of this form until we have determined the true closed form.

Now

$$\bar{E} = -hv\frac{d}{d\alpha}\ln\left(1-e^{-\alpha}\right)^{-(1-\delta)}$$

$$= \frac{-hv}{\left(1-e^{-\alpha}\right)^{-(1+\delta)}}\left[-(1+\delta)\right]\left(1-e^{-\alpha}\right)^{-(2+\delta)}e^{-\alpha}$$

$$= \frac{hv(1+\delta)e^{-\alpha}}{1-e^{-\alpha}} = \frac{hv(1+\delta)}{e^{\alpha}-1}$$

then substituting for α we have the energy density

$$\bar{E} = \frac{hv(1+\delta)}{e^{hv/kT}-1}. \tag{8.22}$$

This is identical in form to Planck's energy density with the exception of the multiplicative term *(1+δ)*. Perhaps this should not come as a s urprise as the phat photon energies are a subset of those Planck used since the phat photon, alone, may only take on limited energy levels. For example, Planck's energy levels are all integer valued levels for which $\varepsilon = 1hv, 2hv, 3hv,...$ while phat photons may only have the values obtained by the square of integers, or $\varepsilon = 1hv, 4hv, 9hv,...$

To obtain the phat blackbody spectrum we must multiply Equation (8.22) with the number, $N(v)dv = \frac{8\pi V}{c^3}v^2 dv$, of waves having this frequency. We find that

$$\rho_T(v)dv = \frac{8\pi v^2}{c^3}\frac{hv(1+\delta)}{e^{hv/kT}-1}dv. \tag{8.23}$$

If one could somehow eliminate all of the other frequencies from a radiating blackbody, the value of δ could be determined.

It was important to learn the form of the energy density of the phat photons, given by Equation (8.22) prior to beginning to look at the development of the notion of a laser using the phat photons so that we are assured the

Boltzmann factor will apply to phat photons as it does for the Einstein, or $N=1$, photons.

The Basis of the Phat Laser

In order to show that there exists a practical use of this new type of photon it may be useful to show that a laser may be developed using the phat photons as an output. Then we may be assured that a source of phat photons is available. The development of the phat laser is done in the same manner as for the usual laser, only using the energies of the phat photon.

The relative number of particles per quantum state at two different energies for a s ystem in thermal equilibrium at temperature T is given, in certain circumstances, by the Boltzmann factor, $e^{-(E_2-E_1)/kT}$. We use this result now to study the behavior of a possibly very important device we will call a Phat Laser.

Consider transitions between two energy states of an atom in the presence of an electromagnetic field. In the spontaneous emission process, the atom is initially in the upper state of energy E_2 and decays to the lower state of energy E_1 by the emission of a photon of frequency

$$v = \left(E_2 - E_1\right) \Big/ N^2 h.$$

Let the spectral energy density of the electromagnetic radiation applied to the atoms be $\rho(v)$. Consider that there are n_1 atoms in energy state E_1 and n_2 in state E_2 where $E_2 > E_1$. The probability per atom per unit time, or transition rate per atom, that an atom in state 1 will undergo a transition to state 2 (stimulated absorption) clearly will be proportional to the energy density $\rho(v)$ of the applied radiation at frequency

$$v = \left(E_2 - E_1\right) \Big/ N^2 h.$$

214

The Dynamic Theory

The transition rate for stimulated emission is also proportional to $\rho(v)$. However, the transition rate for spontaneous emission does not contain $\rho(v)$ because that process does not involve the applied electromagnetic field.

The transition rates also depend on t he detailed properties of the atomic states *1* and *2* through the electric dipole moment matrix element. It has been shown, classically, that an oscillating electric dipole will radiate electromagnetic energy at the average rate, $\overline{R}$ where

$$\overline{R} = \frac{4\pi^3 v^4}{3\varepsilon_o c^3} p^2$$

with p the amplitude of its oscillating electric dipole moment and v the frequency of oscillation. Since we seek to determine the energy carried off by phat photons we must consider the photon energy as $N^2 hv$. The rate of emission of phat photons, R, is

$$R = \frac{\overline{R}}{N^2 hv} = \frac{1}{N^2} \frac{4\pi^3 v^3}{3\varepsilon_o hc^3} p^2$$

Therefore, we expect that the quantum mechanical rate of spontaneous emission of phat photons to be

$$R_{phat} = \frac{R}{N^2} = \frac{1}{N^2} \frac{16\pi^3 v^3 p_{fi}^2}{3\varepsilon_o hc^3} \tag{8.24}$$

where p_{fi} is the matrix element of the electric dipole moment taken between the initial and final states.

Hence, the probability per unit time for a transition from state *1* to state *2* can be written as

$$R_{1\rightarrow 2} = \frac{B_{12}\rho(v)}{N^2}$$

in which B_{12} is a coefficient, calculated by classical quantum mechanics, that includes the dependence on properties of the states *1* and *2*. The total probability per unit time that an atom in state *2* will undergo a transition to state *1* is the sum of two terms, the probability per unit time

Electromagnetic Fields and Wave Particle Duality

A_{21} $/N^2$ of spontaneous emission and the probability per unit time B_{21} $\rho(v)/N^2$ of stimulated emission. Again, A_{21} and B_{21} are coefficients whose values depend on the properties of states 1 and 2, through the appropriate matrix elements. Hence

$$R_{2\to1} = \left[A_{21} + B_{21}\rho(v) \right] / N^2 .$$

If now we consider that the n_1 atoms in state 1 and the n_2 atoms in state 2 of the system are in thermal equilibrium at temperature T with the radiation field of energy density $\rho(v)$, then the total absorption rate for the system $n_1 R_{1\to2}$ and the total emission rate $n_2 R_{2\to1}$ must be equal. That is $n_1 R_{1\to2} = n_2 R_{2\to1}$. Thus we have

$$n_1 \frac{B_{12}\rho(v)}{N^2} = n_2 \frac{\left[A_{21} + B_{21}\rho(v) \right]}{N^2} .$$

If we solve this equation for $\rho(v)$ we obtain

$$\rho(v) = \frac{\dfrac{A_{21}}{B_{21}}}{\dfrac{n_1}{n_2}\dfrac{B_{12}}{B_{21}} - 1} .$$

We now assume we can use the Boltzmann factor, with $N^2 hv = E_2 - E_1$ to obtain $\dfrac{n_1}{n_2} = e^{(E_2 - E_1)/kT} = e^{N^2 hv / kT}$ so that we may write

$$\rho(v) = \frac{\dfrac{A_{21}}{B_{21}}}{\dfrac{B_{12}}{B_{21}} e^{N^2 hv/kT} - 1} . \tag{8.25}$$

This equation, giving the spectral energy density of radiation of frequency v that is in thermal equilibrium at temperature T with atoms of energies E_1 and E_2 must be

216

consistent with the phat photon blackbody spectrum, given in Equation (8.22) above. Hence, we conclude that

$$\frac{B_{12}}{B_{21}} = e^{\left(1-N^2\right)hv/kT} \quad \text{and} \quad \frac{A_{21}}{B_{21}} = \frac{8\pi hv^3}{c^3}. \qquad (8.26)$$

These results are similar to the results first obtained by Einstein in 1917, and therefore the coefficients, for $N=1$ and $\delta=0$ are called the *Einstein A and B coefficients.* Note that the argument does not give us values of the coefficients, but only their ratios. However, if we compute the spontaneous emission coefficient A_{21} from quantum mechanics, we then can obtain the other coefficients from these formulas.

There is much of physical interest here. For one thing, we find from Equation (8.26) that the coefficients of stimulated emission and stimulated absorption are equal only for the Einstein photon for which $N=1$. For another, we see from Equation (8.26) that the ratio of the spontaneous emission coefficient to the stimulated emission coefficient varies with frequency as v^3. This means, for example, that the bigger the energy difference between the two states, the much more likely is spontaneous emission compared to stimulated emission. Still another result is that we can obtain the ratio of the probability A_{21} of spontaneous emission to the probability $B_{21}\rho(v)$ of stimulated emission, namely

$$\frac{A_{21}}{B_{21}\rho\left(v\right)} = e^{hv/kT} - 1 \qquad (8.27)$$

just as it is for the Einstein photon. This shows that, for atoms in thermal equilibrium with phat radiation, spontaneous emission is far more probable than stimulated emission if $hv \prec kT$ regardless of the value of N. Since this condition applies to electronic transitions in both atoms and molecules, stimulated emission can be ignored in such transitions. Stimulated emission can become significant, however, if $hv = kT$, and it may be dominant if $hv \succ kT$, a

Electromagnetic Fields and Wave Particle Duality

condition that applies at room temperature to atomic transitions in the microwave region of the spectrum where v is relatively small.

Phat Lasers

We are now in a position to understand how the concept of phat lasers compares to what we currently understand about ordinary lasers and masers. In general, the ratio of the emission rate to the absorption rate can be written as $n_2 R_{2\to 1}/n_1 R_{1\to 2}$ or

$$\frac{n_2 R_{2\to 1}}{n_1 R_{1\to 2}} = \frac{n_2 A_{21} + n_2 B_{21}\rho(v)}{n_1 B_{12}\rho(v)} = \left[\frac{B_{21}}{B_{12}} + \frac{B_{21} A_{21}}{B_{12} B_{21}\rho(v)}\right]\frac{n_2}{n_1} \quad (8.28)$$

$$= \frac{B_{21}}{B_{12}}\frac{n_2}{n_1} e^{hv/kT} = \frac{n_2}{n_1} e^{N^2 hv/kT}$$

If we have energy states such that $E_2 - E_1 \succ kT$, or $N^2 hv \succ kT$, then Equation (8.28) shows that we obtain

$$\frac{n_2 R_{2\to 1}}{n_1 R_{1\to 2}} \succ \frac{n_2}{n_1}.$$

This result is general in the sense that we have not assumed an equilibrium situation. In situations of thermal equilibrium, where the Boltzmann factor applies, we expect $n_2 \prec n_1$. But in non-equilibrium situations any ratio is possible in principle. If now we have a means of inverting the normal population of states so that $n_2 \succ n_1$ then the emission would exceed the absorption rate. This means that the applied radiation of frequency

$$v = \frac{E_2 - E_1}{N^2 h}$$

will be *amplified* in intensity by the interaction process, more such radiation emerging than entering. Of course, such a process will reduce the population of the upper state until equilibrium is reestablished. In order to sustain the

218

process, therefore, we must use some method to maintain the *population inversion* of the states. Devices that do this are called lasers or masers, depending upon the portion of the electromagnetic spectrum in which they operate.

Practical Applications of Phat Lasers

A photon may be defined as a gauge vector potential with the quantum number set to unity and which satisfies the wave equation. Further, the light quanta come from Weyl's Quantum Principle that also provides the basis for radiation and needs no c onnection to statistics or thermodynamics. Weyl's Quantum Principle provides the logical connection between the wave and particle aspects of light that Einstein was trying to find.

When considering non-weapon applications of laser energy one might look at the transmission of laser light in a light pipe, or optical fiber. Too much energy into the optical fiber destroys the fiber and stops all energy transmission. The photons interact with the atoms of the optical fiber through the electric field of the photon interacting with the orbital electrons of the atoms in the optical fiber. Therefore, the destruction of the optical fiber comes from the strong electric field and not necessarily from the energy itself.

Einstein photons, of energy $\varepsilon = h\nu$ have a one-to-one relation between the electric field and the energy. The quantum number in the phat photon alters this electric field-energy relationship. For phat photons the electric field is proportional to N^2. A given electric field strength may be maintained by reducing the number of phat photons by the factor $1/N$. However, when the electric field strength is held fixed by this means the total energy increases by the factor of N. Therefore, one may send more energy through optical fibers using phat photons than can be done with Einstein photons. One practical result of this would be the reduction of necessary amplifiers in long distance communications systems by the factor of *1/N*.

Rayleigh Scattering for Phat Photons from a constant stimulated energy level

Rayleigh scattering is due to the displacement of bound electrons by the incident electric field. The incident harmonic field induces a dipole in the molecule whose polarizability determines the displacement. The induced dipole oscillates at the same frequency as the incident radiation. This radiation constitutes the scattered light.

We begin the derivation of the Rayleigh scattering cross section for phat photons by assuming that each molecule contains electrons that can be acted on by the oscillating electric field of the laser beam and that the wavelength of the beam is considerably larger than the size of the molecule. This latter assumption permits us to ignore the spatial variation of the electric field $\vec{E}$ over the molecular charge distribution. The beam is furthermore assumed to be incident along the z direction and is polarized along the x direction. The scalar equation of motion of each oscillating electron of such a system is of the familiar form

$$\ddot{x} + \Gamma\dot{x} + \omega_o^2 x = \left(\frac{e}{m}\right)NE_o e^{-i\left(\omega/N^2\right)t}, \qquad (8.29)$$

where N is the quantum number of the phat photon, NE_o is the magnitude of the electric field of the phat photon, and ω/N^2 is the angular frequency of the phat laser beam according to the energy of the phat photon as given by

$$\varepsilon = N^2 h\nu, \qquad (8.30)$$

where ε is the energy step involved and is taken to be a constant in the immediately following investigation.

In Equation (8.29) the Γ is the damping coefficient, $\omega_o = \sqrt{k/m}$ (k here represents the spring constant). If we assume a s teady-state solution of the form $x = x_o \exp\left[-i\left(\omega/N^2\right)t\right]$, then the steady-state displacement as obtained from Equation (8.30) is

$$x = \frac{\left(\frac{e}{m}\right)NE_o e^{\left[-i\left(\omega/N^2\right)t\right]}}{\omega_o^2 - \left(\omega/N^2\right)^2 - i\Gamma\left(\omega/N^2\right)} = x_o e^{\left[-i\left(\omega/N^2\right)t\right]}. \quad (8.31)$$

Because of the much greater mass of the nucleus we may assume that it remains at rest, and the oscillatory motion of the electron is equivalent to an induced electric dipole whose moment is

$$p = ex = p_o e^{\left[-i\left(\omega/N^2\right)t\right]} \quad (8.32)$$

and

$$\left|p_o\right|^2 = \frac{\left(e^2/m\right)^2 \left|NE_o\right|^2}{\left[\omega_o^2 - \left(\omega/N^2\right)\right]^2 + \left(\Gamma\omega/N^2\right)^2}. \quad (8.33)$$

Using the laws of classical electrodynamics it can be shown that the time-averaged net outward flow of energy from this oscillating dipole moment is

$$\left\langle \vec{S}(\theta,r)\right\rangle = \frac{\left(\omega/N^2\right)^4 \left|p_o\right|^2 \sin^2\theta}{32\varepsilon_o \pi^2 c^3 r^2}\hat{r} = \left\langle S\right\rangle\hat{r}, \quad (8.34)$$

where $\langle\ \rangle$ means averaging with respect to time. The average energy flux (or intensity) $\langle S\rangle$ is seen to vary as $1/r^2$, where r is the distance from the center of the dipole to the point of interest, and θ denotes the angle with respect to the x-axis.

When Equation (8.34) is integrated over a sphere of radius r, we obtain the total energy radiated per second by the dipole. Thus,

$$\frac{d\varepsilon}{dt} = \int \langle \vec{S}(\theta, r) \rangle \cdot d\vec{A}$$

$$= \int_0^{2\pi}\int_0^{\pi} Sr^2 \sin\theta \, d\theta \, d\varphi = \frac{\left(\omega/N^2\right)^4 |p_o|^2}{12\pi c^3 \varepsilon_o}, \qquad (8.35)$$

and in terms of Equation (8.33) the total rate becomes

$$\frac{d\varepsilon}{dt} = \frac{\left(\frac{e^2}{m}\right)^2 \left(\omega/N^2\right)^4 N^2 |E_o|^2}{12\pi c^3 \varepsilon_o \left\{\left[\omega_o^2 - \left(\omega/N^2\right)^2\right]^2 + \left[\Gamma\left(\omega/N^2\right)\right]^2\right\}}. \qquad (8.36)$$

Since the scattering cross section σ_{sp} is the ratio of the total scattered energy rate to the total incident beam intensity I_o, it follows that

$$\sigma_{sp} = \frac{1}{I_o}\left(\frac{d\varepsilon}{dt}\right), \qquad (8.37)$$

where

$$I_o = \langle S \rangle = \left(\frac{1}{2}\right) c \varepsilon_o N^2 |E_o|^2. \qquad (8.38)$$

Combining Equations (8.36), (8.37) and (8.38), we find that the scattering cross section per dipole radiator for a phat laser is

$$\sigma_{sp} = \frac{\left(\frac{e^2}{m}\right)^2 \left(\omega/N^2\right)^4}{6\varepsilon_o^2 \pi c^4 \left\{\left[\omega_o^2 - \left(\omega/N^2\right)^2\right]^2 + \left[\Gamma\left(\omega/N^2\right)\right]^2\right\}}. \qquad (8.39)$$

For the special case where the applied frequency ω is much lower than the natural frequency ω_o and the damping coefficient Γ is small, Equation (8.39) reduces to

The Dynamic Theory

$$\sigma_{sp} = \frac{\left(\dfrac{e^2}{m}\right)^2}{6\varepsilon_o^2 \pi c^4 N^8}\left(\frac{\omega}{\omega_o}\right)^4.\qquad(8.40)$$

The derivation of the scattering cross section so far has been entirely classical, and for simplicity we have assumed that each molecule has only one oscillating electron. To generalize our result we need to multiply Equation (8.40) by the so-called oscillator strength f, which is defined as the effective number of electrons per molecule that oscillate at the natural frequency ω_o. The maximum value of the oscillator strength is equal to the total number of electrons in the molecule. Since the more tightly bound inner electrons usually cannot participate in the interaction, the oscillator strength is usually considerably smaller than this limit. Thus the expression for the scattering cross section for phat photons takes on the form

$$\sigma_{sp} = \frac{f e^4 \lambda_o^4}{6\pi\varepsilon_o^2 m^2 c^4 N^8}\frac{1}{\lambda^4},\qquad(8.41)$$

which is the well-known classical scattering formula developed by Rayleigh when $N=1$, or

$$\sigma_{sp} = \frac{\sigma_s}{N^8}.\qquad(8.42)$$

Thus, we see that the scattering cross section for phat lasers is inversely proportional to the phat quantum number to the eighth power.

Figure 8.1 is a plot of the relative Rayleigh scattering cross section for the first four phat photon quantum numbers. As may be seen the Rayleigh scattering cross section drops dramatically as N increases.

Rayleigh Scattering Cross Section

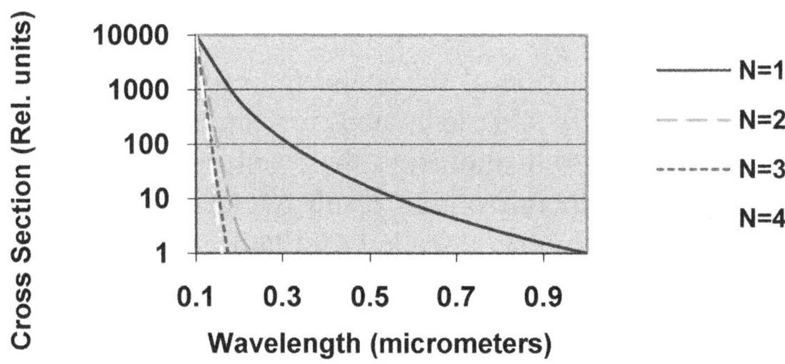

Figure 8.1 Relative Rayleigh Scattering Cross Section as a function of Phat Quantum Number, N, for a constant energy step.

Rayleigh Scattering Cross Section independent of Phat Photon source

In the previous section the energy step was taken to be held fixed throughout the calculation. Now we would like to investigate the comparison of the Rayleigh scattering cross section between phat photons of the same frequency as a function of the photon quantum number. In this case we find that equation of motion for each oscillating electron of the system to be of the form

$$\ddot{x} + \Gamma\dot{x} + \omega_o^2 x = \left(\frac{e}{m}\right)NE_o e^{-i\omega t}, \qquad (8.43)$$

where N and ω determine the energy of the phat photon in accordance with Equation (8.30).

By assuming a steady-state solution of $x = x_o \exp\left[-i\omega t\right]$, we find the solution to Equation (8.43) to be

The Dynamic Theory

$$x = \frac{\left(\frac{e}{m}\right) NE_o e^{[-i\omega t]}}{\omega_o^2 - \omega^2 - i\Gamma\omega} = x_o e^{[-i\omega t]}. \qquad (8.44)$$

Using Equation (8.44) in the same procedure as above we find that the Rayleigh scattering cross section to be the classical cross section, or

$$\sigma_s = \frac{fe^4 \lambda_o^4}{6\pi\varepsilon_o^2 m^2 c^4} \frac{1}{\lambda^4}. \qquad (8.45)$$

Conclusions

A photon may be defined as a gauge vector potential with the quantum number set to unity and which satisfies the wave equation. Further, the light quanta comes from Weyl's Quantum Principle which also provides the basis for radiation and needs no connection to statistics or thermodynamics. Weyl's Quantum Principle provides the logical connection that Einstein was trying to find.

The probability of spontaneous emission of phat photons is inversely proportional to the square of the phat quantum number, or $1/N^2$.

Phat photons may be found to be emitted from all lasers with lower intensities at the phat frequencies with the higher phat quantum numbers. These should be looked for in the output of an existing laser.

Since the scattering cross section compares the energy scattered to the incident energy, the photon quantum number does not enter into the Rayleigh scattering cross section when comparing the scattering of photons of any quantum number. That is; the Rayleigh scattering cross section depends only upon the frequency of the incident beam.

On the other hand, should one desire to compare the Rayleigh scattering cross section for a given laser that may be set up to operate with any photon quantum number, the results are drastically different. In this case the energy of the photon, and the total beam energy, remains the same independent of the quantum number. However, the frequency must decrease as $1/N^2$, or the wavelength must increase as N^2.

This means that for any given laser less energy is lost due to Rayleigh scattering by the atmosphere thereby providing more energy on t arget, though at increased wavelength.

Thermal blooming depends upon the electric field which is proportional to the photon quantum number N. On the other hand the photon energy is proportional to N^2. Therefore, a reduction in the number of phat photons by a ratio of $1/N$ compared to the number of Einstein photons that just lies below the thermal blooming threshold will be able to transport N times the energy possible by the Einstein photons.

Phat Photoelectric Effect

Einstein offered an explanation of the experimentally determined photoelectric effect by offering his equation that gives the maximum kinetic energy of photo electrons freed by photons of energy $h\nu$ should be given by

$$K_{\max} = h\nu - \phi_o. \qquad (8.46)$$

where ϕ_o is the minimum energy required to free the electron.

If we replace Einstein's relation for the energy of the photon with the expression for the phat photon we see that the maximum kinetic energy of the freed electron is given by

$$K_{\max} = N^2 h\nu - \phi_o. \qquad (8.47)$$

In Equation (8.47) the phat photon quantum number shows that if there are phat photons present the maximum kinetic energy would be very different from that predicted by Einstein.

Phat Compton Effect

Compton explained the shifted component of the scattered light when x-rays interact with matter by applying Einstein's energy of the photon to the conservation of momentum and energy relations. He assumed that the scattering process could be treated as an elastic collision between a photon and an electron, governed by the two laws of mechanics, the conservation of energy and the conservation of momentum. We shall do the same here except we will use the phat photon energy expression. Let an incident phat photon of energy $N^2 h\nu$ collide with an electron initially at rest. The photon is scattered through an angle θ, while the electron recoils in a direction φ. The kinetic energy K given to the electron is $(m-m_o)c^2$ by Einstein's special theory of relativity. If $M^2 h\nu_o$ is the frequency of the scattered phat photon, the conservation of energy requires that the sum of the kinetic energy of the electron and the energy of the scattered photon be equal to the energy of the incident photon, or

$$N^2 h\nu = M^2 h\nu_\theta + \left(m - m_o\right)c^2 . \qquad (8.48)$$

Each photon carries momentum equal to its energy $N^2 h\nu$ divided by the speed of light c, or h/λ. Since momentum is a vector quantity and is conserved, the x and y components must obey the equations

$$\frac{N^2 h\nu}{c} = \frac{M^2 h\nu_o}{c}\cos\theta + p\cos\varphi \qquad (8.49)$$

and

$$0 = \frac{M^2 h\nu_o}{c}\sin\theta + p\sin\varphi \qquad (8.50)$$

where p is the momentum of the recoil electron. (The plus sign appears with the $sin\,\varphi$ because it is assumed all angles

Electromagnetic Fields and Wave Particle Duality

are positive when measured counterclockwise form the x axis.)

If we introduce the wavelength of the two photons through $v/c=1/\lambda$ and recall that $p = m_o v / \sqrt{1-(v/c)^2}$, we may rewrite Equations (8.49) and (8.50) as

$$\frac{N^2 h}{\lambda} - \frac{M^2 h \cos\theta}{\lambda_\theta} = \frac{m_o v \cos\varphi}{\sqrt{1-(v/c)^2}}, \text{ and}$$

$$\frac{M^2 h}{\lambda_\theta}\sin\theta = -\frac{m_o v \sin\varphi}{\sqrt{1-(v/c)^2}} \qquad (8.51)$$

Squaring these two equations and adding them eliminates φ leaves

$$\frac{N^4 h^2}{\lambda^2} + \frac{M^4 h^2}{\lambda_\theta^2} - 2\frac{N^2 M^2 h^2}{\lambda\lambda_\theta}\cos\theta = \left(\frac{m_o v}{\sqrt{1-(v/c)^2}}\right)^2$$

$$= \frac{m_o^2 c^2}{1-(v/c)^2} - m_o^2 c^2 \qquad (8.52)$$

Similarly, Equation (8.52) can be divided by c and written

$$\frac{N^2 h}{\lambda} - \frac{M^2 h}{\lambda_\theta} + m_o c = \frac{m_o c}{\sqrt{1-(v/c)^2}} \qquad (8.53)$$

and squared to give

$$\frac{N^4 h^2}{\lambda^2} + \frac{M^4 h^2}{\lambda_\theta^2} - 2\frac{N^2 M^2 h^2}{\lambda\lambda_\theta}$$

$$+2m_o ch\left(\frac{N^2}{\lambda} - \frac{M^2}{\lambda_\theta}\right) + m_o^2 c^2 = \frac{m_o^2 c^2}{\left(1-(v/c)^2\right)} \qquad (8.54)$$

Subtracting Equation (8.52) from Equation (8.54) yields

$$-2\frac{N^2 M^2 h^2}{\lambda\lambda_\theta}(1-\cos\theta) + 2m_o ch\left(\frac{N^2}{\lambda} - \frac{M^2}{\lambda_\theta}\right) = 0 \qquad (8.55)$$

or

228

$$\Delta\lambda = \lambda_\theta - \lambda = \lambda_\theta\left(1 - \frac{N^2}{M^2}\right) + \frac{h}{m_o c}N^2\left(1 - \cos\theta\right)$$

$$= \lambda_\theta\left(1 - \frac{N^2}{M^2}\right) + 0.02426 N^2\left(1 - \cos\theta\right)$$

$$. (8.56)$$

Obviously, Equation (8.56) predicts the wavelength shift of the classical Compton theory when $N=M=1$. Also, it may be seen that when the scattered photon has the same quantum number as the incident photon Equation (8.56) reduces to

$$\Delta\lambda = 0.02426 N^2\left(1 - \cos\theta\right) \qquad (8.57)$$

which argues that for phat photons the wavelength shift differs from the $N=1$ shift by the factor of N^2. Further, since the energy of phat photons depends upon both the frequency and the quantum number, Equation (8.57) reflects the fact that while the wavelength shift is still independent of the incident frequency it is not independent of the quantum number.

Conclusions

Historically, Einstein's light quanta, Schrödinger's wave mechanics, and Maxwell's wave equations for light were all taken to be independent of each other. Indeed, they were all developed at different times with different assumptions and in different fields of physics. Einstein developed his light quanta from statistical thermodynamics. Schrödinger developed his wave mechanics in view of the atomic data that was then being collected. Maxwell had developed his wave equations many years before either Einstein or Schrödinger began their work. Here only Weyl's unity scale factor was employed and it was given as part of his Gauge Principle that has been accepted for the past 75 years. Weyl showed in 1918 that this principle leads to the whole system of Maxwell equations which produce the wave equations. London, in 1927, derived the

Schrödinger wave mechanics from this same principle by restricting his attention to unity scale factors. It is London's work that points out the unity scale factor as the restriction that requires quantization. Without the restriction to a unity scale factor the line integral of the gauge potentials does not have a quantized value. With the restriction to a unity scale factor an assumed gauge potential is required to have quantized paths. Alternatively, an assumed path requires quantized gauge potentials. Quantized vector potentials lead to photons with energies that are dependent upon both frequency and the square of a quantum number.

The quantized energy of the phat photons leads to the prediction of a phat laser. Using the output of a phat laser in communications systems with fiber optics leads to a reduction of needed line amplifiers. Also, the output of phat lasers can deliver more energy through a given atmosphere before thermal blooming takes effect.

The new expression for the energy of phat photons may be applied to both the photoelectric effect and the Compton Effect. In the photoelectric effect the maximum kinetic energy depend upon the phat quantum number. Also we see that the Compton wavelength shift does not depend upon the incident photon frequency, but does depend upon the phat quantum number. The phat photon wavelength shift depends upon the phat quantum number to the fourth power. This argues that the Compton scattering might be used to distinguish phat photons from the Einstein photons.

Chapter 9 Nuclear Physics

In Chapter 6 w e saw that Weyl's Gauge and Quantum Principles require both Maxwell's electromagnetism and quantum mechanics. We also showed that these principles lead to five dimensional quantum mechanics involving a Dirac equation extended into five dimensions. One result of the five dimensional Dirac equation was that there existed other spin vectors which led to the derivation of the requirement that the total spin components can only appear in sets of eight, or octets. The appearance of octets hints at nuclear physics and/or particle physics. However, solving the extended Dirac equation is complicated and time consuming. We can approach nuclear physics by studying the non-singular gauge potential derived in Chapter 7 by first considering Newtonian mechanics and then advancing into non-relativistic quantum mechanics before returning to the Dirac relativistic mechanics.

The appearance of the exponential dependence of the gauge potential for fundamental particles upon space, time and mass was surprising. Yet these dependences were shown in Chapter 7 t o give the correct ratio of the electrostatic to gravitational force.

The radial dependence of the electrostatic gauge potential was surprising and, at first, pleasing. The surprise is that this functional form comes only from the gauge function, f, playing the guiding role in the five dimensional, quantized gauge fields. Thus, the exponential portion of the function comes only from the five dimensional quantization of thermodynamics. This quantum origin of the exponential portion is further supported by the fact that the constant K being a function of the gauge potential quantum numbers.

In the absence of quantum numbers for K, the exponential takes on a value of 1.

The quantization of the electrostatic potential in Chapter 7 is the first time that the electrostatic potential has been quantized theoretically. The story of how this derivation came about may be of some interest. I was making a presentation at the Los Alamos National Laboratory and the then Associate Director for Weapons, Dr. C. Paul Robinson, was in attendance. During the questions at the end of my presentation of Weyl's Quantum Principle, Dr. Robinson suggested that perhaps the WQP would provide the answer to why charged particles only have quantized units of charge. It took me almost six months to develop this derivation and years more to begin to have a slight understanding of its value and finish the presentation in Chapter 7.

The reason for this length of time required for the developing the derivation is that the line integral has a path dependent integrand and I was trying to use the differentials to find what I presumed to be a continuous function. In WQP the line integral stating the scale factor was already known to be path dependent and discontinuous. I should have realized this. The discontinuity shows up because of the quantization and is the reason different energy levels in an atom have different quantum numbers. Therefore, I should not have been surprised to see quantum numbers showing up in the electrostatic potential. After all that was what Dr. Robinson asked me to investigate; quantized electrostatic charge.

However, the fact that the WQP provides a theoretical prediction of the quantization of electrostatic charge that has such extensive experimental observation that it has never been seen to differ from this experimental base is another strong fact supporting the WQP as a principle underlying all of gauge field theory.

Using Newtonian Equations

To arrive at a gauge potential with an additional parameter that depends upon the particle differs from the classical derivation of the radial dependence of the electrostatic potential. However, the classical electrostatic potential is obtained using Gauss' Law by integrating a function with the assumption of a continuous function made implicitly. In the derivation here Weyl's Quantum Principle was used to select gauge potentials from an infinite number of possibilities in a non-integrable manifold, and a differential equation was used for the derivation. Just as in the case of London's display that an infinite number of states are allowed by Weyl's Quantum Principle, here there should be no surprise that an infinite number of particles may exist with different electrostatic gauge potentials where the K and the λ_N are different for different particles. Nor should there be any surprises that the WQP allows a different functional form of the potential when the fact that this potential is selected from functions in a non-integrable manifold, since it may easily contain information not allowed by an integral equation based upon the assumption of continuity.

Forces of interaction between two particles which are the negative slopes of this type of non-singular potential, given the same λ_N for each particle, change their sign when the separation, r, equals λ_N, since

$$force = \frac{N_0 k}{r^2}\left(1 - \frac{\lambda_N}{r}\right)e^{-\frac{\lambda_N}{r}}. \tag{9.1}$$

Therefore, particles that have repulsive long-range interactions must have attractive short-range interactions.

Particles with different λ_N will have forces that do not obey Newton's third law as the force on one particle, due to the presence of a second particle, will vanish when the separation equals λ_2. While the force on the second particle, due to the presence of the first particle, will vanish at a separation of λ_1.

Consequences of such interactions will be considered in this chapter and have no need for renormalization, as might be expected of a non-singular potential. Also, the interactions between unlike particles, $\lambda_1 \neq \lambda_2$, must be solved as a two-particle system that introduces a bilinear form into Dirac's equation for the two-particle system. This reduces to well-known bilinear forms of the Yang-Mills theory of weak nuclear interactions upon a pplying appropriate restrictive assumptions, such as all motion is assumed to be motion *about* the center of mass or all motion is assumed to be motion *of* the center of mass.

Newtonian Mechanics with Non-Singular Potentials

First, let us explore the nature of the non-singular potential using classical Newtonian mechanics.

A non-singular, electrostatic potential of the form

$$\phi = \left(\frac{k}{r} \right) e^{\frac{-\lambda}{r}}, \qquad (9.2)$$

where the λ may be different for each particle, would not satisfy Newton's third law concerning equal and opposite forces for all separations, r, and would necessitate revisiting all of the equations of motion concerning the interaction of unlike particles. An investigation of the equations of motion may start with the well-known change of coordinates which allows one to see the motion about the center of mass,

$$\overline{R} = \frac{m_1 \overline{r_1} + m_2 \overline{r_2}}{m_1 + m_2} \qquad (9.3)$$

$$\overline{r} = \overline{r_1} - \overline{r_2},$$

where the small r's represent vectors to each body or between the bodies, and the capital R represents the vector to the center

The Dynamic Theory

of mass of the two bodies. Of course the inverse transformations associated with the transformation contained in Equations (9.3) exist and are the usual ones associated with the center of mass (COM) approach.

By using the COM approach and writing the low velocity limit of the equations of motion (EOM), we find

$$m_1\ddot{r}_1\hat{r}_1 = \bar{F}_1^i + \bar{F}_1^e$$
$$m_2\ddot{r}_2\hat{r}_2 = \bar{F}_2^i + \bar{F}_2^e.$$

(9.4)

where the superscripts i and e represent internal and external forces respectively and the $^\wedge$ denotes a unit vector. By using the standard definitions for the total mass, $M = m_1+m_2$, and the reduced mass, $\mu=(m_1m_2)/M$ and setting all external forces to zero, Equations (9.4) may be put into the form

$$M\ddot{R}\hat{R} = \bar{F}_1^i + \bar{F}_2^i$$

$$\mu\ddot{r}\hat{r} = \frac{\mu}{m_1}\bar{F}_1^i - \frac{\mu}{m_2}\bar{F}_2^i.$$

(9.5)

The EOM of Equations (9.5) display the effect of Newton's third law on the two-body problem. Should Newton's third law hold in both magnitude and direction then the first equation shows that the force on the COM vanishes while the two forces, which must remain separate when Newton's third law does not hold, becomes a single-force statement without any reference to the mass of the bodies. Further, the first equation gives the motion *OF* the COM while the second equation gives the motion *ABOUT* the COM.

Many other conclusions might be pointed out concerning Equations (9.5). However, it may be of more interest to specialize the forces appearing in the EOM to those of the electric forces between electrically charged particles that exhibit the non-singular potential above. These forces have two traits that have important consequences. First, they depend only upon the separation of the two bodies. That is, the magnitude of the forces depends upon the separation

235

distance, while the direction of the forces is on a line between the bodies. This means that they have "central force" properties. Secondly, they are the result of taking the gradient of a potential. They are not, however, conservative forces in the classical sense for that would include the assumption of Newton's third law.

Though the following presentation will use symbols of general forces it may be useful to display here the functional dependence of the forces that the non-singular potentials produce. The force on body one due to the presence of body two is

$$\bar{F}_1^i = -Z_1 e \bar{\nabla} V_2^i = -\left(\frac{Z_1 Z_2 e^2}{4\pi\varepsilon_0}\right)\frac{1}{r^2}\left(1 - \frac{\lambda_2}{r}\right)e^{\left(-\frac{\lambda_2}{r}\right)}\hat{r}, \quad (9.6)$$

while the force on body two due to the presence of body one is

$$\bar{F}_2^i = -Z_2 e \bar{\nabla} V_1^i = \left(\frac{Z_1 Z_2 e^2}{4\pi\varepsilon_0}\right)\frac{1}{r^2}\left(1 - \frac{\lambda_1}{r}\right)e^{\left(-\frac{\lambda_1}{r}\right)}\hat{r}. \quad (9.7)$$

Equations (9.6) and (9.7) exhibit the property that at large separations they approximately obey Newton's third law, but as the separation approaches the larger of the λ's they begin to severely depart from Newton's third law. Therefore, these two forces cannot be combined into a single central force as is done in classical mechanics, nor can their potentials be combined into a single potential. Therefore, we will proceed to display the method of determining a solution to the equations of motion in the absence of forces that obey Newton's third law.

Additionally, these forces display the attribute of central forces since they each have the form

$$\bar{F} = \hat{r}F(r). \quad (9.8)$$

One of the attributes of central forces is a constant angular momentum since the torque is

$$\bar{N} = \bar{r} \times \bar{F} = (\bar{r} \times \hat{r})F(r) = 0. \quad (9.9)$$

Therefore,

The Dynamic Theory

$$\frac{d\overline{L}}{dt} = \frac{d\left[m(\overline{r} \times \overline{v})\right]}{dt} = 0. \qquad (9.10)$$

From Equation (9.10) it may be seen that since the total angular momentum must be a constant, then both the vectors r and v must lie in a fixed plane perpendicular to the angular momentum, L.

The total angular momentum about the origin is given by

$$\overline{L} = \overline{L}_1 + \overline{L}_2 \qquad (9.11)$$

Then, using the EOM, Equations (9.10), with vanishing external forces,

$$\frac{d\overline{L}}{dt} = \frac{d\overline{L}_1}{dt} + \frac{d\overline{L}_2}{dt} = \overline{r}_1 \times \overline{F}_1^i + \overline{r}_2 \times \overline{F}_2^i. \qquad (9.12)$$

Equation (9.12) may be written in terms of the COM coordinates to find

$$\frac{d\overline{L}}{dt} = \overline{R} \times (\overline{F}_1^i + \overline{F}_2^i) + \overline{r} \times \left(\frac{\mu \overline{F}_1^i}{m_1} - \frac{\mu \overline{F}_2^i}{m_2} \right). \qquad (9.13)$$

By using the forces given by Equations (9.6) and (9.7) then Equation (9.12) becomes

$$\frac{d\overline{L}}{dt} = \frac{k}{r^2}\left[\left(1 - \frac{\lambda_1}{r}\right)e^{-\frac{\lambda_1}{r}} - \left(1 - \frac{\lambda_2}{r}\right)e^{-\frac{\lambda_2}{r}} \right](\overline{R} \times \hat{r}) \qquad (9.14)$$

from which it may be seen that there are two ways for the change in total angular momentum to vanish. The first is that $\lambda_1 = \lambda_2$ and the second is for the COM to be on the line defined by the two bodies. This requires that, for two distinctly different bodies, their motion must have a total angular momentum that is constant with respect to some point in the plane determined by the positions of the two bodies and this point may be taken as the origin. Further, both bodies must be on a l ine with the angle θ, or $\theta + \pi$, in this plane. The significance of this is that Equation (9.14) requires that, not only must the total angular momentum be constant in time, but so must the angular momentum *ABOUT*, and *OF*, the COM.

By using this result together with the EOM, Equations (9.4) and (9.5), with the force laws, Equations (9.6) and (9.7), the Conservation of Energy, and transferring to the cylindrical coordinates typical of motion for central forces, it may be shown that the energy, which is a constant of the low velocity motion, becomes

$$E = K + V + K_c + V_c \qquad (9.15)$$

where $K + V$ is the energy *ABOUT* the COM and $K_c + V_c$ is the energy *OF* the COM. In Equation (9.15), the parts are given by:

$$K = \frac{\mu k}{2r}\left[\left(1-\frac{\lambda_1}{r}\right)\frac{e^{-\frac{\lambda_1}{r}}}{m_2} + \left(1-\frac{\lambda_2}{r}\right)\frac{e^{-\frac{\lambda_2}{r}}}{m_1}\right]$$

$$V = -\frac{\mu k}{r}\left[\frac{e^{-\frac{\lambda_1}{r}}}{m_2} + \frac{e^{-\frac{\lambda_2}{r}}}{m_1}\right] \qquad (9.16)$$

$$K_c = \frac{kR}{2r^2}\left[\left(1-\frac{\lambda_1}{r}\right)e^{-\frac{\lambda_1}{r}} - \left(1-\frac{\lambda_2}{r}\right)e^{-\frac{\lambda_2}{r}}\right]$$

$$V_c = -\frac{k}{r}\left[e^{-\frac{\lambda_1}{r}} - e^{-\frac{\lambda_2}{r}}\right]$$

where $k=(Z_1 Z_2 e^2)/(4\pi\varepsilon_0)$. Now one may assume circular orbits for both the COM and motion about the COM and assume that both orbits have quantized angular momentum to obtain an expanded Bohr model of the two-body problem, which may be shown to have energy states of energies given by

$$E = -\frac{\mu Z^2 e^4 f(\lambda)}{2(4\pi\varepsilon_0)^2 n_\theta^2 \hbar^2}\left(h(\lambda) + k(\lambda)\left[\frac{n_c^2 \mu f(\lambda)}{Mg(\lambda)n_\theta^4}\right]^{\frac{1}{3}}\right) \qquad (9.17)$$

238

with n_θ being the quantum number for the motion about the COM and n_c being the quantum number of the motion of the COM and the functions of lambdas are given by the definitions

$$f(\lambda) = \mu\left[\left(1-\frac{\lambda_1}{r}\right)\frac{e^{-\frac{\lambda_1}{r}}}{m_2}+\left(1-\frac{\lambda_2}{r}\right)\frac{e^{-\frac{\lambda_2}{r}}}{m_1}\right]$$

$$g(\lambda) = \left[\left(1-\frac{\lambda_1}{r}\right)e^{-\frac{\lambda_1}{r}}-\left(1-\frac{\lambda_2}{r}\right)e^{-\frac{\lambda_2}{r}}\right]$$

$$h(\lambda) = \mu\left[\left(1+\frac{\lambda_1}{r}\right)\frac{e^{-\frac{\lambda_1}{r}}}{m_2}+\left(1-\frac{\lambda_2}{r}\right)\frac{e^{-\frac{\lambda_2}{r}}}{m_1}\right]$$

$$k(\lambda) = \left[\left(1-\frac{\lambda_1}{r}\right)e^{-\frac{\lambda_1}{r}}-\left(1-\frac{\lambda_2}{r}\right)e^{-\frac{\lambda_2}{r}}\right].$$

(9.18)

By putting Equations (9.18) in this form, it is easy to see that the vanishing of the lambdas reduces Equation (9.17) to the classical energy equation of Bohr's theory.

The first-order approximation to the Bohr energy levels may be found by assuming that $\lambda_1=0$ and λ_2 is very small compared to r in Equations (9.17) and (9.18). Keeping only terms of the first order of λ_2/r, negative cube roots are avoided and yields the first approximation of the energy as

$$E \sim -\frac{\mu Z^2 e^4}{2(4\pi\varepsilon_0)^2 \hbar^2}\left(1-\frac{\mu^2 \lambda_2 Z e^2}{m_2(4\pi\varepsilon_0)n_\theta^2\hbar^2}\right). \qquad (9.19)$$

For the atomic hydrogen atom ionization energy a λ_2 of 1 Fermi produces a p rediction of the energy level that is approximately $2x10^{-5}$ different from Bohr's model. Comparing this to the statement of experimental error of $5x10^{-5}$ makes it important to redo the ionization energy with reduced experimental error.

Atomic States – Non-relativistic quantum mechanics

An expanded Schrödinger-like wave equation may be developed using similar assumptions that Schrödinger used except that Newton's third law is not assumed nor applied. To be consistent with the energy expression of Equation (9.15), the de-Broglie relations for motion about the COM and for motion of the COM must be used. When this is done, and the required linear behavior and free system limit is met, the resulting wave equation ends up being

$$-\frac{\hbar^2}{\mu}\frac{\partial^2\psi(x,X,t)}{\partial x^2}+V(x,t)\psi(x,X,t)$$

$$-\frac{\hbar^2}{2M}\frac{\partial^2\psi(x,X,t)}{\partial X^2} \qquad (9.20)$$

$$+V_c(x,t)\psi(x,X,t)=i\hbar\frac{\partial\psi(x,X,t)}{\partial t}.$$

The time-independent, expanded wave equation in cylindrical coordinates is then

$$-\frac{\hbar^2}{2\mu}\frac{\partial^2\psi(\overline{r},\overline{R})}{\partial\overline{r}^2}+V(\overline{r},\overline{R})\psi(\overline{r},\overline{R})$$

$$-\frac{\hbar^2}{2M}\frac{\partial^2\psi(\overline{r},\overline{R})}{\partial\overline{R}^2}+V_c(\overline{r},\overline{R})\psi(\overline{r},\overline{R})=E\psi(\overline{r},\overline{R}) \qquad (9.21)$$

where the potentials are those in Equations (9.16).

The estimate of the difference in prediction between the Bohr energy levels and the similar predictions of the non-singular potentials was $2x10^{-5}$. In order to compare the energy levels with the Schrödinger energy levels the potentials of Equations (9.16) should be used in Equation (9.21).

The vector equation of Equation (9.21) may be solved by the method of separation of variables. First, by trying a solution of the form, $\psi(\overline{r},\overline{R})=\psi(\overline{r})\psi(\overline{R})$, one finds that Equation (9.21) separates into two sets of three equations each where one set depends upon (r, θ, ϕ) and the other set depends upon (R, Θ, Φ). The separation constant, E_c, is related to the COM kinetic energy.

The two sets of three equations may be separated further by trying

$$\psi(\bar{r}) = \psi_r \psi_\theta \psi_\phi \ \text{and} \ \psi(\bar{R}) = \psi_R \psi_\Theta \psi_\Phi.$$

The solutions for ψ_Θ and ψ_Φ are identical to the solutions for ψ_θ and ψ_ϕ respectively with the separation constants m_L and L corresponding to the separation constants m_l and l.

The equation for ψ_r becomes

$$\frac{1}{r^2}\frac{d}{dr}\left[r^2\frac{d\psi_r}{dr}\right] + \frac{2\mu}{\hbar^2}\left[E - E_c - V(r) - V_c(r)\right]\psi_r = \frac{\ell(\ell+1)\psi_r}{r^2}.$$

Recognizing the potential difficulties of obtaining solutions with both $V(r)$ and $V_c(r)$ having transcendental terms, an investigation of the influence of the non-singular potentials may be conducted by assuming $\lambda_1 <\!\!< r$ and $\lambda_2 <\!\!< r$.

Following the method of solving Schrödinger's time-independent equation, the following definitions:

$$\rho = 2\beta r; \ \beta^2 = \frac{2\mu(E - E_c)}{\hbar^2}; \ \gamma = \frac{\mu Ze^2}{4\pi\varepsilon_0\hbar^2\beta}$$

and

$$\lambda = \frac{2\mu Ze^2}{4\pi\varepsilon_0\hbar^2}\left[\frac{(2m_1 + m_2)\lambda_1}{(m_1 + m_2)} - \frac{m_1\lambda_2}{(m_1 + m_2)}\right],$$

together with the trial solution,

$$\psi_r(\rho) = e^{-\rho/2}F(\rho) = e^{-\rho/2}\left[\rho_s\sum_{k=0}^{\infty}a_k\rho^k\right], \quad a_0 \neq 0 \ \text{and} \ s \geq 0,$$

requires that two relations must be met; namely

$$s(s+1) = \ell(\ell+1) + \lambda$$

and

$$a_{j+1} = \frac{s+j+1-\gamma}{(s+j+1)(s+j+2) - [\ell(\ell+1) + \lambda]}a_j.$$

In order to satisfy $s \geq 0$, $s = l + \delta$ where

$$\delta = -\left(\ell + \frac{1}{2}\right)\left[1 - \sqrt{1 + \frac{\lambda}{\left(\ell + \frac{1}{2}\right)^2}}\right].$$

Further, the series will terminate if $\gamma - \delta = n$ is an integer. The solution may then be put into the form

$$E - E_c = \frac{E_n}{\left(1 + \frac{2\delta}{n}\right)} ; \text{ where } E_n = \frac{-\mu Z^2 e^4}{(4\pi\varepsilon_0)^2 2\hbar^2 n^2}.$$

The COM radial equation has no potential terms and is given by the solution of

$$\frac{1}{R^2}\frac{d}{dr}\left(R^2 \frac{d\psi_R}{dR}\right) + \frac{2ME_c}{\hbar^2}\psi_R = L(L+1)\psi_R.$$

This equation has a Bessel function solution of the form

$$\psi_R = R^{-\frac{1}{2}}[C_1 I_v(\alpha R)] \text{ where}$$

$$v = \frac{1}{2}\sqrt{1 + 4L(L+1)} \text{ and}$$

$$\alpha = \sqrt{\left|\frac{2ME_c}{\hbar^2}\right|}.$$

This solution does not fully establish the value of E_c. This value can be determined, but is not done here, by considering that Equation (9.14) requires that

$$\overline{R} \times \hat{r} = 0.$$

Nuclear Forces – Relativistic quantum mechanics

To obtain Lorentz covariant EOMs, one must begin with a Lorentz covariant relation between energy and momentum. In special relativity, the energy, E, momentum, p, and rest mass, m, of a free particle are connected by the relation

$$E^2 = p^2 c^2 + m^2 c^4. \tag{9.22}$$

Defining the four quantities

$$P^\mu = \left(\frac{E}{c}, \overline{p} \right)$$

and the corresponding quantities

$$P_\mu = \left(\frac{E}{c}, -\overline{p} \right)$$

shows that the quantity $P_\mu P^\mu$ is a Lorentz scalar. Using the summation convention, Equation (9.22) becomes

$$P_\mu P^\mu = \frac{E^2}{c^2} - p^2 = m^2 c^2. \tag{9.23}$$

We now need to apply these definitions and relations to the system of two particles in order to display the solution for the relativistic quantum mechanical system when the forces do no satisfy Newton's third law.

For two free particles taken together as a system, the energy, momentum and mass should all be additive that is

$$E_s = E_1 + E_2; \quad \overline{P}_s = \overline{P}_1 + \overline{P}_2; \quad M = m_1 + m_2.$$

For the two-particle, free system, Equation (9.23) becomes

$$P_{s\mu} P_s^\mu = \frac{E_s^2}{c^2} - P_s^2 = M^2 c^2$$

or

$$(p_{1\mu} + P_{2\mu})(p^{1\mu} + p^{2\mu}) = (m_1 + m_2)^2 c^2. \tag{9.24}$$

Equation (9.24) leads to a free system Klein-Gordon equation of the form

$$\left[h^2(\partial_{1\mu} + \partial_{2\mu})(\partial_1^\mu + \partial_2^\mu) + (m_1 + m_2)^2 c^2 \right]\psi(x) = 0 \quad (9.25)$$

as the 2-particle, free system Klein-Gordon equation.

Equation (9.25) allows us to use natural units where the units of action and the speed of light may be set to unity. [It must be pointed out that the unit of action depends upon the gauge function, but is held fixed here to allow comparison to current work.]

Standard procedure of relativistic quantum mechanics allows one to seek a wave equation of the form

$$i\frac{\partial}{\partial t}\psi(x) = H\psi(x) \quad (9.26)$$

where H is the system Hamiltonian.

A necessary (but not sufficient) condition that Equation (9.26) be Lorentz covariant is that the Hamiltonian, H, be linear in the spatial derivatives. When the Hamiltonian of the two-body system is

$$H = H_1 + H_2 = -(\alpha \cdot \overline{P}_1 + \beta_1 m_1) - (\overline{\alpha}_2 \cdot \overline{P}_2 + \beta_2 m_2).$$

For free particles, there is no distinction between the α's and β's, and the system's Hamiltonian would be given by

$$H = -[\overline{\alpha} \cdot (\overline{P}_1 + \overline{P}_2) + \beta(m_1 + m_2)] \quad (9.27)$$

when there is no interaction between the particles.

The properties of the α's and β would then be the same as developed classically or

$$\alpha_1^2 = \alpha_2^2 = \alpha_3^2 = 1$$

$$\beta^2 = 1$$

$$\{\alpha_1, \alpha_2\} = \{\alpha_2, \alpha_3\} = \{\alpha_3, \alpha_1\} = \alpha_3\alpha_1 + \alpha_1\alpha_3 = 0$$

$$\overline{\alpha}\beta + \beta\overline{\alpha} = 0.$$

In the x-representation, the Hamiltonian equation becomes

$$i\frac{\partial}{\partial t}\psi(x) = [i\overline{\alpha} \cdot (\overline{\nabla}_1 + \overline{\nabla}_2) - \beta(m_1 + m_2)]\psi(x) \quad (9.28)$$

where $\psi(x)=<x|>$. The two, independent particle, free system Hamiltonian obeys the free system relativistic energy-

The Dynamic Theory

momentum equation, but the requirement of Lorentz covariance has not yet been imposed on the system.

Multiply Equation (9.28) on the left by the matrix β and remembering that the total Hamiltonian in Equation (9.27) is the sum of the Hamiltonian of each particle, results in

$$[i(\partial_{1\mu} + \partial_{2\mu})\gamma^{\mu} + m_1 + m_2]\psi(x) = 0. \qquad (9.29)$$

Equation (9.29) may also be written as

$$[i\partial_{\theta\mu}\gamma^{\mu} + m_{\theta}]I^{\theta}\phi(x) = 0 \quad where \quad I^{\theta} = \begin{pmatrix} 1 \\ 1 \end{pmatrix}. \qquad (9.30)$$

All inertial observers must agree on Equation (9.30). Now the properties of the matrices are the same as in the classical case in that

$$[\gamma^{\mu}, \gamma^{\nu}] = 2g^{\mu\nu} = \begin{cases} 1, \ \mu = \nu = 0 \\ -1, \ \mu = \nu = 1, 2, 3 \ as \ well \ as \ 0, \ \mu \neq \nu \end{cases}$$

and

$$\gamma^{0^{+}} = \gamma^{0} \quad (Hermitian)$$

$$\gamma^{i^{+}} = -\gamma^{i} \quad (anti - Hermiian).$$

Just as under the Lorentz transformation, $x^{\mu'} = L^{\mu}_{\nu}x^{\nu}$, each component of a vector transforms into a linear combination of all components, also require that each component of the wave function transforms into a linear combination of all four components; i.e.,

$$\psi(x)\vec{L}T\psi'(x') = S\psi(x)$$

Where the symbol $\vec{L}T$ means "transform under a Lorentz transformation into…". S must be a 4x4 matrix. An inverse transformation is required to exist or $S^{-1}S = SS_{-1} = I.$

Now if Equation (9.30) is to be Lorentz covariant we must require that

$$[i\partial'_{\theta\mu}\gamma^{\mu} + m_{\theta}]I^{\theta}\phi'(x') = 0 \quad where \quad \partial'_{1\mu} = \frac{\partial}{\partial x_1^{\mu_1}} \qquad (9.31)$$

so that the primed and unprimed inertial observers agree on the form of the Hamiltonian.

To obtain Equation (9.31) from (9.30) one must write $\psi(x) = S^{-1}\psi'(x')$ and

$$\partial_\mu = \frac{\partial x^{\nu'}}{\partial x'^\mu}\partial_{\nu'} = L^\nu_\mu \partial_{\nu'}.$$

Putting these into Equation (9.31) gives

$$[iL^\nu_{\theta\nu}\partial_{\theta\nu}\gamma^\mu + m_\theta]I^\theta S^{-1}\psi'(x') = 0.$$

Multiplying on the left by S then gives

$$[iL^\nu_{\theta\mu}\partial_{1\nu'}S\gamma S^{-1} + m_\theta]I^\theta\psi'(x') = 0,$$

which would be identical to Equation (9.30) if $L^\nu_{1\mu}S\gamma^\mu S^{-1} = \gamma^\nu$ and $L^\nu_{2\mu}S\gamma^\mu S^{-1} = \gamma^\nu$.

The Lorentz transformation should be the same for both particles so that $L^\nu_{1\mu}{=}L^\nu_{2\mu}$ and then the spinor transformation is identical to the classical case for each particle and each leads to the same condition that must be imposed on the S matrix in order that the two particle Dirac equation be Lorentz covariant.

Notice that, by considering the two particles as a system, the total system angular momentum includes the total angular momentum of both particles and is to be conserved.

Weak Nuclear Forces – Un-like particles

Now consider the situation where the two particles making up the system interact with each other with forces which do not only obey Newton's third law yet neither body interacts with the system's surroundings.

Therefore, consider the substitution

$$P_\mu \to \tilde{P}_\mu - eA_\mu(x)$$

in the two particle, non-interacting, free system Hamiltonian. In the x-representation, this becomes

$$i\partial_\mu \to i\partial_\mu - eA_\mu(x),$$

so that the Equation (9.30) becomes

The Dynamic Theory

$$[(i\partial_{1\mu} - e_1 A_{2\mu}(x) + i\partial_{2\mu}$$
$$- e_2 A_{1\mu}(x))\gamma^\mu + m_1 + m_2]\psi(x) = 0. \qquad (9.32)$$

Equation (9.32) may also be written

$$[(i\partial_{\theta\mu} I^\theta - e_\theta A_\mu^{T\theta}(x))\gamma^\mu + m_\theta I^\theta]\psi(x) = 0, \qquad (9.33)$$

where $A^{T\theta}_{\ \mu}$ is the transpose of $A^\theta_{\ \mu}$.

When e_1 is the electric charge of the first particle, e_2 is the electric charge of the second particle, A_1 is the potential of the first particle, and A_2 is the potential of the second particle, then Equations (9.32) and (9.33) should describe the interaction between the two bodies making up the system which does not have any interactions with its surroundings.

Consider the problem to be separated into the motion **of** the COM and the motion **about** the COM. Then subscript 1 $\Rightarrow r, \theta, \phi$ with $A_{1\mu}(r, \theta, \phi)$ and $A_{2\mu}(r, \theta, \phi)$ and subscript $2 \Rightarrow R, \Theta, \Phi$.

Recalling that one of the requirements on the free system Hamiltonian was that its solutions also be solutions of the free system Klein-Gordon equation, it should be of some interest to see whether or not the solutions of Equation (9.33) are also solutions of the corresponding Klein-Gordon equation. Since the Klein-Gordon equation is a second-order differential equation, whereas the Hamiltonian is of first-order, it will be necessary to operate on the latter with first order differential operators in order to cast it into a form similar to the Klein-Gordon equation. If we operate on the left of Equation (9.33) with the operator

$$[(i\partial_{\theta\mu} I^\theta - e_\theta A^{T\theta\mu}(x))\gamma^\mu - (m_\theta I^\theta)],$$

we obtain

$$[(i\partial_{\theta\mu} I^\theta - e_\theta A^{T\theta\mu})\gamma^\mu (i\partial_{\theta\nu} I^\theta - e_\theta A^{T\theta\nu})\gamma^\nu$$
$$- (m_\theta I^\theta)^2]\psi(x) = 0. \qquad (9.34)$$

The corresponding Klein-Gordon equation comes from Equation (9.25) with the substitution $\partial_\mu \to \partial_\mu - eA_\mu$. Then we would have

$$[(i\partial_{\theta\mu}I^\theta - e_\theta A_\mu^{T\theta})(i\partial_\theta^\mu I^\theta - e_\theta A^{T\theta\mu}) - (m_\theta I^\theta)^2]\psi(x) = 0. \quad (9.35)$$

We notice that in Equation (9.35) we have

$$\gamma^\mu\gamma^\nu = \frac{1}{2}[\gamma^\mu,\gamma^\nu] + \frac{1}{2}[\gamma^\mu,\gamma^\nu] = g^{\mu\nu} = \sigma^{\mu\nu}$$

$$where \ g^{\mu\nu} = \frac{1}{2}[\gamma^\mu,\gamma^\nu]$$

which may be shown to be $\sigma^{0i}=(1/2)[\gamma^0\gamma^i\text{-}\gamma^i\gamma^0]$. But

$$i\frac{d\alpha}{dt} = [\alpha, H] \ and \ H$$

$$= -[\bar{\alpha}\cdot\bar{P}_\theta + \beta m_\theta]I^\theta \ then \ \gamma^0 S$$

$$= \gamma^0[\bar{\gamma}\cdot\bar{P}_\theta - m_\theta]I^\theta.$$

Now let $\bar{\alpha} = \bar{x}_1^i + \bar{x}_2^i = \bar{x}_\theta I^\theta$, then look at

$$\dot{x}_\theta^i I^\theta = -i[x_1^i + x_2^i), \gamma^0 H]$$

$$= -i[x_1^i + x_2^i), \gamma^0\bar{\gamma}\cdot(\bar{P}_1 + \bar{P}_2) - \gamma^0(m_1 + m_2)]$$

$$= -i[x_1^i + x_2^i), \gamma^0(-\gamma^j P_{1j} - \gamma^j \bar{P}_{2j})]$$

$$= i\{\gamma^0\gamma^j[x_1^i, P_{1j}] + \gamma^0\gamma^j[x_2^i, P_{2j}]\}$$

$$= i\{\gamma^0\gamma^j(-i\hbar\delta_{ij}) + \gamma^i\gamma^j(-i\hbar\delta_{ij})\}$$

$$= 2\gamma^0\gamma^j$$

$$(9.36)$$

Therefore, Equation (9.36) gives us

$$\sigma^{0i} = \frac{1}{2}(\dot{x}_1^i + \dot{x}_2^i) = \frac{1}{2}\dot{x}_\theta I^\theta.$$

On the other hand $\sigma^{ij}=-2iS^k$ $(i,j,k$ cyclic), but with two particles in the system this should be written

$$\sigma^{ij} = -i(S_1^k + S_2^k) = -iS_\theta^k I^\theta. \quad (9.37)$$

Putting Equations (9.35)-(9.37) into Equation (9.34) we find

$$[(i\partial_{1\mu} - e_1 A_{2\mu} + i\partial_{2\mu} - e_2 A_{1\mu})(i\partial_{1\mu} - e_1 A_{2\mu} + i\partial_{2\mu} - e_2 A_{1\mu})(g^{\mu\nu} + \sigma^{\mu\nu})$$

$$- (m_1 + m_2)^2]\psi(x)$$

$$= [(i\partial_{1\mu} - e_1 A_{2\mu} + i\partial_{2\mu} - e_2 A_{1\mu})(i\partial_1^\mu - e_1 A_2^\mu + i\partial_2^\mu - e_2 A_1^\mu) - (m_1 + m_2)^2$$

$$+ (i\partial_{1\mu} - e_1 A_{2\mu} + i\partial_{2\mu} - e_2 A_{1\mu})(i\partial_{1\nu} - e_1 A_{2\nu} + i\partial_{2\nu} - e_2 A_{1\nu})\sigma^{\mu\nu}]\psi(x) = 0$$

or

$$\left[\begin{array}{l}(i\partial_{0\mu}I^{0}-e_{0}A_{\mu}^{T0})(i\partial_{0}^{\mu}I^{0}-e_{0}A^{T0\mu})-(m_{0}I^{0})\\+(i\partial_{0\mu}I^{0}-e_{0}A_{\mu}^{T0})(i\partial_{0\nu}I^{0}-e_{0}A_{\mu}^{T0})(i\partial_{0\nu}I^{0}-e_{0}A_{\nu}^{T0})\sigma^{\mu\nu}\end{array}\right]\psi(x)=0. \quad (9.38)$$

Most of the terms involving the matrices $\sigma^{\mu\nu}$ vanish. This is because the scalar product of a symmetric and an anti-symmetric tensor vanishes. That is, if $S^{\mu\nu}$ and $T^{\mu\nu}$ satisfy $S^{\mu\nu}=S^{\nu\mu}$, $T^{\mu\nu}=-T^{\nu\mu}$ then $S^{\mu\nu}T^{\mu\nu}=0$.

In Equation (9.38), the quantity $\sigma^{\mu\nu}$ is anti-symmetric under interchange of μ and v. The terms involving $-ie_1\partial_{1\mu}A_{2\nu}\sigma^{\mu\nu}\psi(x)=-ie_1\sigma^{\mu\nu}\psi(x)(\partial_{1\mu}A_{2\nu})-ie_1A_{2\nu}\partial_{1\mu}\sigma^{\mu\nu}\psi(x)$ are also anti-symmetric. Now let us write out the coefficient of $\sigma^{\mu\nu}$ in Equation (9.38) :

$$(-\partial_{1\mu}\partial_{1\nu}-i\partial_{1\mu}e_1A_{2\nu}-\partial_{1\mu}\partial_{2\nu}-i\partial_{1\mu}e_2A_{1\nu}$$
$$-ie_1A_{2\mu}\partial_{1\nu}+e_1A_{2\mu}e_1A_{2\nu}-ie_1A_{2\mu}\partial_{2\nu}+e_1A_{2\mu}e_2A_{1\nu}$$
$$-\partial_{2\mu}\partial_{1\nu}-i\partial_{2\mu}e_1A_{2\nu}-\partial_{2\mu}\partial_{2\nu}-i\partial_{2\mu}e_2A_{1\nu}$$
$$-ie_2A_{1\mu}\partial_{1\nu}+e_2A_{1\mu}e_1A_{2\nu}-ie_2A_{1\mu}\partial_{2\nu}+e_2A_{1\mu}e_2A_{1\nu}).$$

Obviously the terms $-\partial_{1\mu}\partial_{1\nu}$, $-\partial_{2\mu}\partial_{2\nu}$, $e^2{}_2A_{1\mu}A_{1\nu}$, and $e^2{}_1A_{2\mu}A_{2\nu}$ are symmetric so they may be dropped. Combining terms leads to

$$[i\partial_{1\mu}(e_1A_{2\nu}+e_2A_{1\nu})-(\partial_{1\mu}\partial_{2\nu}+\partial_{2\mu}\partial_{1\nu})$$
$$-ie_1A_{2\mu}(\partial_{1\nu}+\partial_{2\nu})+2e_1e_2(A_{2\mu}A_{1\nu})$$
$$-i\partial_{2\mu}(e_1A_{2\nu}+e_2A_{1\nu})-ie_2A_{1\mu}(\partial_{1\nu}+\partial_{2\nu})].$$

The terms $-(\partial_{1\mu}\partial_{2\nu}+\partial_{2\mu}\partial_{1\nu})$ and $e_1e_2(A_{2\mu}A_{1\nu}+A_{1\mu}A_{2\nu})$ are symmetric under an interchange of μ and v. The following is left

$$[ie_1(\partial_{1\mu}+\partial_{2\mu})A_{2\nu}$$
$$-ie_2(\partial_{1\mu}+\partial_{2\mu})A_{1\nu}$$
$$-i(e_1A_{2\mu}+e_2A_{2\mu})(\partial_{1\nu}+\partial_{2\nu})].$$

Using

$$-ie_1\partial_{1\mu}A_{2\nu}\sigma^{\mu\nu}\psi = -ie_1\sigma^{\mu\nu}\psi(\partial_{1\mu}\partial_{2\nu}) - ie_1A_{2\nu}\partial_{1\mu}\sigma^{\mu\nu}\psi$$

$$-ie_1\partial_{2\mu}A_{2\nu}\sigma^{\mu\nu}\psi = -ie_1\sigma^{\mu\nu}\psi(\partial_{2\mu}\partial_{2\nu}) - ie_1A_{2\nu}\partial_{2\mu}\sigma^{\mu\nu}\psi$$

$$-ie_2\partial_{1\mu}A_{1\nu}\sigma^{\mu\nu}\psi = -ie_2\sigma^{\mu\nu}\psi(\partial_{1\mu}\partial_{1\nu}) - ie_2A_{1\nu}\partial_{1\mu}\sigma^{\mu\nu}\psi$$

$$-ie_2\partial_{2\mu}A_{1\nu}\sigma^{\mu\nu}\psi = -ie_2\sigma^{\mu\nu}\psi(\partial_{2\mu}\partial_{1\nu}) - ie_2A_{1\nu}\partial_{2\mu}\sigma^{\mu\nu}\psi$$

leads to

$$[-ie_1\sigma^{\mu\nu}\psi(\partial_{1\mu}A_{2\nu} + \partial_{2\mu}A_{2\nu}) - ie_2\sigma^{\mu\nu}\psi(\partial_{1\mu}A_{1\nu} + \partial_{2\mu}A_{1\nu})$$

$$-ie_1A_{2\nu}(\partial_{1\mu} + \partial_{2\mu})\sigma^{\mu\nu}\psi - ie_2A_{1\nu}(\partial_{1\mu} + \partial_{2\mu})\sigma^{\mu\nu}\psi$$

$$-ie_1A_{2\mu}(\partial_{1\nu} + \partial_{2\nu})\sigma^{\mu\nu}\psi - ie_2A_{1\mu}(\partial_{1\nu} + \partial_{2\nu})\sigma^{\mu\nu}\psi].$$

In the last four terms the coefficients of $\sigma^{\mu\nu}$ are symmetric in an interchange of μ and ν so only the first two terms remain. These may be written in symmetric and anti-symmetric form since

$$\partial_\mu A_\nu = \frac{1}{2}(\partial_\mu A_\nu + \partial_\nu A_\mu) + \frac{1}{2}(\partial_\mu A_\nu - \partial_\nu A_\mu).$$

Only the anti-symmetric part will survive. We may rewrite Equation (9.38) as

$$[(i\partial_{1\mu} - e_1A_{2\mu} + i\partial_{2\mu} - e_2A_{1\mu})(i\partial_1^\mu + e_1A_2^\mu + i\partial_2^\mu - e_2A_1^\mu) - (m_1 + m_2)^2$$

$$-ie_1\sigma^{\mu\nu}(\partial_{1\mu}A_{2\nu} - \partial_{1\nu}A_{2\mu}) - ie_1\sigma^{\mu\nu}(\partial_{2\mu}A_{2\nu} - \partial_{2\nu}A_{2\mu})$$

$$-e_2\sigma^{\mu\nu}(\partial_{1\mu}A_{1\nu} - \partial_{1\nu}A_{1\mu}) - ie_2\sigma^{\mu\nu}(\partial_{2\mu}A_{1\nu} - \partial_{2\nu}A_{1\mu})]\psi = 0.$$

Defining $B^\beta_{\theta\mu\nu} = \partial_{\theta\mu}A^{T\beta}{}_\nu - \partial_{\theta\nu}A^{T\beta}{}_\mu$ this may be written as

$$[(i\partial_{\theta\mu}I^\theta - e_\theta A_\mu^{T\theta})(i\partial_\theta^\mu I^\theta - e_\theta A^{T\theta\mu})$$

$$-(m_\theta I^\theta)^2 - i\sigma^{\mu\nu}e_\beta B^\beta_{\theta\mu\nu}I^\theta]\psi(x) = 0. \tag{9.39}$$

Equation (9.39) appears to be a very interesting equation. Typically the fields $F_{\mu\nu} = \partial_\mu A_\nu - \partial_\nu A_\mu$ are identified as the electromagnetic field and the result is the Klein-Gordon equation with fields. Here the derivatives and gauge potentials are mixed so care must be taken when interpreting the terms. Also remember

$$
\sigma^{\mu\nu} = \begin{pmatrix}
0 & \dot{x}_1 + \dot{x}_2 & \dot{y}_1 + \dot{y}_2 & \dot{z}_1 + \dot{z}_2 \\
-\dot{x}_1 - \dot{x}_2 & 0 & -2i(S_1 + S_2)^3 & 2i(S_1 + S_2)^2 \\
-\dot{y}_1 - \dot{y}_2 & 2i(S_1 + S_2)^3 & 0 & -2i(S_1 + S_2)^1 \\
-\dot{z}_1 - \dot{z}_2 & -2i(S_1 + S_2)^2 & 2i(S_1 + S_2)^1 & 0
\end{pmatrix}
$$

or this may be written as $\sigma^{\mu\nu} = \sigma^{\mu\nu}{}_\theta I^\theta$.

Notice that should both particles have spin $\pm 1/2$ the total system spin, which is $S = S_1 + S_2 = S_\theta I^\theta$, must take on the values -1, 0, or 1, similar to the values allowed for isospin. A system with three spin $1/2$ particles would take on values of $-3/2$, $-1/2$, $1/2$, or $=3/2$.

Perhaps Equation (9.39) may best be understood by considering the motion *ABOUT* the COM and the motion *OF* the COM as being our two motions. Further, consider the two-particle system for which the potentials of Equations (9.16), hold. Now consider the two extremes of, first, the motion *OF* the COM is small compared to the motion *ABOUT* the COM. This would be the case when the particle separation is large compared to the larger of the λ's. In this case it may be seen that, in Equation (9.16) , the COM potential vanishes, and the remaining potential about the COM approaches the classical electrostatic potential. Further, Equation (9.33) collapses to the classical Dirac equation for the motion of a particle in the field of another. Next, consider the case when the particle separation is of the order of the two λ's and the motion *ABOUT* the COM may be neglected. In this case the potential *ABOUT* the COM is being driven to zero by the exponents while the potential of the COM is becoming dominant.

Notice that the potentials at large separations produce a dominant negative potential, as in the hydrogen atom. Separations of the order of the λ's produce a very sharp positive potential which returns to zero as the separations become less than the smaller of the λ's. This sets up the condition whereby the COM may be trapped in a positive energy well from which it may escape through tunnelling.

In order to see this conclusion, consider the plot of the potential of the center of mass in Equation (9.16), V_c in Figure 9.1.

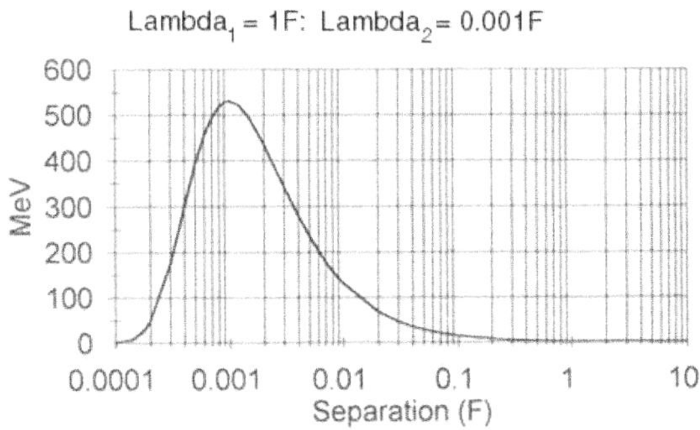

Figure 9.1: COM Potential for $\lambda_1 >> \lambda_2$

Equation (9.39), the Klein-Gordon equation with fields, describes phenomena usually referred to as weak nuclear forces. The $B^{\beta}{}_{\theta\mu\nu}$ are defined as are the Yang-Mills fields. The potential plotted in Figure 9.1 is the positive potential that traps the center of mass of the neutron wherein a proton is in orbit around an electron.

Before the reader begins to claim that the Heisenberg Uncertainty Principle denies that an electron can ever be within the confines of a neutron, please recall that in Chapter 7 we showed that the unit of action, or uncertainty, depends upon the gauge function. In the case of the neutron we will see later that the weak force gauge fields significantly decrease the uncertainty of the electron and moderately decrease the uncertainty of the proton.

Three-body Systems

Currently various types of quarks are thought to be the constituents of protons, neutrons and other nuclear particles. Scattering data conclusively show three entities within a proton. These are interpreted as three quarks. However, the preceding display that the neutron may be a proton in orbit around an electron introduces a new concept into the possible structure of a proton. If a neutron is a single orbiting proton around a single electron, what would be the result of two positrons in orbit around a single electron? Though our first reaction may be the old one due to memories of the uncertainty principle that argues one should not even ask such a question, suppose we use the methodology just developed and investigate what may be learned about a three body system.

Consider the extension of the above to a system of three particles interacting with each other yet having no outside fields influencing their motions. This would lead to the three components of momentum

$$\tilde{P}_{1\mu} = P_{1\mu} - e_1(A_\mu^2 + A_\mu^3)$$

$$\tilde{P}_{2\mu} = P_{2\mu} - e_2(A_\mu^3 + A_\mu^1)$$

$$\tilde{P}_{3\mu} = P_{3\mu} - e_3(A_\mu^1 + A_\mu^2)$$

or

$$\tilde{P}_{\theta\mu} = P_{\theta\mu} - e_\theta C_\mu^{\beta\gamma}$$

with the definition $C_\mu^{\beta\gamma} = A_\mu^\beta + A_\mu^\gamma$, where θ, β, and γ are cyclic. Then Equation (9.33) may be written

$$[(i\partial_{\theta\mu}I^\theta - e_\theta C_\mu^{\beta\kappa})\gamma^\mu + m_\theta I^\theta]\psi(x) = 0 \qquad (9.40)$$

where θ, β, γ are cyclic. By defining $B^{\eta\beta\kappa}_{\theta\mu\nu} = \partial_{\theta\mu}C^{\eta\beta\kappa}_\nu - \partial_{\theta\nu}C^{\eta\beta\kappa}_\mu$ Equation (9.40) becomes

$$\left[\begin{array}{l}\left(i\partial_{\theta\mu}I^\theta - e_\theta C_\mu^{\beta\kappa}\right)\left(i\partial_\theta^\mu I^\theta - e_\theta C^{\beta\kappa\mu}\right) \\ -\left(m_\theta I^\theta\right) - i\sigma^{\mu\nu}e_\eta B^{\eta\beta\kappa}_{\theta\mu\nu} I^\theta\end{array}\right]\psi(x) = 0 \qquad (9.41)$$

where θ, β, κ, and η range from 1 to 3 and θ, β, and η are cyclic.

Equation (9.41) represents the Klein-Gordon equation for three particles when one particle is different from the other two. However, for three like particles it remains valid, but has some terms that vanish due to the equal and opposite forces.

Also, Equation (9.41) is the equation describing the strong nuclear forces that bind two protons to a single electron to form the deuterium nucleus. Of course this occurs when the separation of all three particles is approximately of nuclear separations. It may also be seen that the three-body system of two protons in orbit around a single electron is a symmetrical system wherein the COM has no motion.

Returning to consideration of the three body system being two positrons in orbit around a single electron, we see that the solution would be virtually the same as for two protons in orbit around an electron. However, in the case of positrons the orbits would be much smaller due to the fact that the lambda's of the positron are expected to be much smaller than the lambda's of the protons. For the system of two positrons in orbit around an electron Equation (9.41) would provide the details. Such a system would satisfy the scatter data that requires three entities within a proton. The entities in this case would be positrons and an electron rather than three quarks.

We have now shown how the WQP requires the electrostatic potential to be quantized and non-singular. Further, we have shown that nuclear forces are derived when the relativistic equations of motion are expressed for the non-singular electromagnetic forces that do not satisfy Newton's third law. The short-range weak nuclear forces are primarily due to this violation of Newton's third law. It is no wonder then that the decay of a neutron is accompanied by motion of the center of mass. It must necessarily move in the decay.

On the other hand the non-singular electromagnetic forces between like particles obey Newton's third law at all separations. However, the non-singular character of the forces requires repulsion between the like particles for all separations greater than the λ of the particles. When the separation is less than λ the force between the particles is attractive just as the strong nuclear forces are seen to be experimentally.

Yes, there is plenty of room for a discussion of quarks that may be taken up later.

The extension to N particles should follow readily and would provide for solving the internal motions of a system made up of N particles interacting relativisticly with each other in the absence of external forces. The discussion of the two-particle system above led to the conclusion that when, all motion is taken to be motion OF the COM, the potential well formed was a positive energy well which would allow the calculation of the lifetime of the state. This would argue that an N-particle system, for which the COM is required to move, that any bound state of the COM might also be trapped in a positive energy well. The lifetimes of these states would then be subject to calculation.

The Neutron

When unlike particles are considered care must be taken to keep the lambdas in the forces straight. The force on any charged particle due to the presence of another, second particle, is the product of the charge of the first particle and the field of the second particle. Thus, the force on the first particle goes to zero at the lambda of the second. For the force on a proton due to the field of an electron

$$F_p = q_p E_e = \left(-\frac{k}{r^2}\right)\left(1 - \frac{\lambda_e}{r}\right)e^{-\frac{\lambda_e}{r}} \qquad (9.42)$$

while the force on an electron due to the field of a proton is

$$F_e = q_e E_p = \left(\frac{k}{r^2}\right)\left(1 - \frac{\lambda_p}{r}\right)e^{-\frac{\lambda_p}{r}}. \qquad (9.43)$$

By looking at proton and electron like-particle scattering data it would appear that the lambda of the proton must be much larger than that of the electron. If this is the case then the force on the electron due to the near presence of the proton goes to zero while the proton is still attracted to the electron. Any further decrease in the separation causes the electron to experience a repulsive force; although the proton is still attracted to the electron. This immediately raises the eyebrows. C an it be that Newton's third law, concerning the equal and opposite forces, does not hold in Nature? The answer is, certainly. Newton's third law does not hold in high-speed electromagnetic interactions when viewed by the retarded potentials; it was found to be violated during beta decay until the hypothesis of the neutrino reinstated the summation of particle spins. Should one then throw out the unlike-particle forces because they violate Newton's third taw without seeing what predictions these forces might lead to?

If one proceeds with the unlike-particle forces, he finds very quickly that it appears possible that the proton might find a very close orbit, at a separation from the electron by a distance lambda, in which it could settle down into a Bohr orbit around the electron. On the other hand the electron would experience no force from the orbiting proton. Such a state might cause one to think of the neutron. Here one runs into the question of particle spins that beta decay brought out and which led to the hypothesis of the neutrino. Also, the argument is offered that Heisenberg's Uncertainty Principle requires that the electron could never be in an orbit so tightly bound that the orbit is less than nuclear separations. This argument hinges upon the unit of action being Planck's constant. But remember the dependence of the Poisson brackets upon the geometry?

Another argument against the neutron being an electron and proton in nuclear sized orbits is based on an argument that the principle of angular momentum cannot be conserved. The non-singular forces, which require that the force between the electron and proton be directed on a line between them, also require that the angular momentum be conserved, as was shown earlier. However, the unit of action depends upon the gauge function and this requires

256

that, when Bohr-type orbits are considered, there is an effective unit of action for the electron orbit and a different effective unit of action for the proton orbit. Thus, the effective unit of action for the electron orbit requires that in the neutron the orbital angular momentum would be given by $\hbar_e$ and its intrinsic spin angular momentum would be $(1/2)\hbar_e$. Similarly, for the proton the orbital angular momentum would be $\hbar_p$ and the spin $(1/2)\hbar_p$.

After the neutron decays, the angular momentum is the sum of the two particles' intrinsic spin angular momenta, which is given by $\hbar$ because both particles are free and therefore, each has an intrinsic spin angular momentum of $(1/2)\hbar$. Thus, the conservation of angular momentum is expressed as

$$\frac{1}{2}(\pm\hbar_e \pm \hbar_p) + \hbar_e + \hbar_p = \hbar. \qquad (9.44)$$

Experimental evidence of orbital and/or spin angular momentum is contained in the experimental magnetic moments. If one equates the intrinsic and orbital magnetic moments of the electron and proton while they are in the orbital configuration to the experimental value of the neutron's magnetic moment they have

$$\pm\frac{1}{2}\left(\frac{\hbar_e}{\hbar}\right)\mu_B \pm \frac{1}{2}\left(\frac{\hbar_p}{\hbar}\right)\mu_n$$

$$-\left(\frac{\hbar_e}{\hbar}\right)\mu_B + \left(\frac{\hbar_p}{\hbar}\right)\mu_n = -1.91315\mu_B \qquad (9.45)$$

where μ_B is the Bohr magneton and μ_n is the nuclear magneton. Equations (9.44) and (9.45) require that $\hbar_e = 8.0517 \times 10^{-4}\hbar$ and $\hbar_p = 0.66585\hbar$. Thus, within the proposed theory the neutron appears to be a proton in orbit around an electron.

Not surprisingly then, it is possible to build a nuclear model of the protons-around-electrons, and electrons-around-positrons, states that allow one to predict the masses of the nuclei which have a mass number less than 10 a mu with better RMS error than the best of the semi empirical mass formulas have for mass numbers greater than oxygen. This should possibly be considered all the more significant since the semi empirical mass formulas have ever increasing errors for the low mass numbers and are not even used below an amu of 16. It remains for this nuclear model to be extended to the higher mass numbers, but it appears from the work done thus far that one can only expect that the correspondence with experiment will improve with increasing mass numbers.

Is it p ossible that the non-singular forces can explain the phenomena associated with the weak forces? Certainly the nuclear mass predictions argue that a nuclear model based upon these orbits does not miss far and is a much cleaner model than currently used. Initial looks at the neutrino experiments using the proposed theory offer other explanations for these experimental results but are too lengthy to include here. It should be remembered that these experiments must be explained by the proposed theory if the unlike forces are to fully account for phenomena that the weak nuclear forces are now thought to explain. We will see later how the neutrino enters into this new scheme of things.

Nuclear Model

Another way of visualizing the neo-coulombic force is to make a plot of it and compare it with a plot of the columbic force. Figure 9.2 compares these two forces plotted with the separation variable in fermions and normalized so that the columbic force at one Fermi separation is unity. Note that this plot compares the forces for like particles to ensure that λ is the same for both particles. Figure 9.2 shows that the neo-coulombic force is virtually indistinguishable from the Coulomb force for separations greater than approximately *10λ*. However, at a separation of exactly λ, the force is identically zero. In terms of the classical notion of nuclear forces, we would say that at separations greater than *10λ*, the nuclear force is negligible, whereas at a separation of λ the magnitude of the nuclear force was equal to the magnitude of the coulombic force. The neo-coulombic force becomes an attractive force for separations less than λ. This is exactly the behavior to be expected of a non-singular potential. For a potential to be non-singular it must tend to zero as r goes to zero. Such a potential which tends to zero for r tending to zero and for r tending to 4 must have a maximum absolute value in between. At that maximum the force, being determined by the slope of the potential, will go to zero and will be of the opposite signs on each side of the zero.

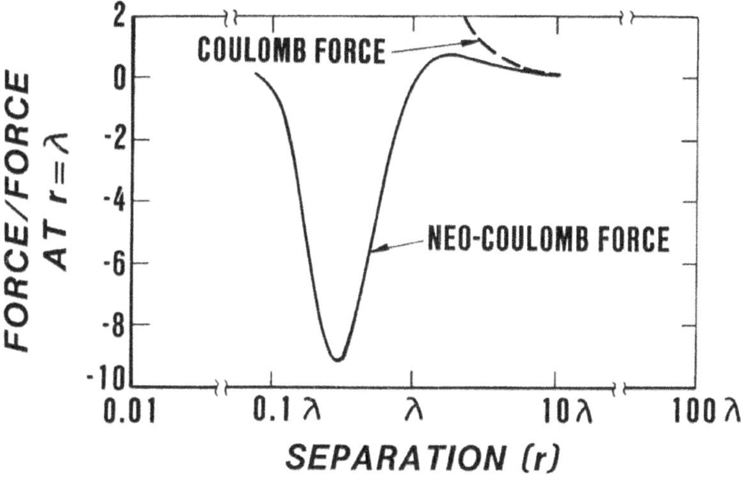

Figure 9.2. Comparison of coulomb and non-coulombic forces at short range.

Now let us look at the force between unlike particles, say a proton and an electron. Consider the electron and proton to be placed on a horizontal surface separated by a distance, *r*, with the proton to the right of the electron. Thus, the long range attractive forces between these two particles will cause the proton to experience a force to the left while the electron will experience a force to the right. We may than write the force on the proton that is due to the positive charge of the proton being in the electron field as

$$F_p = q_p E_e = \frac{-e^2}{4\pi\varepsilon_o r^2}\left(1 - \frac{\lambda_e}{r}\right)e^{\frac{-\lambda_e}{r}} \tag{9.46}$$

where the electron field involving the electron lambda has been accounted for. The electron force owing to the electron charge being in the proton field is given by

$$F_e = q_e E_p = \frac{e^2}{4\pi\varepsilon_o r^2}\left(1 - \frac{\lambda_p}{r}\right)e^{\frac{-\lambda_p}{r}} \tag{9.47}$$

Figure 9.3 plots both these forces as a function of the separation, *r*, where, $\lambda_p = 10^{-15}$ *m*, or $\lambda_p = 1$ Fermi has been assumed. The electron-electron scattering data show that the electron-electron interaction behaves in a coulombic manner even when separations are approximately *0.01-0.1* Fermi. To be consistent with this data, we have assumed $\lambda_e = 10^{-3}$ Fermi.

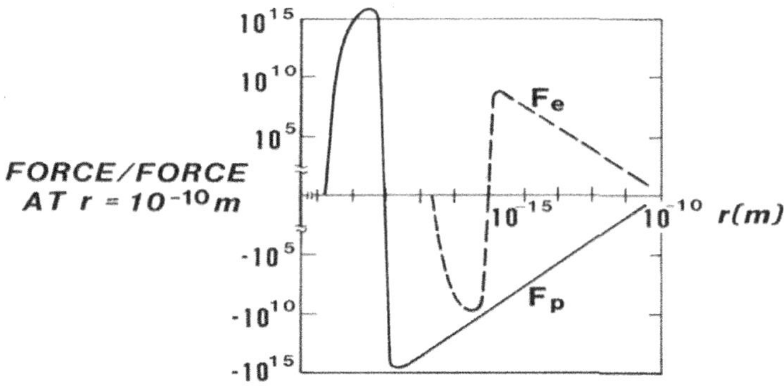

Figure 9.3. Unlike particle forces

The Dynamic Theory

From this plot of the force on the proton and the force on the electron, we see that for separations less than about 10 Fermi's the forces become extremely unsymmetrical. This immediately and visually demonstrates that the neo-coulombic exponential force violates Newton's third law requiring that the force on the proton be equal in magnitude and opposite indirection to the force on the electron. The question arises whether or not a violation of Newton's third law has ever been seen as the result of an interaction between an electron and a proton? The answer, based on a neutron disintegration from which a proton and electron emerge, is definitely yes; Newton's third law was seen to be violated. To reinstate Newton's third law in neutron disintegration and all other beta decay, Pauli postulated the existence of the neutrino. Fermi later developed his theory of weak interactions, from which appeared the necessity to talk of a fourth force in nature.

It was previously shown that the neo-coulombic force, which requires distinct λ for distinct fundamental particles, leads to the Yan-Mills equations for describing the weak forces. Does this mean the neutrino does not exist? What about the experimental evidence submitted in support of the capture of a free neutrino? We will return to the neutrino later.

If we again consider the plots of the proton and electron forces in Figure 9.3 we see that, at atomic separations and greater distances, the forces obey Newton's third law and the difference between the neo-coulombic and columbic forces is so small that it could not be detected in atomic or macroscopic phenomena. But as the separation becomes smaller, the picture begins to change. When the r approaches λ_p, the electron is no l onger attracted to the proton as strongly as the proton is attracted to the electron. If the separation is exactly λ_p, then the electron is indifferent to the proton's presence. The proton, on t he other hand, is still very much attracted to the electron. If for the moment, we ignore the interpretation of Heisenberg's uncertainty principle that would say it cannot be, then we could easily imagine a circular proton orbit around a stationary electron, during which the proton stays at a radius of λ_p from the electron. The electron should be stationary during such motion because it would experience no force.

We now consider a separation between the electron and proton, which is some simple fraction of λ_p. Here, we find the electron repulsed by the proton, but the proton is still attracted to the electron. Notice that the force on both particles, from our initial positioning of the proton on the right, is to the left. If both particles were given an angular momentum such that they were placed into synchronized circular orbits, then because their synchronous motion always results in the force on both particles being directed along the line separating them and from the proton toward the electron or from the electron away from the proton then, again ignoring arguments from the uncertainty principle, circular orbits in which the electron is in a small orbit about a space point could be imagined, where the proton is in a much larger orbit about the same space point.

Let us follow this picture a little farther and write simple Newtonian like force laws for this situation. The situation envisioned is presented in Figure 9.4. The electron position is given by r_e from the origin, and the position of the proton is given by r_p. The separation between them is

$$r = r_p - r_e . \tag{9.48}$$

Because the force is always directed along the line separating the two particles, we may write the radial equation of motion for the proton as

$$m_p \left(\frac{v_p}{r_p} \right) = \frac{-e^2}{4\pi\varepsilon_o r^2} \left(1 - \frac{\lambda_e}{r} \right) e^{\frac{-\lambda_e}{r}} \tag{9.49}$$

where the assumed circular motion has been taken into account and v_p is the tangential proton velocity. The electron equation of motion is given by

$$m_e\left(\frac{v_e}{r_e}\right) = \frac{e^2}{4\pi\varepsilon_o r^2}\left(1-\frac{\lambda_p}{r}\right)e^{\frac{-\lambda_p}{r}} \qquad (9.50)$$

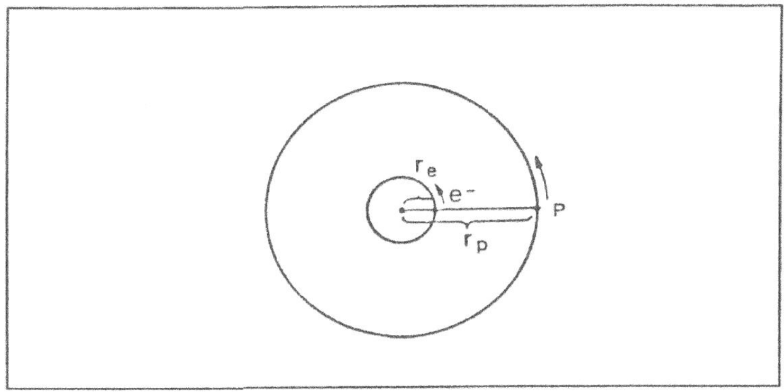

Figure 9.4. Circular orbits for the electron and proton in a neutron.

The right-hand side of Equations (9.49) and (9.50) are both functions of the separation, r, whereas the two left-hand sides are individually functions of r_e and r_p. A solution is possible only when the three equations, Equations (9.48), (9.49), and (9.50), are solved simultaneously.

An alternative approach is to add Equation (9.49) and (9.50) then obtain the equation of motion for the center of mass,

$$m_p\left(\frac{v_p^2}{r_p}\right) + m_e\left(\frac{v_e^2}{r_e}\right) = \frac{e^2}{4\pi\varepsilon_o r^2}\left[\left(1-\frac{\lambda_p}{r}\right)e^{\frac{-\lambda_p}{r}} - \left(1-\frac{\lambda_e}{r}\right)e^{\frac{-\lambda_e}{r}}\right] \qquad (9.51)$$

or

$$M\left(\frac{V^2}{R}\right) = \frac{e^2}{4\pi\varepsilon_o r^2}\left[\left(1-\frac{\lambda_p}{r}\right)e^{\frac{-\lambda_p}{r}} - \left(1-\frac{\lambda_e}{r}\right)e^{\frac{-\lambda_e}{r}}\right] \qquad (9.52)$$

where $\quad R = \left(m_p r_p + m_e r_e\right)/\left(m_p + m_e\right)$ and $\quad M = m_p + m_e$.

From Equation (9.52) we see that bound states, where the center of mass is in motion as the result of the asymmetrical force, may only occur when the separation is less than λ_p.

All of these equations of motion exhibit a feature not usually found in equations of motion. That is, because the force depends on the separation, r, between the particles and not strictly on the position, r_e, for the election, then the usual integration of the force over a change of position, which produces the potential energy, cannot be readily done because,

$$V\left(r_e\right) = -\int F_e \, dr_e$$

$$= -\int \left[\frac{e^2}{4\pi\varepsilon_o r^2}\left(1 - \frac{\lambda_p}{r}\right)e^{\frac{-\lambda_p}{r}}\right] dr_e \qquad (9.53)$$

However, we may now use the process of determining the motion of the center of mass and the motion about the center of mass rather than the motion of the individual particles. We have already obtained this solution above and found that the energies were given by

$$E = K + V + K_c + V_c, \qquad (9.54)$$

with

$$K = \frac{\mu k}{2r}\left[\left(1 - \frac{\lambda_p}{r}\right)\frac{e^{-\frac{\lambda_p}{r}}}{m_e} + \left(1 - \frac{\lambda_e}{r}\right)\frac{e^{-\frac{\lambda_e}{r}}}{m_p}\right]$$

$$V = -\frac{\mu k}{r}\left[\frac{e^{-\frac{\lambda_p}{r}}}{m_e} + \frac{e^{-\frac{\lambda_e}{r}}}{m_p}\right]$$

$$K_c = \frac{kR}{2r^2}\left[\left(1 - \frac{\lambda_p}{r}\right)e^{-\frac{\lambda_p}{r}} - \left(1 - \frac{\lambda_e}{r}\right)e^{-\frac{\lambda_e}{r}}\right]$$

$$V_c = -\frac{k}{r}\left[e^{-\frac{\lambda_p}{r}} - e^{-\frac{\lambda_e}{r}}\right],$$

(9.55)

In Equation (9.54) the subscript c refers to the center of mass value while the absence of a subscript refers to motion about the center of mass.

Heisenberg's Uncertainty Principle and the Gauge Function.

The suggestion that bound states of electrons and protons might exist where the orbits are of the approximate order of magnitude of nuclear dimensions, is essentially a return to the notion that a neutron might be such a state. This idea gave way under arguments of conservation of momentum and Heisenberg's Uncertainty Principle to the view that electrons are forbidden to be found within the nucleus. Therefore, let us take another look at those fundamental tenets of quantum mechanics, the Poisson brackets.

The classical Poisson bracket is defined by

$$\{F,G\} = \sum_j \left(\frac{\partial F}{\partial q_j}\frac{\partial G}{\partial p_j} - \frac{\partial F}{\partial p_j}\frac{\partial G}{\partial q_j}\right)$$

(9.56)

where F and G are any two functions of the canonically conjugate variables q_j and p_j. The special relations that occur when F and G are q_j and p_j, respectively, are especially important in quantum mechanics; these are, classically:

$$\{q_j, q_k\} = 0$$

$$\{p_j, p_k\} = 0 \tag{9.57}$$

$$\{p_j, p_k\} = \delta_{jk},$$

where δ_{jk} is the Kronecker delta. The classical Poisson brackets of Equation (9.56) are obtained when Euclidean spaces are assumed. However, the definition of Poisson brackets remains valid for general metric spaces, when the notion of covariant differentiation is used. If we now consider the momenta, expressed in a general coordinate system, the covariant components, $p_j = mg_{ij}\dot{x}^i$ and

$p_j = g^{ji} p_i = m\dot{x}_j$ are the contravariant components. Covariant differentiation must be carried out with respect to contravariant vector components. There, in a general space the canonically conjugate variables to be considered are x^j and p^k, and the Poisson bracket of the position and momenta becomes

$$\{x^j, p^k\} = \left[\frac{\partial x^j}{\partial x^l} + \left\{{}^{\,j}_{s\,l}\right\} x^s\right]\frac{\partial p^k}{\partial p^l} - \frac{\partial x^j}{\partial p^l}\left[\frac{\partial p^k}{\partial x^l} + \left\{{}^{\,k}_{n\,l}\right\} p^n\right] \tag{9.58}$$

$$= \left[\delta_{il} + \left\{{}^{\,j}_{s\,l}\right\} x^s\right]\delta_{lk}$$

or

$$\{x^j, p^k\} = \delta_{ik} + \left\{{}^{\,j}_{s\,k}\right\} x^s \tag{9.59}$$

Quantum mechanics adopts the operator,

$$\left(\frac{\hbar}{i}\right)\frac{\partial}{\partial x^j} \rightarrow p_j \tag{9.60}$$

for the momentum. This, in general case, becomes the covariant operator

The Dynamic Theory

$$\left(\frac{\hbar}{i}\right)(\ )_{,j} \to p_j. \tag{9.61}$$

The operator for the contravariant momentum components is then

$$\left(\frac{\hbar}{i}\right)g^{jl}(\ )_{,l} \to p^j \tag{9.62}$$

Now if we look at the quantum Poisson bracket, where the operators are operating on a scalar ψ, then

$$\left[x^j, p^k\right]\Psi = \left[x^j\left(\frac{\hbar}{i}\right)g^{kl}\frac{\partial\Psi}{\partial x^l} - \left(\frac{\hbar}{i}\right)g^{kl}\left(x^j\Psi\right)_{,l}\right]$$

$$= x^j\left(\frac{\hbar}{i}\right)g^{kl}\frac{\partial\Psi}{\partial x^l} - g^{kl}\left(\frac{\hbar}{i}\right)\left[\frac{\partial x^j}{\partial x^l} + \left\{{}_{s}{}^{j}{}_{l}\right\}x^s\right]\Psi$$

$$\qquad\qquad - \left(\frac{\hbar}{i}\right)g^{jl}x^j\frac{\partial\Psi}{\partial x^l} \tag{9.63}$$

$$= i\hbar\left[\delta_{jl} + \left\{{}_{s}{}^{j}{}_{l}\right\}x^s\right]\Psi$$

This may be written in terms of the classical Poisson bracket, Equation (9.56), as

$$\left[x^j, p^k\right]\Psi = i\hbar g^{kl}\left\{x^j, p^l\right\}\Psi. \tag{9.64}$$

If the space is Euclidean, then the g^{kl} become the Kronecker delta and the Christoffel symbols vanish and the quantum Poisson bracket of Equation (9.64) becomes

$$\left[x^j, p_k\right]\Psi = i\hbar\delta_{jk}\Psi \tag{9.65}$$

because $p^k = g^{kl}p_l = \delta^{kl}p_l = p_k$. However, from Equations (9.59) and (9.63), we see that the metric does play a role in the quantum operators. This should also be seen in the use of the operators in the Schrödinger Hamiltonian operator, because

$$p_j p^j = \frac{\hbar}{i}\frac{1}{\sqrt{g}}\frac{\partial}{\partial x^j}\left(\sqrt{g}\,p^j\right)$$

$$= \frac{\hbar}{i}\frac{1}{\sqrt{g}}\frac{\partial}{\partial x^j}\left(\sqrt{g}\left(\frac{\hbar}{i}\right)g^{jl}\frac{\partial}{\partial x^l}\right) \qquad (9.66)$$

$$= \frac{\hbar^2}{\sqrt{g}}\frac{\partial}{\partial x^j}\left(\sqrt{g}\,g^{jl}\frac{\partial}{\partial x^j}\right)$$

becomes the operator to be used in a general space and, of course, is the operator currently used in applying Schrodinger's equation to the hydrogen atom. The geometrical effect may be seen also in Dirac's equation by using the same procedure.

Now, of what benefit is this discussion of geometrical effect upon quantum mechanics in considering the neo-coulombic force? Recall that the neo-coulombic force came from a gauge function in a Weyl space. A gauge function has a geometrical effect that could be thought of as effectively changing the unit of action in quantum mechanics. To see the basis for this statement, let us recall the quantum Poisson bracket operations on a scalar, Equation (9.64),

$$\left[x^j,p^k\right]\Psi = i\hbar\left[\delta_{jl} + \left\{{}_{s}^{\ j}{}_{l}\right\}x^s\right]\Psi \qquad (9.67)$$

and let us define

$$\hbar'\delta_{jk} = \hbar g^{kl}\left[\delta_{jl} + \left\{{}_{s}^{\ j}{}_{l}\right\}x^s\right] \qquad (9.68)$$

then we can write

$$\left[x^j,p^k\right]\Psi = i\hbar'\delta_{jk}\Psi \qquad (9.69)$$

which has the same form as now used in quantum mechanics but the effective unit of action $\hbar'$ depends on the geometry as seen by Equation (9.68).

Neutron units of action

While we have not yet displayed an analytical expression for $\hbar'$, this absence of an analytical expression

268

for the effective unit of action does not completely stop us from considering the possibility that a neutron may be a proton in a large orbit about an electron in a small orbit. We may, for the moment, acknowledge the difficulty of obtaining an analytical expression for $\hbar'$ by allowing the $\hbar'$, or the unit of action, for the proton and electron to be a function of their orbit, and we may designate $\hbar'_e$ to be the effective unit of action for the electron orbit in a neutron and $\hbar'_p$ to be the unit of action for the neutron's proton orbit. If the effective unit of action depends upon the orbit, as it appears here that it must, then the interpretation that Heisenberg's Uncertainty Principle rules out the possibility of an electron being contained within nuclear discussions is inapplicable.

Another argument against the neutron being an electron and proton in nuclear sized orbits is based on an argument that the principle of angular momentum cannot be conserved. The neo-coulombic forces, which require that the force between the electron and proton be directed on a line between them, require that angular momentum be conserved as was shown earlier. However, the effective unit of action for the electron orbit requires that, in the neutron the orbital angular momentum would be given by $\hbar'_e$ and its intrinsic spin angular momentum would be $\hbar'_e/2$. Similarly, for the proton the orbital angular momentum would be $\hbar'_p$ and the spin $\hbar'_p/2$.

After the neutron decays, the angular momentum is the sum of the two particles' intrinsic spin angular momenta, which is given by $\hbar$ because both particles are free and, therefore, each has an intrinsic spin angular momentum of $\hbar/2$. Therefore, the conservation of angular momentum is expressed as

$$\frac{1}{2}\left(\pm\hbar_e \pm \hbar_p\right)+\hbar_e+\hbar_p = \hbar. \tag{9.70}$$

Nuclear Physics

Experimental evidence of orbital and/or spin angular momentum is contained in the experimental magnetic moments. If we equate the intrinsic and orbital magnetic moments of the electron and proton while they are in the orbital configuration to the experimental value of the neutron's magnetic moment we have

$$\pm\frac{2}{2}\frac{\hbar_e}{\hbar}\mu_B \pm\frac{2}{2}\frac{\hbar_p}{\hbar}\mu_n + \frac{\hbar_e}{\hbar}\mu_B + \frac{\hbar_p}{\hbar}\mu_n = -1.91315\mu_n \quad (9.71)$$

where μ_B is a Bohr magneton and μ_n is a nuclear magneton.

Equations (9.70) and (9.71) represent two equations in the two unknowns, $\hbar'_e$ and $\hbar'_p$, which may be solved to obtain the effective units of action for the electron and proton orbits making up a neutron such that angular momentum in conserved during neutrons' decay and that the correct magnetic moment of the neutron is ensured. Substituting the experimentally measured values of intrinsic magnetic moments for the electron and proton into Equation (9.71) would produce a m ore accurate solution because it would then contain the anomalous magnetic moments. Then we would have

$$\pm 2.002319\frac{\hbar_e}{\hbar}\mu_B \pm 2.79275\frac{\hbar_p}{\hbar}\mu_n + \frac{\hbar_e}{\hbar}\mu_B + \frac{\hbar_p}{\hbar}\mu_n = -1.91315\mu_n. \quad (9.72)$$

The only simultaneous solution of Equations (9.70) and (9.72), for which $\hbar'_e$ and $\hbar'_p$ are both positive, are

$$\hbar'_e = 8.0517x10^{-4}\hbar$$
$$\hbar'_p = 0.66586\hbar \quad (9.73)$$

The values of the effective units of action for the proton and electron given in Equation (9.73) show that angular momentum is conserved during the decay of a neutron when the neutron is considered to be a proton in orbit around an electron under the neo-coulombic force.

The third major argument against a neutron being states of the electron and proton orbits stems from the experimental evidence on the violation of Newton's third law during decay. That is, the energy of the electron emerging after decay is inconsistent with the equal and opposite columbic forces between an electron and a proton. Here, we find that the neo-coulombic forces are unequal in magnitude and opposite in direction; thus the energy of an electron emerging as the result of crossing from such an orbit cannot be consistent with Newton's third law. There is a fourth argument against this picture of a neutron: the possible existence of the neutrino. The above picture of the neutron produces no ne ed to postulate the existence of neutrinos. What then can be said about the experimental evidence that has been put forward in support of the capture of free neutrinos? The answer to this question will be presented in the later section on f ive dimensional wave solutions.

Nuclear Masses

The difficulty produced by the asymmetry of forces that arises in the interaction of an electron with a proton may be avoided if two protons are considered to be in orbit about the single electron. If we think of a snapshot of such a case we would find that the situation depicted in Figure 9.5 allows us to visualize the forces.

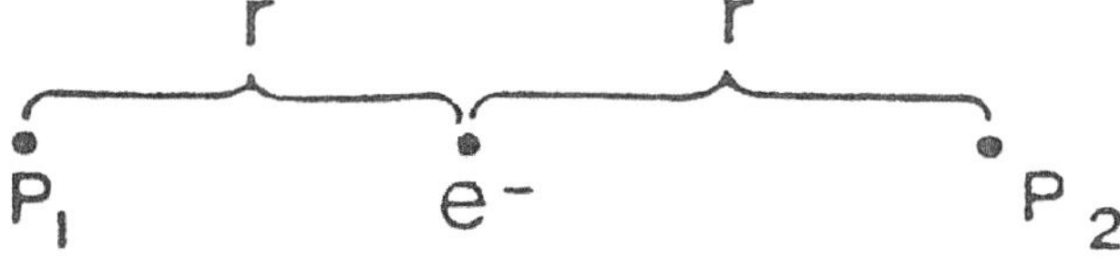

Figure 9.5. Two protons in orbit around an electron.

The force on the electron would be zero because it has a proton on each side diametrically opposed to one another. The force on each proton will be made up of two parts; one, the force that is due to the presence of the electron, and the other, owing to the other proton. The symmetry guarantees that each proton will experience an identical force, if circular orbits are assumed, toward the center of rotation. The force on the proton on the left would be

$$F_{p1} = \frac{-e^2}{4\pi\varepsilon_o r^2}\left[\left(1-\frac{\lambda_e}{r}\right)e^{\frac{-\lambda_e}{r}} - \frac{1}{4}\left(1-\frac{\lambda_p}{(2r)}\right)e^{\frac{-\lambda_p}{2r}}\right]. \quad (9.74)$$

To be sure, quantum mechanical procedure should be used; however, it may be beneficial to begin by assuming circular orbits similar to Bohr's initial approach to atomic structure. This may indicate the potential utility of the force in Equation (9.74), as well as perhaps identifying procedures to be used later.

Any nuclear orbits should probably be relativistic; therefore, in cylindrical coordinates, where the velocity for motion in a plane is given by

$$\overline{v} = \dot{r}\hat{r} + r\dot{\theta}\hat{\theta},$$

then we have

$$\gamma = \sqrt{1-\frac{v^2}{c^2}} = \sqrt{1-\frac{\left(\dot{r}^2+r^2\dot{\theta}^2\right)}{c^2}} .$$

For circular orbits, this becomes

$$\gamma = \sqrt{1-\frac{r^2\dot{\theta}^2}{c^2}} . \quad (9.75)$$

Thus, the relativistic equations of motion for the proton become

$$\frac{d}{dt}\left[\frac{m_p\left(\dot{r}\hat{r}+r\dot{\theta}\hat{\theta}\right)}{\gamma}\right] = \frac{-e^2}{4\pi\varepsilon_o r^2}\left[\left(1-\frac{\lambda_e}{r}\right)e^{\frac{-\lambda_e}{r}} - \frac{1}{4}\left(1-\frac{\lambda_p}{(2r)}\right)e^{\frac{-\lambda_p}{2r}}\right] \quad (9.76)$$

Equation (9.76) separates into two equations

$$\frac{d}{dt}\left[\frac{m_p\dot{r}}{\gamma}\right] = \frac{-e^2}{4\pi\varepsilon_o r^2}\left[\left(1-\frac{\lambda_e}{r}\right)e^{\frac{-\lambda_e}{r}} - \frac{1}{4}\left(1-\frac{\lambda_p}{(2r)}\right)e^{\frac{-\lambda_p}{2r}}\right] (9.77)$$

272

and

$$\frac{d}{dt}\left[\frac{m_p r\dot{\theta}}{\gamma}\right] = 0.$$

The second of these equations says that the angular momentum is given by

$$\frac{m_p r^2 \dot{\theta}}{\gamma} = L_p = n\hbar'_p \tag{9.78}$$

where $\hbar'_p$ indicates that whereas the unit of angular momentum will be a constant for a given orbit, it may be different for different orbits.

The first of the equations, Equation (9.77) is

$$\frac{d}{dt}\left[\frac{m_p \dot{r}}{\gamma}\right] = \frac{\left(m_p \ddot{r} - m_p r\dot{\theta}^2\right)}{\gamma} - \frac{\left(m_p \dot{r}\frac{d\gamma}{dt}\right)}{\gamma}$$

$$= \frac{-e^2}{4\pi\varepsilon_o r^2}\left[\left(1 - \frac{\lambda_e}{r}\right)e^{\frac{-\lambda_e}{r}} - \frac{1}{4}\left(1 - \frac{\lambda_p}{(2r)}\right)e^{\frac{-\lambda_p}{2r}}\right] \tag{9.79}$$

but for circular motion $\dot{r} = 0$, therefore,

$$\frac{m_p r\dot{\theta}^2}{\gamma} = \frac{e^2}{4\pi\varepsilon_o r^2}\left[\left(1 - \frac{\lambda_e}{r}\right)e^{\frac{-\lambda_e}{r}} - \frac{1}{4}\left(1 - \frac{\lambda_p}{(2r)}\right)e^{\frac{-\lambda_p}{2r}}\right] \tag{9.80}$$

Substituting from Equation (9.78) into Equation (9.80) we have

$$\frac{\left(n\hbar'_p\right)^2 \gamma}{m_p r^3} = \frac{e^2}{4\pi\varepsilon_o r^2}\left[\left(1 - \frac{\lambda_e}{r}\right)e^{\frac{-\lambda_e}{r}} - \frac{1}{4}\left(1 - \frac{\lambda_p}{(2r)}\right)e^{\frac{-\lambda_p}{2r}}\right] \tag{9.81}$$

The potential energy for one of the protons can be found by integrating the force and is

$$V(r) = -\int F(r)\, dr$$

$$= \int \frac{e^2}{4\pi\varepsilon_o r^2}\left[\left(1-\frac{\lambda_e}{r}\right)e^{\frac{-\lambda_e}{r}} - \frac{1}{4}\left(1-\frac{\lambda_p}{(2r)}\right)e^{\frac{-\lambda_p}{2r}}\right] dr \quad .(9.82)$$

$$= \frac{e^2}{4\pi\varepsilon_o r}\left[\frac{1}{4}e^{\frac{-\lambda_p}{2r}} - e^{\frac{-\lambda_e}{r}}\right]$$

Then the total energy of the three body system, including rest energy, would be

$$E_T = \frac{2e^2}{4\pi\varepsilon_o r}\left[\frac{1}{4}e^{\frac{-\lambda_p}{2r}} - e^{\frac{-\lambda_e}{r}}\right] + \frac{2m_p c^2}{\gamma} + m_e c^2 . \qquad (9.83)$$

However, by substituting Equation (9.78) into Equation (9.75) and solving for γ, we find

$$\gamma = \frac{1}{\sqrt{1+\left(\dfrac{n\hbar'_p}{m_p rc}\right)^2}} \qquad (9.84)$$

Thus, substituting Equation (9.84) into Equation (9.81) produces a transcendental equation whose solution gives $r(n)$, which may then be used in Equation (9.83) to obtain the total energy of the system. The mass of the system should then be found from

$$M = \frac{E_T}{c^2} . \qquad (9.85)$$

Because this system has one electron and two protons, it has a total electric charge of $+1$ and would have a mass of approximately 2 amu. This is the same characteristic exhibited by the deuterium nucleus. If this is the structure of the H_2 nucleus, then the mass given by Equation (9.85) should correspond to the mass of the ground state nuclear mass for $n = 1$. If the $1e^-$, $2p^+$ case existing where $n = 1$ is the ground state H_2 nucleus, then is the excited state represented by two protons in the $n = 2$ state or can it be represented by one proton in an $n = 1$ orbit and one in an n

= 2 orbit? The equations developed here consider only the case when both protons are in the same orbit. Any consideration of the protons being in different orbits introduces an asymmetry in the forces and a similar difficulty faced in the neutron case. Therefore, for the moment we will consider only the simpler cases, where symmetry reduces the complexity of the solution. Notice, though, that even in the simpler symmetric case, no analytical solution exists of Equation (9.81) for $r(n)$ because the force contains a transcendental function.

In my Los Alamos National Laboratory report titled, "The Unifying Effect of the Dynamic Theory" I explored the potential of modeling the nuclei using a model of a sub nuclear core made up of electrons or electrons and positrons which had an overall excess of negative electrical charge. This negative sub nuclear core was surrounded by shells of orbiting protons much like a reversal of the atomic structure of electrons orbiting around excess positive charge on the nucleus. Without exploring the sub nuclear core construction we can look at a structure for the nucleus by denoting the excess electron charge of the core by the integer Y, by which we mean the total number of core electrons less the number or core positrons, then by denoting the number of shell protons in orbit around this nuclear core with the currently used mass number, A, we find that the charge on the nucleus, Z, is given by

$$Z = A - Y.$$
(9.86)

Equation (9.86) indicates that the excess core electron number behaves identically with the neutron numbers in current nuclear theory, although there are no neutrons as such in this nuclear model. Indeed, the neutron, in this picture, is simply another state, namely Y = 1 and A = 1.

This suggests a picture of the nucleus in which there are protons in orbits about a nuclear core. The number of protons are given by the current mass number, A. The radii of the proton shell orbits are approximately the value of λ_p;

that is, about 1 Fermi. The core may be made up of electrons in orbit about positrons and is sized approximately the same as λ_{e-}, which is much, much less than λ_p. This view of the nucleus is similar to that of the atomic view, but here the nuclear core plays the role of the atomic electrons. The force law for the shell proton orbits might then be given for the two proton symmetric case by

$$F = \frac{-e^2}{4\pi\varepsilon_o r^2}\left[\left(1-\frac{\lambda_e}{r}\right)e^{\frac{-\lambda_e}{r}} - \frac{1}{4}\left(1-\frac{\lambda_p}{(2r)}\right)e^{\frac{-\lambda_p}{2r}}\right]. \qquad (9.87)$$

The equations specifying the proton shell orbits are given by Equation (9.87) and the total energy of the nuclei would be given by

$$E(A,Y) = \sum_n \left\{ \frac{A(n)}{R(n)}\left[\frac{1}{4}e^{-\left[\frac{\lambda_p}{R(n)}\right]} - Ye^{-\left[\frac{\lambda_e}{R(n)}\right]}\right] + \frac{A(n)m_p^2}{\gamma(n,R(n))} \right\} + E_c(Y) \quad (9.88)$$

where $A(n)$ is the number of protons with the quantum number, n; $R(n)$ is the radius of the proton orbit with the number n; $E_c(Y)$ is the energy of the nuclear core for which Y is the excess electron charge; and $\gamma(n,R(n))$ is the relativistic γ evaluated for n and $R(n)$. The mass of the nuclei with energies given by Equation (9.85) would then be

$$M(A,Y) = \frac{E(A,Y)}{c^2}. \qquad (9.89)$$

This approach has a simple look to it. For instance, if $Y = 1$ then $E_c = 0.511$ MeV, the rest energy of the electron. Then the ground state for ^{2}H would be

$$E(2,1) = 2E_1 + E_c \qquad (9.90)$$

whereas the excited state is

$$E(2,1)^* = 2E_2 + E_c. \qquad (9.91)$$

Using $E_c = 0.511$ MeV we find that the energy of a single proton in the $n = 2$ orbit would be

$$E_2 = \frac{\left[E(2,1)^* - E_c\right]}{2} \qquad (9.92)$$

276

The Dynamic Theory

Thus, the ^{3}He nuclei energy would be

$$E(3,1) = E(2,1) + \frac{\left[E(2,1)^* - E_c\right]}{2}. \qquad (9.93)$$

Using this approach I put together the binding energies for nuclei up t o an A=10 by determining the average energy of each level from the experimental data. The resulting predicted nuclei masses had a 2.9 MeV RMS error. The binding energy per mass number is plotted versus mass number in Figure 9.6.

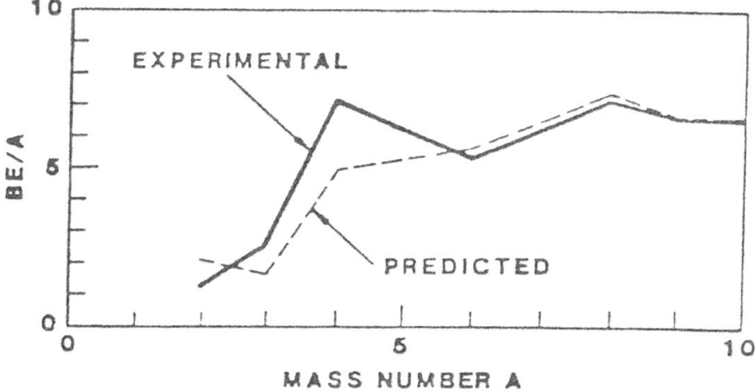

Figure 9.6. Binding Energy pre mass number versus mass number.

Deuterium-Deuterium Fusion into Helium

The suggestion that a neutron consists of a proton in orbit around an electron is very different from the standard nuclear model and raises many questions such as, "Isn't this a violation of the Heisenberg uncertainty principle?" and "Doesn't such motion violate spin conservation?" These questions have already been answered. However, if a neutron is a proton in orbit around an electron, then an additional orbiting proton leads to a deuterium nucleus that provides for a very different prediction of a fusion reactor.

277

Each deuteron has a magnetic moment whose spin axis is normal to the plane of the three particles and the axis of the deuterons may be aligned end to end by the use of a magnetic field. Once this preconditioning has been done, if the deuterons are nudged together (see Figure 9.7), the long range repulsive interaction between protons cause the protons of the approaching deuterons to stay as far from each other as far as possible. The protons, therefore, self align to establish the minimum threshold energy for fusion.

On the other hand, the electrons repel other electrons though they may be attracted to protons. A very stable configuration may be obtained by the fusion of two deuterons. In this stable configuration the four protons are in orbit in a plane about an orbit spin axis while the two electrons are located on the spin axis equal distances above and below the plane of the protons orbit (see Figure 9.8).

The forces of the interactions between all six particles of the two deuterons have been written down and studied using a Runga-Kutta integration method in a spreadsheet. This study showed that this alignment of the deuterium nuclei spin axis provides a much lower threshold for fusion than all other orientations. Further, the deuterium nuclei in this orientation preferentially fuse to the helium nucleus without emitting any particle radiation.

However, since the forces and the resulting potentials are transcendental equations, they have not yet been analytically integrated and must be submitted to computer solution to determine the numerical value of the fusion threshold for different methods of nudging the deuterons together. Further, in order to complete the study of the deuterium-deuterium to helium reaction the relativistic solutions of the equations must be obtained.

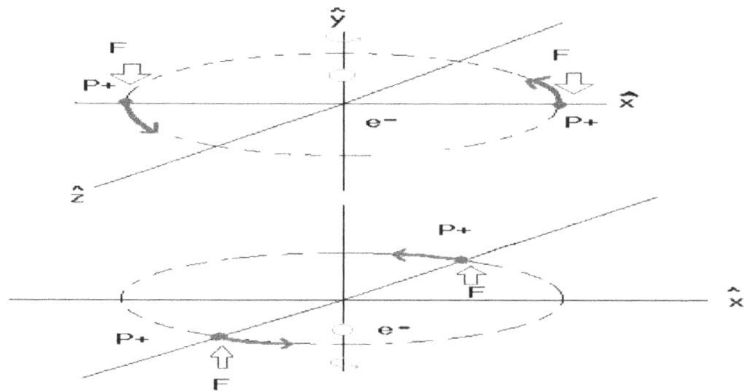

FIGURE 9.7. Two Deuterium Nuclei Being Nudged Together.

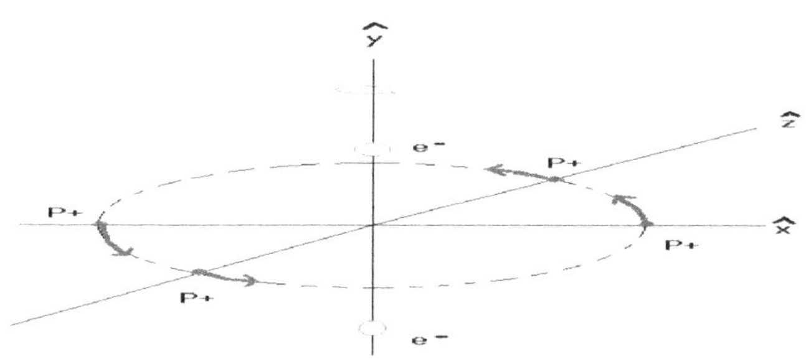

FIGURE 9.8. Helium Nucleus Formed From Two Deuterium Nuclei.

Weyl's Gauge Principle and the restriction to a unity scale factor not only require the electromagnetic gauge fields, but also require that the electrostatic potential be quantized and non-singular. The non-singular character of the potentials violates Newton's third law of equal and opposite forces. However, when quantum mechanical procedures are applied in the absence of Newton's third law the relativistic quantum mechanical equations of the motion OF and ABOUT the center of mass contain the Yang-Mills fields and the SU3 group fields of the standard model. Thus, while the new view and approach are completely different from the standard model, the standard model is not violated for its equations are contained within the new view. Plus, the new view adds predictions not possible in the standard model.

The equations of motion for the center of mass for the two-particle system of a proton and an electron show that a bound orbit of the proton around the electron exists at nuclear separations. The center of mass of the neutron is bound inside a positive energy well from which it may escape by quantum tunneling. The rate of neutron disintegration may be calculated and shown to compare with the experimental half-life.

At nuclear separations another proton may be added to the neutron to form the deuterium nucleus with two protons in orbit around a single electron. This configuration for the deuterium nucleus shows that should one deuterium nucleus approach another at high energy and with a random orientation with respect to the first deuterium there will be very high threshold for fusion and a high probability that a tritium nucleus, wherein three protons orbits around a single electron, will be formed and a neutron released. However, when the spin axis of the two deuterium nuclei are aligned and the nuclei are placed so that their spin axis are aligned the natural tendency of protons long range repulsion causes the two spinning deuterium nuclei to lock into phase with each other where the lines between the protons within each deuterium nucleus are at ninety degrees to each other.

One means of holding two deuterium nuclei in such an orientation might be to place them into the crystal lattice of a metal under a high magnetic field that locks up their spin axis in the correct orientation. The crystal structure should be such that each deuterium nucleus should have a clear view of a neighboring deuterium nucleus along it spin axis without another nucleus in the path. Temperature vibrations such as to provide motion of the deuterium nuclei toward each other will set up the quantum tunneling that will result in the two deuterium nuclei preferentially fusing into a helium nucleus without any attendant radiation.

Virtual Particle Masses of the Proton

Today the particle physics community is abuzz with the possibility of learning more about the universe we live in due to the fact that the Large Hadron Collider is now coming on line and will provide new results for scientists to analyze.

The non-singular electrostatic potential offers an alternate nuclear model to that presented by the standard model. This new theory describes nuclear interactions as those due to a non-singular electrostatic potential. This provides a functional form for the nucleon potential for nucleons of interest. The LHC will cause protons to collide at high energy (some 7,000 MeV) so it would seem appropriate to look at the possible virtual particles that may be associated with protons.

It was shown that the non-singular potential leads to a nuclear model wherein the deuterium nucleus consists of two protons in a nuclear orbit around a single electron. This is where we would like to go since the LHC uses protons to collide into protons. First, we should work ourselves up from two-particle interactions to the three-particle interactions.

A two-particle system with un-like particles was analyzed using the motion about the center of mass and the motion of the center of mass as the two entities in the system. The neutron was shown to be a proton in a nuclear orbit around an electron. The difference in the λ_p of the proton and the λ_e of the electron established this nuclear orbit in which the center of mass is in motion but trapped in a positive energy well from which it must tunnel out by quantum tunneling. The potential keeping the center of mass in motion in a neutron was shown to be

$$V_c = -\frac{k}{r}\left[e^{\frac{-\lambda_p}{r}} - e^{\frac{-\lambda_e}{r}} \right].$$

(9.94)

It has been shown that the unit of action depends upon the gauge function and the experimental spins require that the effective unit of action for the proton in the neutron be $0.66586\hbar$ while the unit of action for the electron is $8.0517 \times 10^{-4}\hbar$. The mass of the neutron and these values argue the proton orbital radius is about 1.5×10^{-19} m which also provides a prediction of a half-life of something over 600 seconds for the neutron using the positive potential well shown in Figure 9.9.

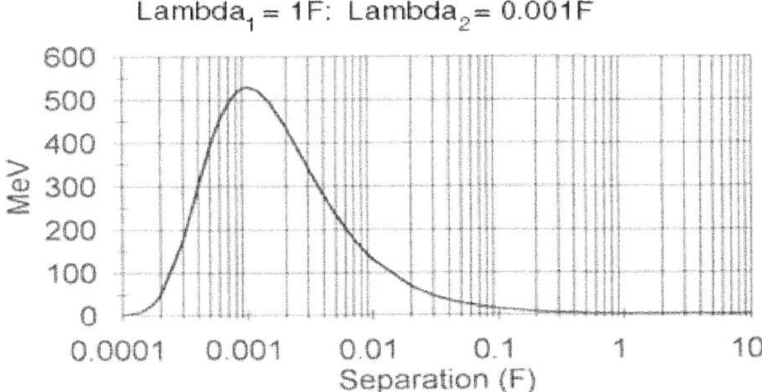

Figure 9.9. Positive potential well of the Center of Mass of a neutron.

Previously it was shown that a deuterium nucleus consisted of relativistic circular orbits of two protons around an electron. These orbits established energy levels given by

$$E_T = \frac{2k}{r}\left[\frac{e^{\frac{-\lambda_p}{2r}}}{4} - e^{\frac{-\lambda_e}{r}}\right] + 2m_p c^2 \sqrt{1 + \left(\frac{n\hbar'}{m_p rc}\right)^2} + m_e c^2 \quad (9.95)$$

where $\hbar'$ is the effective unit of action. Experimental magnetic moments of the deuteron set the effective unit of action at $\hbar' = 0.25012\hbar$. Using this value of the effective unit of action for the protons in the deuteron and the deuteron mass in Equation (9.95) gives $r = 3.05$ Fermi.

This configuration for the deuterium nucleus lead to the concept of a compact fusion reactor in which two deuterium nuclei preferentially fused into a helium nucleus and liberate energy.

The virtual particle masses may be predicted using the static solution of the Klein Gordon equation, which are solutions of

$$\nabla^2 \Psi = \frac{m^2 c^2}{\hbar'^2}\Psi \quad (9.96)$$

where Ψ is the potential, $\hbar'$ is the effective unit of action and m is the mass of the virtual particle. Solutions for spherically symmetric potentials may be found using the fact that

$$\nabla^2 \Psi = \frac{1}{r^2}\frac{d}{dr}\left(r^2 \frac{d\Psi}{dr}\right). \quad (9.97)$$

The masses of the photon and the virtual particles within the neutron and the deuterium nucleus will be calculated before addressing the mass of the virtual particles within the proton. The progression will be seen in that the neutron is one proton in nuclear orbit around an electron, the deuterium nucleus is two protons in orbit around an electron and the proton is two positrons in orbit around an electron.

The classical electrostatic potential is given by

$$V \cong \left(\frac{Ze}{4\pi\varepsilon_o r} \right) e^{-\left(\frac{\lambda_N}{r}\right)} \qquad (9.98)$$

when $\lambda_N = 0$. When the classical electrostatic potential is used in Equation (9.97) it is easy to verify that the mass of the virtual particle associated with the classical electrostatic potential is zero. This is the mass of the photon. The non-singular potential obtained when $\lambda_N \neq 0$ also shows a virtual particle mass of zero.

The mass of the virtual particle associated with the neutron is determined when the potential holding the center of mass of the neutron in place is used. This potential is given by Equation (9.94). Using this potential in Equation (9.97) gives a mass specified by

$$m_n = \frac{\hbar'}{rc} \sqrt{ \frac{1}{r} \frac{ \left[\frac{\lambda_p e^{\frac{-\lambda_p}{r}}}{m_e}\left(2-\frac{\lambda_p}{r}\right) + \frac{\lambda_e e^{\frac{-\lambda_e}{r}}}{m_p}\left(2-\frac{\lambda_e}{r}\right) \right] }{ \left[\frac{e^{\frac{-\lambda_p}{r}}}{m_e}\left(1-\frac{\lambda_p}{r}\right) + \frac{e^{\frac{-\lambda_e}{r}}}{m_p}\left(1-\frac{\lambda_e}{r}\right) \right] } } . \qquad (9.99)$$

When $\lambda_e = \frac{m_e}{m_p}\lambda_p$ and $\lambda_e \ll r$ Equation (9.99) simplifies

to

$$m_n = \frac{\hbar'}{rc} \sqrt{ \frac{\lambda_p}{r} \frac{ \left[e^{\frac{-\lambda_p}{r}}\left(2-\frac{\lambda_p}{r}\right) + \frac{2m_e^2}{m_p^2} \right] }{ \left[e^{\frac{-\lambda_p}{r}}\left(1-\frac{\lambda_p}{r}\right) + \frac{m_e}{m_p} \right] } } . \qquad (9.100)$$

Equation (9.100) now needs to be evaluated using the previously determined values for the neutron. The proton orbital radius in the neutron was found to be approximately 10 Fermi while the effective unit of action was 0.66586 $\hbar$.

The Dynamic Theory

Putting these values into Equation (9.100) gives a virtual particle mass of

$$m_N \approx 3.28x10^3 \, GeV/c^2 . \qquad (9.101)$$

The mass of the virtual particle associated with the deuterium nucleus may be found using the potential for the proton orbit. This potential was determined to be given as the potential energy portion of Equation (9.95). By putting the deuterium's proton potential into Equation (9.97) the virtual particle mass is found to be given by

$$m_D = \frac{\hbar'}{rc} \sqrt{\frac{1}{r} \frac{\left[\dfrac{-\lambda_p^2 e^{\frac{-\lambda_p}{2r}}}{16r} + \lambda_e e^{\frac{-\lambda_e}{r}} \left(2 - \dfrac{\lambda_e}{r} \right) \right]}{\left[\dfrac{e^{\frac{-\lambda_p}{2r}}}{4} - e^{\frac{-\lambda_e}{r}} \right]}} . \qquad (9.102)$$

Again, as above, when $\lambda_e << r$ Equation (9.102) reduces to

$$m_D = \frac{\hbar'}{rc} \sqrt{\frac{\lambda_p}{r} \frac{\left[\dfrac{-\lambda_p e^{\frac{-\lambda_p}{2r}}}{16r} + 2 \dfrac{m_e}{m_p} \right]}{\left[\dfrac{e^{\frac{-\lambda_p}{2r}}}{4} - 1 \right]}} . \qquad (9.103)$$

Now the values of the proton's orbits in the deuterium nucleus was found to be $\hbar' = 0.25012\hbar$ and $r = 3.25$ Fermi. These values in Equation (9.103) give the mass of the deuterium nucleus proton orbits to be

$$m_D \approx 30.5 \, MeV/c^2 . \qquad (9.104)$$

The proton has been viewed by the standard model as a nucleon consisting of three quarks. The non-singular potential and the theory that produced it do not need quarks to describe the effects currently ascribed to the weak and the strong nuclear forces. Indeed the proton may be described by two positrons in very tight, sub-nuclear, orbits around an electron in the same fashion as two protons in a deuterium nucleus. The equation giving the positron orbits within the proton is the same as the equation for the proton orbits in the deuterium nucleus except the values of $\lambda_{pos} \neq \lambda_p$ and the effective unit of action is different. Therefore, the mass of the virtual particle associated with the positron orbits within the proton may be found using Equation (9.102) before any approximations have been made.

To be accurate the formal solution of the relativistic quantum description of the three body problem must be found. However, an estimate may be made if both the effective unit of action and the lambda of the positron are both scaled using ratio of the positron to proton masses. This scaling would produce $\hbar' = 0.0001362\hbar$ and $\lambda_{pos} = 0.00177$ Fermi.

Using these values in Equation (9.95) for the total energy of the proton and setting the total energy to the experimental mass of the proton yields a radius for the positrons within the proton of $r_{pos} \approx 5.728x10^{-5}$ Fermi. These values give a virtual particle mass (using Equation (9.102)) for the positrons within the proton of

$$m_p \approx 7,254 MeV \, / \, c^2 . \qquad (9.105)$$

The non-singular potential leads to a very different model of nuclear physics and particles than does the standard model. In particular, the non-singular potential provides a potential for various particles making up t he different nucleons and, thereby, provides the means to predict the virtual masses associated with these particle orbits.

The Dynamic Theory

The virtual particle associated with the proton orbits within the neutron was predicted to be 3.28×10^3 GeV/c^2. While this seems very massive the values of the effective unit of action agrees with the experimental magnetic moments and the bound radius predicts the correct neutron mass and half-life. These agreements are too good to question the virtual mass prediction too much.

The virtual particle associated with the proton orbits within the deuterium nucleus was predicted to be 30.5 MeV/c^2.

The virtual particle associated with the positron orbits within the proton was predicted to be 7,254 MeV/c^2. This mass is right at the energy claimed for the LHC. This argues that it may be possible for the LHC to break the colliding protons into their constituent parts, the positrons and electrons.

These predicted virtual particle masses should be compared with data that will be generated by the Large Hadron collider. In particular, since the analyses of data from colliders typically involve guessing a form factor to use in the fitting of the data and then determining how well the guess fits the data, guesses that provide alternates to those suggested by the standard model should be available in order to support an objective analysis.

Nuclear Physics

Chapter 10 Gauge Gravitational Fields

General Theory of Relativity

In the preceding chapters we have seen how the Weyl Quantum Principle has required Maxwell's electromagnetism, quantum mechanics, wave-particle duality, a variable unit of action, quantized photon energy, weak nuclear forces for un-like particles and strong nuclear forces for particles with the same non-singular electrostatic potential. However, in doing this Newton's first and second law were shown to hold while his third law only resulted from the forces themselves and was not specified beforehand. Now we turn to Newton and Einstein's work toward explaining the effects of gravity. Newton used an assumed force law for gravity and Einstein put the effects of gravity into the curvature of four-dimensional space-time. Can the WGP, which has yielded electromagnetism and both nuclear forces, also specify gravitational forces?

Several researchers used an extension into five dimensions in their search to unify the electromagnetic, nuclear and gravitational forces. However, in each of these attempts, the real world was thought to be described in the four dimensions of space and time and the fifth dimension was used in order to get additional degrees of freedom to work with in their search for the unifying field. On the other hand, numerous researchers used the basis of quantum mechanics in their research into the unification of the various forces of nature. Neither approach has produced the desired unification.

Consider combining these two areas of research in the following manner. Let a five-dimensional manifold have some physical reality in all five dimensions. The first four are the four of space and time as used in Einstein's relativity. The physical manifestation of the fifth dimension is to be learned during this investigation. Whatever the fifth dimension turns out to be, requiring it to be conserved will embed a four-dimensional hyper-surface into the five-dimensional manifold.

The starting point of this investigation is the exponent of the non-integrable Weyl scale factor

$$l = l_0 \exp\left[\frac{1}{\gamma}\int \phi_j \, dx^j\right].$$

The Weyl Quantum Principle requires this exponent to be quantized as in

$$\int \phi_j \, dx^j = i2\pi N$$

where the i on the right hand side indicates an imaginary value and the summation convention is to be applied to the j's. Here the indices are to be considered to range from 0 to 4, with 0, (1,2,3) and 4 representing time, space and the new fifth dimension respectively. Just as Weyl showed the gauge fields may be derived from these gauge potentials and the components of the 5-dimensional field tensor may be written in matrix form as

$$F_{ij} = \begin{bmatrix} 0 & E_1 & E_2 & E_3 & V_4 \\ -E_1 & 0 & B_3 & -B_2 & V_1 \\ -E_2 & -B_3 & 0 & B_1 & V_2 \\ -E_3 & B_2 & -B_1 & 0 & V_3 \\ -V_4 & -V_1 & -V_2 & -V_3 & 0 \end{bmatrix}. \qquad (10.1)$$

We shall now show that conservation of the fifth dimension embeds a four-dimensional hyper-surface within the five dimensional manifold and that the equations giving the curvature of this hyper-surface are Einstein's field equations of his general theory of relativity.

For simplicity, all the field components may be set to zero except for the V_4 component. The surface field tensor will be given by

The Dynamic Theory

$$F_{\alpha\beta} = F_{ij} y_\alpha^i y_\beta^j \qquad (10.2)$$

where

$$y_\alpha^i = \frac{\partial y^i}{\partial x^\alpha} \delta_\alpha^i$$

for $i=0, 1, 2, 3$ and

$$y_\alpha^4 = \frac{\partial y^4}{\partial x^\alpha}.$$

The space indices, i, j, k range over $0, 1, 2, 3, 4$ while the surface indices α, β, η, ν only take on values of $0, 1, 2, 3$. The surface metric is found from

$$g_{\alpha\beta} = a_{ij} y_\alpha^i y_\beta^j = a_{\alpha\beta} + h_{\alpha\beta}$$
$$\textit{where} \quad h_{\alpha\beta} = 2a_{\alpha4} y_\beta^4 + a_{44} y_\alpha^4 y_\beta^4. \qquad (10.3)$$

The space energy-momentum tensor for matter under the influence of gauge fields is given by

$$T_{sp}^{ij} = \gamma u^i u^j + \frac{1}{c^2}\left[F_k^i F^{kj} + \frac{1}{4} a^{ij} F^{kl} F_{kl} \right] \qquad (10.4)$$

and may be written in terms of the surface metric as

$$T_{sp}^{\alpha\beta} = \gamma u^\alpha u^\beta + \frac{1}{c^2}\left[\begin{array}{c} F_k^\alpha F^{k\beta} + F_4^\alpha F^{4\beta} \\ + \frac{1}{4}(g^{\alpha\beta} - h^{\alpha\beta})(F^{\mu\nu} F_{\mu\nu} + F^{4\nu} F_{4\nu}) \end{array} \right]$$

(10.5)

since

$$u^4 \equiv \frac{dy^4}{dt} = \frac{\partial y^4}{\partial t} + \overline{\nabla} \bullet (y^4 \overline{u}) = 0 \qquad (10.6)$$

is the statement required by the conservation of the fifth dimension.

The surface energy-momentum tensor may now be found within the space tensor and written

$$T^{\alpha\beta} = T_{sp}^{\alpha\beta} - \frac{1}{c^2}\left[F_4^\alpha F^{4\beta} - \frac{1}{2} h^{\alpha\chi} F^{4\nu} F_{4\nu} \right]. \qquad (10.7)$$

The form of this expression for the surface energy-momentum tensor suggests writing

$$CT^{\alpha\beta} \equiv G^{\alpha\beta} \equiv R^{\alpha\beta} - \frac{1}{2}g^{\alpha\beta}R \qquad (10.8)$$

which, of course, look like the field equations of Einstein's general theory of relativity, when Equations (10.8) are taken to be Einstein's field equations,

$$G^{\mu\nu} = -\frac{8\pi K}{c^2}T^{\mu\nu} \qquad (10.9)$$

where K is the gravitational constant.

Then the field V_4, which is the only non-zero field component considered, must be related to the gravitational field and the fifth gauge potential must be related to the gravitational potential. Therefore, the physical reality of the fifth dimension is gravitating mass or its equivalent, mass.

Charge-to-Mass Ratio and Magnetic Moments

Before we plunge into the derivation of predictions to compare with the general theory of relativity let us first consider a question that arises from the necessity of keeping the units consistent among the five-dimensional field quantities E, B, V, and V_4 when they are all to be considered as components of the five dimensional gauge field. By considering the units of these field components it is soon found that a charge-to-mass ratio is needed in order that the units of the gravitational field components may be compared, or put in the same equation, with the electric and magnetic components. Let us first see if we can determine this ratio.

When the derivation of the fields allowed for fundamental particles is to be considered, these field expressions give rise to the specification of a charge-to-mass ratio which allows conversion of classical gravitation field units to electromagnetic field units. The gravitational field component in the system of field equations with the electric and magnetic components brings up the requirement for a gravitation-to-electromagnetic unit

292

conversion. Looking at the different field quantities may show this need. First consider the electric case. The field units are given by $[E]$ = volt/meter, while the expression for the electric force density is $F_e=\rho E$ with units of Newton/meter.

For the gravitational field the units are $[V]=$ webers/meter squared, while the gravitational "current" density has units given by $[J_4]$ = ampere/meter^2. Thus, the gravitational force density is given by $F_g=(J_4/c)V$ where again the units are Newton/meter^3. In order to compare this system to the classical gravitational system we need to be able to go from a gravitational field with units of Newton/kilogram to units of volt/meter. Now

$$\frac{nt}{kg} = \frac{volt\text{-}coul}{m\text{-}kg} = \left(\frac{volt}{m}\right)\left(\frac{coul}{kg}\right).$$

Thus, if β is a quantity with units of coul/kg, then $(1/\beta)$ is the conversion factor we seek. Similarly, we need to convert the gravitational mass density, (J_4/c), with units of coul/m^3 to units of kg/m^3. Obviously β will also be the conversion factor for this also. The question is; "How do we determine this charge-to-mass ratio and is it unique?"

If we consider the fields to be fields of fundamental particles we may determine β. Thus, let us look at the solution of the gauge function for fundamental particles given previously, that is

$$ln\, F^{\frac{1}{2}} = f_r f_\theta f_\phi f_\gamma f_t \, .$$

We showed that the mass, time and radial dependences led to a solution for Weyl Unitary Particles of the form

$$\ln f^{\frac{1}{2}} = \left(\frac{c}{H_o}\right)\left(\frac{1}{r}\right)e^{-\left(\frac{\lambda_N}{r}\right)}\sin^{\lambda_0}\theta\, e^{-H_o t}e^{-\left(\frac{H_o}{a_o c}\right)\gamma}$$

$$\lambda_o \ll 1 \Rightarrow \sin^{\lambda_0}\theta \cong 1$$

$$N_0 = N_4$$

$$Z = \left(N_0 - N_1\right) \tag{10.10}$$

$$q = Zee^{-H_o t}e^{-\left(\frac{H_o}{a_o c}\right)\gamma}$$

$$J_4 = c\rho$$

and $N_1 = N_2 = N_3$

The field tensor was given by

$$\frac{F'_{ij}}{\left(\dfrac{e}{4\pi\varepsilon_o}e^{-H_o t}e^{-\left(\frac{H_o}{a_o c}\right)\gamma}\right)} \approx \begin{vmatrix} 0 & Z\left(\frac{1}{r^2}\right)\left(1-\frac{\lambda_N}{r}\right)e^{-\left(\frac{\lambda_N}{r}\right)} & Zr_o\lambda_0\left(\frac{1}{cr^2}\right)e^{-\left(\frac{\lambda_N}{r}\right)}\cot\theta & 0 & 0 \\ -Z\left(\frac{1}{r^2}\right)\left(1-\frac{\lambda_N}{r}\right)e^{-\left(\frac{\lambda_N}{r}\right)} & 0 & 0 & 0 & -Z\left(\frac{1}{r^2}\right)\left(1-\frac{\lambda_N}{r}\right)e^{-\frac{\lambda_N}{r}} \\ -Zr_o\lambda_0\left(\frac{1}{cr^2}\right)e^{-\left(\frac{\lambda_N}{r}\right)}\cot\theta & 0 & 0 & 0 & -Z\left(\frac{1}{r^2}\right)e^{-\left(\frac{\lambda_N}{r}\right)}\cot\theta \\ 0 & 0 & 0 & 0 & 0 \\ 0 & Z\left(\frac{1}{r^2}\right)\left(1-\frac{\lambda_N}{r}\right)e^{-\frac{\lambda_N}{r}} & Z\left(\frac{1}{r^2}\right)e^{-\left(\frac{\lambda_N}{r}\right)}\cot\theta & 0 & 0 \end{vmatrix} \tag{10.11}$$

The electric field, when $r \gg \lambda$, is

$$E_r \approx Z\left(\frac{e}{4\pi\varepsilon_o r^2}\right). \tag{10.12}$$

The electrical force on a charge of e would then be given by

$$F_e = eE_r \approx Z\left(\frac{e^2}{4\pi\varepsilon_o r^2}\right). \tag{10.13}$$

The gravitational field when $r \gg \lambda$, is

$$V_r \approx -Z\left(\frac{e}{4\pi\varepsilon_o r^2}\right). \tag{10.14}$$

However the units of the gravitational field are volt/m here. We need to look at the units conversions. To convert the units of the gravitational field from volts/m to N/kg we need to multiply by a charge to mass ratio, β, to get

$$V_g = V_r\beta = -Z\beta\left(\frac{e}{4\pi\varepsilon_o r^2}\right)\left(1-\frac{\lambda_N}{r}\right)e^{-\left(\frac{\lambda_N}{r}\right)}e^{-H_o t}e^{-K_\gamma \gamma} .\,(10.15)$$

To get the gravitational mass we need to divide the gravitational charge by the charge to mass ratio, β, to get

$$m = \frac{M}{c\beta} \approx \frac{e}{\beta}. \qquad\qquad (10.16)$$

The gravitational field then is given by

$$V_g = V_r\beta = -Z\beta\left(\frac{m\beta}{4\pi\varepsilon_o r^2}\right)\left(1-\frac{\lambda_N}{r}\right)e^{-\left(\frac{\lambda_N}{r}\right)}e^{-H_o t}e^{-K_\gamma \gamma} .\,(10.17)$$

Now we can form the gravitational force by multiplying Equation (7.74) by the gravitational mass of our second particle which is also m and obtain

$$F_g = mV_g = -Zm^2\beta^2\left(\frac{1}{4\pi\varepsilon_o r^2}\right)\left(1-\frac{\lambda_N}{r}\right)e^{-\left(\frac{\lambda_N}{r}\right)}e^{-H_o t}e^{-K_\gamma \gamma} . \,(10.18)$$

The only charge to mass ratio that will convert the units to the to the well known Newton gravitational force is

$$\beta = \sqrt{4\pi\varepsilon_o G} = 2.4296 \times 10^{-11}\ coul\,/\,kg . \qquad (10.19)$$

Of course this is the charge to mass ratio that was used in Chapter 7.

The surprising, and pleasing, thing about the charge to mass ratio given by Equation (10.19) is that it is formed as the product of two known physical quantities rather than depending upon new quantities, such as a_0, whose value is not well known. Further, in retrospect, it appears to be the simplest, if not the only, combination of an electromagnetic parameter and a gravitational parameter whose units are coul/kg. It is worthwhile to point out that the dependence of the fields upon time and mass density is extremely small but is essential in establishing β, and the inductive coupling between the electromagnetic and gravitational fields.

Magnetic Moments

In Chapter 6 one result of the five dimensional quantum mechanics was a p rediction that even an electrically neutral body that was spinning must also have a magnetic moment. The charge-to-mass ratio completes the notion that a rotating gravitational, electrically neutral body should have a magnetic moment stemming from the effective electric charge associated with the gravitational mass. Given the charge-to-mass ratio we may quickly look at its prediction for the earth's magnetic moment.

Using β, the "effective" charge associated with a gravitational mass is given by $q_{eff} = \beta M$. For the earth this effective charge would be $q_{eff} = 1.454 \times 10^{14} \, coul$. Thus, if the magnetic moment of the earth is given by

$$\mu = \left(q_{eff}/2M \right) A ,$$

where A is the earth's angular momentum then we have

$$\mu = \left(\frac{q_{eff}}{2M} \right) I \omega$$

$$= \frac{(1.454 \times 10^{14})(9.71 \times 10^{37})(7.29 \times 10^{5})}{2(5.983 \times 10^{24})}$$

or $\mu = 8.6 \times 10^{22} \, amp\text{-}m^{2}$.

This predicted value of the Earth's magnetic moment compares very well with the experimental value of 8.1×10^{22} amp- $m^{2.}$

Perihelion Advance

No serious suggestion that the additional vector field in the five-dimensional gauge equations be the gravitational field can be made without giving due consideration to the explanation of the planetary perihelion advance provided by Einstein's general theory of relativity. Though several attempts have been made to explain the perihelion advance by other means none has succeeded in casting much doubt on Einstein's explanation.

Let us recall some of the main features of the classical problem of planetary orbits. Kepler's first law states that a planet describes a closed elliptical orbit with the sun at a focal point. However, the presence of such small influences as other planets moving in the Suns' field causes a perturbation in the motion of a given planet, and the resulting orbit is not precisely elliptic. Indeed, one may think of the actual orbit as a slightly bumpy ellipse that may precess in the plane of motion; that is, the perihelion shifts about and does not always occur at the same angular position.

The fact that the idealized classical orbit is a closed ellipse is a result peculiar to the Newtonian inverse-square law; in fact, Newton himself found that, if the force of gravity were proportional to $1/r^{(2+\delta)}$ instead of $1/r^2$, then a planetary orbit would not be closed and a perihelia shift of order δ would occur. Indeed, this result was taken to indicate that, since planetary orbits are very nearly closed, the Newtonian inverse-square law must be very accurate, as in fact it is.

Let us now ask where may there be room for differences between the predictions of classical celestial mechanics and the celestial mechanics of the general theory of relativity or the non-singular potential presented here. Since Kepler's first law is experimentally verified to be correct to a high accuracy, we might expect that non-Newtonian theories may merely add a few bumps to the nearly elliptic orbits and contribute somewhat to perihelia motion. Since angles are much more conveniently measured in astronomy than are distances, it is natural to concentrate on pe rihelia motion. Conveniently enough, there is, in fact, a well-known discrepancy in classical mechanics concerning the perihelia motion of the planet Mercury. Because of Mercury's high velocity and eccentric orbit, the perihelion position can be accurately determined by observation. The difference between the classically predicted perihelia shift (due to perturbations by other planets) and the observed perihelia shift is 43 seconds of arc per century. Even though this is a very small difference, it is about a hundred times the probable observational error and represents a true discrepancy from the very precise predictions of celestial mechanics that has bothered astronomers since the middle of the nineteenth century.

The first attempt to explain this discrepancy consisted in hypothesizing the existence of a new planet, Vulcan, inside the orbit of Mercury, and much theoretical work was done to predict the position of Vulcan, using the known perturbation on Mercury's orbit. However, careful observation failed to discover the hypothetical planet, and the hypothesis was finally abandoned in 1915 when Einstein used general relativity theory to explain the observed effect. Earlier in this chapter the five dimensional manifold was shown to yield Einstein's general theory of relativity when mass, the fifth dimension, was conserved. This argues that there are two different, but mathematically equivalent, ways to describe gravitational effects when mass is conserved. One, Einstein's general theory, has already been applied to the description of the advance of the perihelion of planetary orbits. The second description may be done from the force approach within the five dimensional manifold.

Now let us look at what the non-singular gravitational potential offers as an explanation for the perihelia advance and then compare it to the predictions of the general relativity theory.

The classical equations of motion are

The Dynamic Theory

$$m\ddot{r} - mr\dot{\theta}^2 = F(r)$$

$$mr\ddot{\theta} + 2m\dot{r}\dot{\theta} = 0 . \qquad (10.20)$$

The second of Equations (10.20) has the solution

$$\dot{\theta} = \frac{L}{mr^2}, \qquad (10.21)$$

where L is the angular momentum.

Using Equation (10.21), the first of Equations (10.20) may be written

$$M\ddot{r} = F(r) + \frac{L^2}{Mr^3} \qquad (10.22)$$

or

$$M\ddot{r} \equiv -\frac{\partial}{\partial r}\left[v'(r)\right], \qquad (10.23)$$

where

$$v'(r) = v(r) + \left(\frac{L^2}{2Mr^2}\right). \qquad (10.24)$$

We are seeking the prediction of the non-singular gravitational potential with respect to the perihelion advance. This may be found by comparing the frequency of small radial oscillations about steady circular motion for the effective potential given by Equation (10.24) for the non-singular potential with the frequency of revolution.

By considering the non-singular potential, Equation (10.24) becomes

$$v'(r) = \left(\frac{K}{r}\right)e^{\frac{-\lambda}{r}} + \left(\frac{L^2}{2Mr^2}\right), \qquad (10.25)$$

with $K = -GMm$, where G is the gravitation constant, M is the mass of the sun, and m is the mass of the planet of interest.

Equation (10.21) gives the frequency of revolution. To determine the frequency of small radial oscillations about steady circular motion we need to evaluate the second derivative of the effective potential, v, the radius for which the first derivative is zero.

The first derivative of the effective potential is obtained by differentiating Equation (10.25) with respect to r. This may be found to be

$$\frac{\partial}{\partial r}[v'(r)] = \left(\frac{-K}{r^2}\right)\left(1-\frac{\lambda}{r}\right)e^{\frac{-\lambda}{r}} - \left(\frac{L^2}{mr^3}\right). \tag{10.26}$$

The second order derivative of the effective potential may be found to be approximately

$$\frac{\partial^2 v'(r)}{\partial r^2} \cong \left(\frac{2K}{r^3}\right)\left(1-\frac{2\lambda}{r}\right)e^{\frac{-\lambda}{r}} + \left(\frac{3L^2}{mr^4}\right), \tag{10.27}$$

when terms involving λ^2/r^2 are considered negligible with respect to terms involving λ/r.

We may determine r_0 from the condition

$$\frac{\partial[v'(r)]}{\partial r} = 0 = \left(\frac{GMm}{r_0^2}\right)\left(1-\frac{\lambda}{r_0}\right)e^{\frac{-\lambda}{r_0}} - \left(\frac{L^2}{mr_0^3}\right). \tag{10.28}$$

The radius, r_0 is the radius of near circular orbit and the effect of the exponential factor and $(1-\lambda/r)$ factor will be negligible for $\lambda<<r$. Thus, we may approximate Equation (10.28) by

$$0 = \left(\frac{GMm}{r_0^2}\right) - \left(\frac{L^2}{mr_0^3}\right)$$

so that

$$r_0 \cong \left(\frac{L^2}{GMm^2}\right). \tag{10.29}$$

If we approximate the exponential factor in Equation (10.27) by its power series expansion and retaining only those terms whose dependence upon λ/r are linear or less, then Equation (10.27) is approximated by

$$\frac{\partial^2 v'(r)}{\partial r^2} \cong \left(\frac{2K}{r^3}\right)\left(1-\frac{2\lambda}{r}\right)\left(1-\frac{\lambda}{r}\right)+\left(\frac{3L^2}{mr^4}\right)$$
$$=\left(\frac{2K}{r^3}\right)\left(1-\frac{3\lambda}{r}\right)+\left(\frac{3L^2}{mr^4}\right).$$

(10.30)

Now the frequency of small radial oscillations about steady circular motion may be found from

$$\omega^2 \equiv \left(\frac{1}{m}\right)\left[\frac{\partial^2 v'(r)}{\partial r^2}\right]_{r=r_0},$$

Thus, we have

$$\omega^2 \cong \left(\frac{1}{m}\right)\left\{2GMm\left(\frac{GMm^2}{L^2}\right)^3\left[1+\left(\frac{3\lambda GM m^2}{L^2}\right)\right]+\left(\frac{3L^2}{m}\right)\left(\frac{GMm^2}{L^2}\right)^4\right\}$$
$$=\left(\frac{2G^4M^4m^6}{L^6}\right)\left[1+\left(\frac{3\lambda GM m^2}{L^2}\right)\right]+\frac{3G^4M^4m^6}{L^6}$$
$$=\left(\frac{G^4M^4m^6}{L^6}\right)\left[1+\left(\frac{6\lambda GM m^2}{L^2}\right)\right].$$

(10.31)

Taking the square root of Equation (10.31) may now make an approximation for the frequency of small radial oscillations about steady circular motion. By considering the second term of the second factor as small compared to one, we have

$$\omega \cong \left(\frac{G^2M^2m^3}{L^3}\right)\left[1+\left(\frac{3\lambda GM m^2}{L^2}\right)\right].$$

(10.32)

The perihelion advance per revolution may now be found as the difference between Equations (10.32) and Equation (10.21) evaluated at r_0, divided by the orbital frequency, or

$$\delta\theta = 2\pi\left(\frac{\omega - \dot\theta}{\dot\theta}\right)$$

$$= 2\pi\left\{ \frac{\left(\dfrac{G^2 M^2 m^3}{L^3}\right)\left[1 + \left(\dfrac{3\lambda G M m^2}{L^2}\right)\right]}{-\dfrac{G^2 M^2 m^3}{L^3}} \right\}\left(\frac{L^3}{G^2 M^2 m^3}\right) \qquad (10.33)$$

so that

$$\delta\theta \cong 2\pi\left(\frac{3\lambda G M m^2}{L^2}\right). \qquad (10.34)$$

The perihelion advance predicted by Einstein's general theory of relativity is given by

$$\delta\theta_{GiR} = 2\pi\left(\frac{3 G^2 M^2 m^2}{c^2 L^2}\right).$$

(10.35)

If λ were to be such as to provide an identical prediction as the general theory then λ would have to satisfy

$$\lambda = \frac{GM}{c^2}. \qquad (10.36)$$

For G = 6.7 x 10^{-8} gr^{-1}cm^3/sec^2, M = 1.98 x 10^{33}gr, and c = 3 x 10^{10} cm/sec,

$$\lambda = \frac{(6.7\ x\ 10^{-8})(1.98\ x\ 10^{33})}{(3\ x\ 10^{10})^2}\,cm = 1.47\ x\ 10^5\ cm \qquad (10.37)$$

or

$$\lambda = 1.47\ x\ 10^3 m. \qquad (10.38)$$

This is an extremely small value compared to the radius of the sun but is sufficient to provide the same prediction of perihelion advance as the general theory of relativity.

We may be tempted here to be surprised at the fact that the non-singular gauge gravitational potential yields the same prediction of the perihelion advance as Einstein's general theory of relativity. But suppose we reflect a little on what we have just done. We supposed there might be a physical fifth dimension in Weyl's gauge principle so that we had a five-dimensional gauge field. We then imposed conservation of the fifth dimension and found that this conservation requirement embedded a four-dimensional hyper-surface into the five-dimensional manifold. We found the curvature of the four-dimensional hyper-surface to be given be Einstein's field equations of his general theory of relativity. We then recognized that the statement of conservation of mass gave us two equivalent ways of describing gravitational phenomena. First, we could use Einstein's curved space representing the four-dimensional curve hyper-surface. On the other hand, we could, just as equivalently, use the gauge forces given by the non-singular gauge potential in the five-dimensional manifold. Should we now be surprised that both methods produce the same prediction of perihelion advance?

We should note here that the resulting linear dependence of λ upon the mass is the reason λ of the positron was scaled down from the lambda of the proton in Chapter 9 during our prediction of the virtual mass of the particles within the proton.

Cosmology

The hot big bang model of the Universe is the model that is in vogue now. Virtually all the journals print numerous articles relating to some aspect of the hot big bang model. The model is based upon t he Newtonian gravitational force and the notion of a scale of the universe that is changing with time. This notion is borrowed from Einstein's general theory of relativity; however, Einstein's theory is not used in the hot big bang model itself. It would seem a shame to discuss a new, non-singular, gravitational potential such as presented in this book without some discussion of its possible impact upon t he hot big bang model. It would have been preferable to wait until the entire solution could be presented. However, this is not possible now so this presentation will include a discussion of how one might expect the new potential to impact the hot big bang model and the problems that render the solution illusive. Further, this analysis ignores the time dependence that was just shown to be necessary. This aspect will be addressed in a l ater chapter. Here we seek only to explore the difference between the standard model and the non-singularity of the gravitational potential.

The development of the standard big bang model begins with considering a spherical piece of the universe with an observer at the center. This sphere is considered to be filled with "dust" of density $\rho(t)$ with a galaxy of interest placed at the outer boundary of the sphere which has a radius denoted by x.

When Gauss's law and Newton's laws of motion and gravitation are used one arrives at

$$m_g\frac{d^2x}{dt^2} = \frac{4\pi}{3}\frac{x^3\rho(t)Gm_g}{x^2} = \frac{4\pi}{3}G\rho(t)m_g x. \quad (10.39)$$

But consider what happens if one wishes to compare this with the cosmology produced by the non-singular, gravitational gauge potential. Then Equation (10.39) becomes

$$m_g\frac{d^2x}{dt^2} = \frac{4\pi}{3}\frac{x^3\rho(t)Gm_g}{x^2}\left(1-\frac{\lambda}{x}\right)e^{-\frac{\lambda}{x}}$$
$$= \frac{4\pi}{3}G\rho(t)x\left(1-\frac{\lambda}{x}\right)e^{-\frac{\lambda}{x}}. \quad (10.40)$$

Now let us replace x with the co-moving coordinate $x=R(t)r$ where $R(t)$ is the scale factor of the universe and r is the co-moving distance coordinate as is done in the standard model. When we also normalize the density to its value at the present epoch, ρ_o, by $\rho(t)=\rho_o R^{-3}(t)$ we obtain

$$\frac{d^2R}{dt^2} = \frac{4\pi G\rho_o}{3}\left(1-\frac{\lambda}{Rr}\right)\frac{e^{-\frac{\lambda}{Rr}}}{R^2}. \qquad (10.41)$$

We can begin to see the trend to be expected from the universe from Equation (10.41) by noting that should we look back in time to the point when $R=\lambda/r$ then we would have a point in time, say T_1, when the acceleration of the universe would have been zero. At times before T_1 there would have been an acceleration outward while for times after T_1, such as the current epoch, the rate of the expansion of the universe is slowing down. This is a very different story than is told by the standard model. But how is it different? It is the same as the standard model in that from Equation (10.41) one sees that the universe was forced into expansion at early times and is now slowing down its rate of expansion. One big difference between the story to be told by Equation (10.41) and the standard model is that Equation (10.41) gives the reason for the initial expansion and it denies that the universe was ever collapsed to a singular point as supposed by the standard model. To better see the first contention we should proceed a little further.

A plot of the dynamics of the universe might be helpful in understanding the predictions of Equation (10.41). First Figure 10.1 shows the deceleration expected in the past. It shows that at times prior to roughly 10 billion years ago the universe was in a period of very rapid expansion. Around 10 billion years ago there arrived a point of transition from an accelerated expansion in the universe to an expanding universe wherein the expansion was slowing down, or decelerating.

Figure 10.1 Cosmological Deceleration

Next Figure 10.2 s hows the velocity of expansion that accompanies the cosmological deceleration of Figure 10.1. The early rise in expansion velocity echo's the early inflation shown in Figure 10.1.

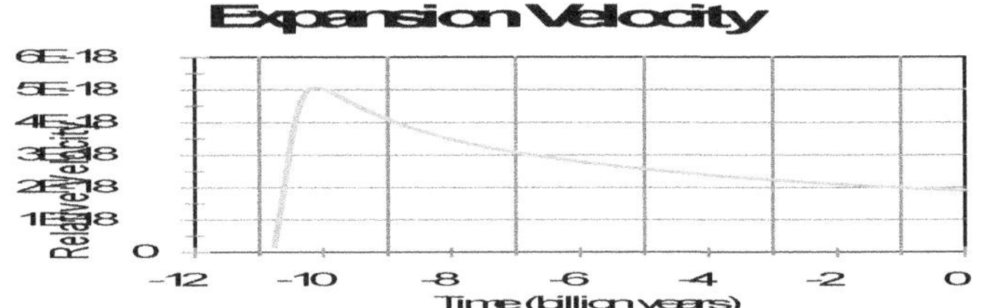

Figure 10.2 Cosmological Expansion Velocity

In Figure 10.3 t he evolution of the universe is displayed, however, the evolution does not show the early inflation as well as the deceleration and expansion velocity graphs do.

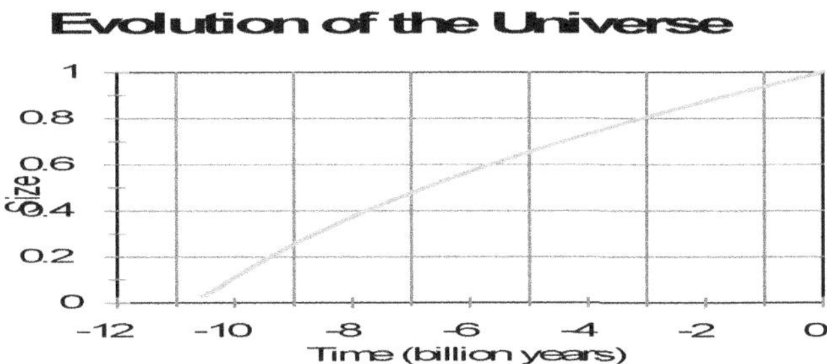

Figure 4.3 Cosmological Evolution

If we multiply Equation (10.41) by dR/dt and integrate with respect to time we find

$$\dot{R}^2 = \frac{8\pi G}{3}\left[\rho(t)R^2 e^{\frac{\lambda}{Rr}} + \frac{\varepsilon(t)R^2}{2c^2}\right] kc^2. \qquad (10.42)$$

The Dynamic Theory

In Equation (10.42) we have included the radiation term for completeness. Now let us evaluate the constant of integration, k, by setting the values of R, dR/dt, ρ, and ε at their present day values of 1, H_o, ρo, and ε_o. Then Equation (10.42) becomes

$$H_o^2 = \frac{8\pi \rho_o G}{3} e^{\frac{\lambda}{r}} + \frac{4\pi G \varepsilon_o}{3c^2} kc^2. \tag{10.43}$$

If we now make the definitions

$$\rho_c \equiv \frac{3H_o^2}{8\pi G} \quad and \quad \Omega \equiv \frac{\rho_o e^{\frac{\lambda}{r}}}{\rho_c}, \tag{10.44}$$

Equation (10.44) may be put into Equation (10.43) to obtain

$$\dot{R}^2 = H_o^2 \left(\frac{\Omega e^{\frac{\lambda}{Rr}}}{Re^{\frac{\lambda}{r}}} + 1 - \Omega \right) + \frac{H_o^2 \varepsilon_o}{2c^2 \rho_c} \left(\frac{1}{R^2} \right). \tag{10.45}$$

We may now take a look at some of the implied dynamics from Equation (10.45). First look at the dynamics as R tends to infinity and there is no radiation. For this case we would have

$$\dot{R}_\infty^2 = H_o^2 (1 - \Omega), \tag{10.46}$$

which is the same as in the standard model.

Now suppose we look backward in time to the time when dR/dt was zero? Then Equation (10.45) becomes

$$0 = \rho_o Re^{\frac{\lambda}{Rr}} \; \rho_o R^2 e^{\frac{\lambda}{r}} + \rho_c R^2 + \frac{\varepsilon_o}{2c^2} (1 - R^2). \tag{10.47}$$

This is a transcendental equation that could be solved for R if we knew λ, r and the density of the dust and radiation at the current epoch. It may be seen from Equation (10.47) that if there is no radiation and R does not equal zero then

$$R = \frac{e^{\frac{\lambda}{Rr}}}{\left[e^{\frac{\lambda}{r}} \frac{\rho_c}{\rho_o} \right]}. \tag{10.48}$$

There is also a trivial solution at $R=0$ in Equation (10.47), but for this case the acceleration is also zero and, therefore, no dynamics are allowed.

While at first glance it may appear that we have as good a solution as is arrived at in the standard model, consider the following points. Our galaxy was considered to be on t he outside limit of a sphere of dust. For the current epoch the density of the dust is very small locally and Gauss's law for considering the total mass of the sphere of dust to be placed at the center should hold very well. But what about when we are looking back in time when the density was a l ot greater? The transcendental nature of the potential argues that our solution may be more difficult to come by.

A second point concerns the fact that we have developed the gauge function in prior sections. If this is the scale of the universe as the gauge function is supposed to be, then why are we again trying to solve for it here? If the scale of the universe is given by the gauge function then the dynamics may be over specified if we put the radiation into the equation for the acceleration such as Equation (10.42). On the other hand, what is the source for the radiation? If there is no hot big bang for the radiation to come from where might it o riginate? The five-dimensional gauge fields that include both gravitational and electromagnetic fields imply an inductive coupling between the electromagnetic and the gravitational fields. Could the radiation be due to the expansion of the gravitating mass of the universe? If so then knowledge of the gauge function might turn the equation of motion for our galaxy into a prediction of the radiation required at the present time. This prediction might then be compared to the measured radiation. However, there is the necessity to have a λ for the universe. It may be obtained from the gauge function also as GM/c^2. But how is M determined?

The Dynamic Theory

Perhaps this is sufficient to point out that the overall picture of cosmology to be given by the non-singular, gravitational gauge potential is not yet complete, but in any event will likely be very different from the hot big bang model of the universe. Will it allow for high temperatures needed for accounting for the abundances of the elements? Since it allows for the universe to be much smaller in the past it would have the associated high temperatures. Yet it should not have the infinite temperatures associated with a singular universe.

Quantum Gravity

We have seen previously how the WQP required quantum mechanics and how allowing a fifth dimension introduced a five dimensional gauge field with gravitational components. Now we wish to show how these two combine to provide us with quantum gravity. We shall suppose that we have a non-singular gravitational potential in the five-dimensional manifold of space, time and mass density and require that mass be conserved and a unity scale factor. These requirements yield a gravitational Schrödinger wave equation similar to the atomic one, or

$$-\frac{\hbar_g^2}{\mu}\frac{\partial^2 \psi(x,X,t)}{\partial x^2} + V(x,t)\psi(x,X,t) - \frac{\hbar_g^2}{2M}\frac{\partial^2 \psi(x,X,t)}{\partial X^2}$$
$$+V_c(x,t)\psi(x,X,t) = i\hbar_g \frac{\partial \psi(x,X,t)}{\partial t} \tag{10.49}$$

where the subscript, g, reminds us that the unit of action depends upon the gravitational gauge function and may be different from Planck's constant. We have retained the two-body approach that was used in the electrostatic case because the non-singular potential for two different masses have different λ as seen from $\lambda = Gm/c^2$. Therefore, for different masses Newton's third law is not satisfied just as in the electrostatic case. The time independent, expanded

309

gravitational wave equation in cylindrical coordinates then becomes

$$-\frac{\hbar_g{}^2}{2\mu}\frac{\partial^2\psi(\bar{r},\bar{R})}{\partial\bar{r}^2}+V(\bar{r},\bar{R})\psi(\bar{r},\bar{R})-\frac{\hbar_g{}^2}{2M}\frac{\partial^2\psi(\bar{r},\bar{R})}{\partial\bar{R}^2} \quad (10.50)$$

$$+V_c(\bar{r},\bar{R})\psi(\bar{r},\bar{R})=E\psi(\bar{r},\bar{R})$$

where the potentials are the gravitational potentials and have the same form as the electrostatic potentials for the COM motion and the motion about the COM. These may be recalled as

$$K=\frac{\mu k}{2r}\left[\left(1-\frac{\lambda_1}{r}\right)\frac{e^{-\frac{\lambda_1}{r}}}{m_2}+\left(1-\frac{\lambda_2}{r}\right)\frac{e^{-\frac{\lambda_2}{r}}}{m_1}\right]$$

$$V=-\frac{\mu k}{r}\left[\frac{e^{-\frac{\lambda_1}{r}}}{m_2}+\frac{e^{-\frac{\lambda_2}{r}}}{m_1}\right] \quad (10.51)$$

$$K_c=\frac{kR}{2r^2}\left[\left(1-\frac{\lambda_1}{r}\right)e^{-\frac{\lambda_1}{r}}-\left(1-\frac{\lambda_2}{r}\right)e^{-\frac{\lambda_2}{r}}\right]$$

$$V_c=-\frac{k}{r}\left[e^{-\frac{\lambda_1}{r}}-e^{-\frac{\lambda_2}{r}}\right]$$

with the k being the gravitational constant.

The vector equation of Equation (10.50) may be solved as before, by the method of separations of variables. First, by trying a solution of the form, $\Psi(\bar{r},\bar{R})=\Psi(\bar{r})\Psi(\bar{R})$, one finds the Equation (10.50) separates into two sets of three equations each where one set depends upon (r,θ,ϕ) and the other set depends upon (R,Θ,Φ). The separation constant, E_c, is related to the COM kinetic energy.

The Dynamic Theory

The two sets of three equations may be separated further by trying $\Psi(\bar{r}) = \Psi_r \Psi_\theta \Psi_\phi$ and $\Psi(\bar{R}) = \Psi_R \Psi_\Phi \Psi_\Theta$

The solutions for Ψ_Θ and Ψ_Φ are identical to the solutions for Ψ_θ and Ψ_ϕ respectively with the separation constants m_L and L corresponding to the separation constants m_l and l.

The equations for Ψ_r becomes

$$\frac{1}{r^2}\frac{d}{dr}\left[r^2\frac{d\psi_r}{dr}\right] + \frac{2\mu}{\hbar^2}\left[E - E_c - V(r) - V_c(r)\right]\psi_r = \frac{\ell(\ell+1)\psi_r}{r^2}.$$

Recognizing the possible difficulties of obtaining solutions with both $V(r)$ and $V_c(r)$ having transcendental terms, an investigation of the influence of the non-singular gravitational potential may be conducted by assuming $\lambda_1 << r$ and $\lambda_2 << r$.

Following the method previously used for solving Schrödinger's time–independent equation, the following definitions:

$$\rho \equiv 2\beta r; \ \beta^2 \equiv \frac{2\mu(E - E_c)}{\hbar_g^2}; \gamma \equiv \frac{\mu G m_1 m_2}{\hbar_g^2 \beta}$$

and

$$\lambda \equiv \frac{2\mu G m_1 m_2}{\hbar_g^2}\left[\frac{(2m_1 + m_2)\lambda_1}{(m_1 + m_2)} - \frac{m_1 \lambda_2}{(m_1 + m_2)}\right],$$

together with the trial solution,

$$\Psi_r = e^{\frac{-\rho}{2}} F(\rho) = e^{\frac{-\rho}{2}}\left[\rho^s \sum_{k=0}^{\infty} a_k \rho^k\right],$$

$$a_0 \neq 0 \ and \ s \geq 0,$$

requires that two relations must be met: namely

$$s(s+1) = l(l+1) + \lambda$$

and

311

$$a_{j+1} = \frac{s+j+1-\gamma}{(s+j+1)(s+j+2)-\left[l(l+1)+\lambda\right]}a_j.$$

In order to satisfy $s \geq 0, s = l + \delta$ where

$$\delta = -\left(l+\tfrac{1}{2}\right)\left[1 - \sqrt{1 + \frac{\lambda}{\left(l+\tfrac{1}{2}\right)^2}}\right].$$

Further, the series will terminate if $\gamma - \delta = n$ is an integer. The solution may then be put into the form

$$E - E_c = \frac{E_n}{\left(1 + \frac{2\delta}{n}\right)} ; where \ E_n = \frac{-\mu\left(Gm_1 m_2\right)^2}{2\hbar_g^2 n^2}.$$

The COM radial equation has no potential terms and is

$$\frac{1}{R^2}\frac{d}{dr}\left(R^2\frac{d\psi_R}{dR}\right) + \frac{2ME_c}{\hbar_g^2}\psi_R = L(L+1)\psi_R.$$

This equation has a Bessel function solution of the form

$$\Psi_R = R^{-\frac{1}{2}}\left[C_1 I_v\left(\alpha R\right)\right]$$

$$where \ v = \tfrac{1}{2}\sqrt{1+4L\left(L+1\right)} \ and$$

$$\alpha = \sqrt{\left|\frac{2ME_c}{\hbar_g^2}\right|}.$$

This solution does not fully establish the value of E_c. This value can be determined, but is not done here, by considering that $\overline{R} \times \hat{r} = 0$.

Limiting case solutions

In the limit where $\lambda_1 = \lambda_2 = 0$, it may easily be seen that the above solution collapses back to a solution for the gravitational problem just as the solution for the electrostatic case. That is, in solving for the $\Psi_\phi\left(\phi\right)$ equation, we shall find that the equation has acceptable solutions only for certain values of m_l. Using these values of m_l in the equation for $\Psi_\theta(\theta)$, it turns out that this equation has acceptable solutions only for certain values of l. With these values of l in the equation for $\Psi_r(r)$, this equation is found to have acceptable solutions only for certain values of the total energy, E; that is, the energy of the gravitational orbits is quantized! In exactly the same manner as in the atomic case, the angular momentum and the energy are both quantized functions. This fact allows one to write

$$mr^2 \tfrac{d\theta}{dt} = angular\ momentum$$

$$= n\hbar_g\ and$$

$$\tfrac{1}{2}m\left(\tfrac{dr}{dt}\right)^2 + \tfrac{n^2 h_g^2}{2mr^2} + V(r) = E.$$

These equations provide the elliptical solutions familiar to planet orbits. There the semi-major axis is given by $a = \left|\tfrac{-GMm}{2E}\right|$ and the eccentricity may be found as

$$\xi = \sqrt{1 + \tfrac{2E_n^2 h_g^2}{m(GMm)^2}}\ .$$

For the quantum numbers n, l, and m_l, there are many possible quantum states to find the planetary orbits to be in. However, not all integers are allowed as quantum numbers. The solar system with all the planets and comets may also be considered as a multi-planet system with all the similar potential relations between the planets as has been seen for the multi-electron atoms. Here there is an increased complexity and richness of possibilities in the differing properties of each planet. These solutions may allow classifying stars in the universe as the atoms are now classified; that is, there may be a periodic table of stars.

Those who have not read of the data supporting the quantization of the solar system may wish to look in Dr. Halton Arp's book "Seeing Red" published by Apeiron (1998). The quantized orbits of the planets, taken individually as a single planet system, would show quantized semi-major axis as $a_n = \left| \frac{GMmh_g^2 n^2}{\mu(GMm)^2} \right| = \left| \frac{h_g^2 n^2}{\mu GMm} \right|$. If the experimental value of a_n for the Earth is taken with the Earth's quantum number, 5, then the value of the gravitational unit of action for the solar system would be $\hbar_g = \sqrt{\frac{a_5 \mu_E GMm_E}{25}} \approx 5.32 \times 10^{39} \, J-\sec$.

Red Shifts

The long range $1/r^2$ dependence of the new three-dimensional vector gauge field component and the work above suggests that these components are the components of the gravitational field. If this is to be the case the proposed theory must then explain the same phenomena that the general theory of relativity predicts. First, note that the gravitational field components in the gauge field tensor must have units equivalent to the electric field components. Following up on t his, we showed that a charge-to-mass ratio is needed to convert the gravitational field units from the familiar units of acceleration to the volts/meter units used in the gauge field tensor. By considering the new fields and comparing them to the currently used fields we found this ratio to be given by the square root of the product of the gravitational constant and the dielectric constant, or

$$\beta = \sqrt{4\pi\varepsilon G} = 8.6127 x 10^{-11} \quad coul / kg. \qquad (10.53)$$

An interesting result follows immediately. If the fundamental charge-to-mass ratio works as it appears to, and electrically neutral spinning bodies have magnetic moments, then the predictions of magnetic moments for electrically neutral bodies may be made by determining the effective charge density of the rotating gravitating body using the charge-to-mass ratio and the spin of the body. We presented a s imple calculation of the earth's magnetic moment, assuming uniform mass distribution, and by this method produced a prediction of the magnetic moment 1.06

314

times the actual value. This prediction seems surprisingly close considering the uncertainties in the density measurements of the mass distribution of the earth.

One of the predictions of Einstein's general theory of relativity concerns the tendency of light from stars and other objects in the heavens to be shifted towards the red color end of the spectrum. Looking at the emission and reception of light within the framework of the proposed theory one finds that the unit of action, which establishes the energy of any state, depends upon both the relative time and the gravitational field at the time and place of the emission and reception. This is so because the theory holds the gravitational field to be a gauge field and it is the gauge function that determines the applicable unit of action. It was previously shown that

$$[x^j, p^k]\psi = i\hbar g^{kl}\left[\delta_{jl} + \binom{j}{sl}x^s\right]\psi. \tag{10.54}$$

Thus, for a metric with only a gauge function the effective unit of action would be given by

$$\hbar' = \hbar\exp[2f_{,t}f_{,r}f_{,\gamma}]. \tag{10.55}$$

By recalling the form of the gauge function we previously developed and integrating it over the volume of the gravitating body so that it depends upon the total mass of the body, one may use Equation (10.55) to find the first order expression for the unit of action for emission of a photon to be

$$\hbar_e \cong \hbar\exp\left[\frac{W_e}{R_e}e^{-\left(\frac{\lambda_e}{R_e}\right)}e^{-H_o t_e}\right] \cong \hbar\exp\left[\frac{W_e(1-H_o t_e)}{R_e}e^{-\left(\frac{\lambda_e}{R_e}\right)}\right] \tag{10.56}$$

where the subscript, e, denotes emission. Similarly, the unit of action for the reception of a photon can be found to be

$$\hbar_r \cong \hbar\exp\left[\frac{W_r}{R_r}e^{-\left(\frac{\lambda_r}{R_r}\right)}e^{-H_o t_r}\right] \cong \hbar\exp\left[\frac{W_r(1-H_o t_r)}{R_r}e^{-\left(\frac{\lambda_r}{R_r}\right)}\right]. \tag{10.57}$$

If photon energy is conserved between emission and reception then

$$\hbar_e \nu_e = \hbar_r \nu_r. \tag{10.58}$$

If one sets $t_e = 0$, $t_r = L/c$, and $W = (-GM/c^2)$, then they find that, to the first order with respect to time, the shift in frequency is given by

$$z \cong \frac{\Delta\lambda}{\lambda_e} = \exp\left\{ \left(\frac{-G}{c^2}\right)\left[\frac{M_r e^{\frac{-\lambda_r}{R_r}}}{R_r} - \frac{M_e e^{\frac{-\lambda_e}{R_e}}}{R_e}\right] + \left(\frac{H_o L}{c}\right)\frac{\left(\dfrac{M_r}{R_r}\right)}{\left(\dfrac{M}{R}\right)} \right\} - 1. \tag{10.59}$$

In Equation (10.59) the M and R without subscripts refer to the mass and radius of the Earth. If the reception is done at the Earth's surface the ratio of the gravitational potential at reception and the Earth's surface cancels each other. This factor occurs in the expression for the red shift because our standard for comparison is the frequencies measured at the Earth's surface.

By looking at the first order approximations of this prediction one finds that the time dependence of the gravitational field produces the linear dependence and is given by Hubble's constant while the gravitational potential produces the same prediction that comes from Einstein's theory.

Looking a little closer one finds that the time dependence of the red shift produces an experimental number, $H_o^{-1} = (5.6+0.6)\times10^{17}$ sec. (1.61×10^{-18} sec^{-1} < H_o < 2.0×10^{-18} sec^{-1}), that corresponds to the same time dependence that has been measured and reported for the moon's orbit ($H_o=1.9\times10^{-18}$sec^{-1}), which is well within experimental error. It is somewhat pleasing that a prediction coming from the same time dependence originating in the gauge function leads to a comparison of phenomena involving cosmological distances agrees with phenomena involving the much shorter distance involved in the moon's orbit. Another possible plus to this prediction is that, because the prediction involves an exponential dependence upon time and gravitational potential between the emission and reception of the light, then the distances that are currently ascribed to distant bodies by their red shifts may be much greater than the actual distances. Also, the possible red shifts from dense gravitating bodies may be much greater than is now believed possible thereby removing the mystery from many objects. This is may be especially so when one uses the full expression for the effective unit of action rather than making the first order approximations we did here.

The time dependence of the gravitational field stems from the five dimensionality of classical thermodynamics just as the exponentials that enter into the equation.

Because of the distances involved the approximation $\delta \ll R$ may be used for both the emitter and the receiver. Then the approximation of Equation (10.59) may be written as

$$\ln(z+1) \cong -\frac{G}{c^2}\left[\frac{M_r}{R_r} - \frac{M_e}{R_e}\right] + \left(\frac{HL}{c}\right)\frac{\left(\dfrac{M_r}{R_r}\right)}{\left(\dfrac{M}{R}\right)} \tag{10.60}$$

so that on the Earth's surface we would have the approximation

$$\ln(z_{ES}+1) \cong -\frac{G}{c^2}\left[\frac{M}{R} - \frac{M_e}{R_e}\right] + \left(\frac{HL}{c}\right) \tag{10.61}$$

while if we were to obtain experimental redshifts using the Hubble Telescope while in orbit at an orbital height of h we would have

$$\ln\left(z_{HT}+1\right) \cong -\frac{G}{c^2}\left[\frac{M}{(R+h)}-\frac{M_e}{R_e}\right]+\left(\frac{HL}{c}\right)\frac{R}{(R+h)}. \quad (10.62)$$

Equation (10.61) and (10.62) represent two equations in the two unknowns, L and M_e/R_e. The solution of these equations is given by the equations

$$\frac{M_e}{R_e} \cong \left(\frac{c^2}{hG}\right)\left[(R+h)\ln\left(z_{HT}+1\right)-R\ln\left(z_{ES}+1\right)\right] \quad (10.63)$$

and

$$L \cong \left(\frac{c}{H}\right)\left\{\ln\left(z_{ES}+1\right)-\ln\left(z_{HT}+1\right)\left(1+\frac{R}{h}\right)+\frac{GM}{C^2R}\right\}. \quad (10.64)$$

One may then see how comparing the red shifts obtained from the Earth's surface with those taken at a height above it will allow the determination of distance to, and the gravitational potential of, a cosmological object. The ability to obtain solutions from the two equations for red shifts at different receiving gravitational potentials does not exist in other predictions of red shifts. It is because of the appearance of the ratio of gravitational potentials in Equation (10.59) that allows the Dynamic Theory to make the distance and potential predictions. For objects with a large gravitational potential compared to that of the Earth the major change in $ln(z+1)$ comes from the ratio of h to $R+h$. For an orbit height of 380 miles this ratio is 0.0876. This means that the expected change in the measured redshirt from the Earth's surface to the Hubble Telescope orbit is of the order of a few percent. A student survey of books reporting red shifts in the optical range puts the experimental error between a few percent and near 30 percent. Therefore, care needs to be taken or frequencies sought which have less experimental error.

When I first made this prediction while I was still at Los Alamos National Laboratory, Dr. Dan Ross still worked with me as a collaborator and on one of his trips to Washington DC he stopped by the offices at the Hubble Telescope center and talked with some folks there about this prediction and if they

318

had seen any such deviation from the red shifts measured at the Earth's surface. They indicated that since there was no reason to believe such a difference should exist all measurements made by the Hubble Telescope was calibrated to the Earth's surface so any difference would be lost. They were programmatically driven and could not recalibrate the data. It would be interesting to be able to compare the actual data to the red shift data at the Earth's surface. This probably would entail a rewrite of the computer analysis software currently used on the Hubble Telescope data.

Dark Matter

Much has been written, hypothesized, and calculated on the subject of Dark Matter and Dark Energy. However, none consider a time dependent gravitational field. A gravitational field that gets weaker with time will display galaxy dynamics responding to a much stronger field before sending light from space towards the Earth that can only be received many light years later. The theoretical basis for such a time dependent gravitational field has already been presented. Three elements of this theory apply to the potential explanation of Dark Matter and Dark Energy. These elements include:

1. The theory is a five dimensional gauge theory with Weyl geometry. This means that the fields within the theory are gauge fields. However, the theory is not another Kalusa-Klein type of theory in that the fifth dimension describes a real physical property, mass density, and, therefore, is not hidden or obscured by some mathematical technique. The five dimensionality of the gauge theory requires that the gravitational field be time dependent.

2. Quantum Mechanics is required by restricting the Weyl scale factor within the gauge theory to have only a value of unity. This was noted by

Schrödinger before he published his wave equations and later it was shown by London that this restriction required Schrödinger's wave equations. This quantization requires that the gauge potentials be quantized and non-singular.

3. The fundamental Weyl geometry requires that the Poisson brackets and the unit of action be dependent upon the gauge function. This variable unit of action leads to a relation determining the red shift of light coming to Earth from distant stars.

These three aspects of the new theory suffice to offer a different view of the data from which the hypothesis of dark matter and dark energy have evolved.

Data wherein the tangential velocities of stars in the arms of a spiral galaxy differed from Newtonian predictions were first reported nearly seventy years ago. A fundamental theory supporting these data has not heretofore been given, though empirical theories have been presented. The best of these theories is the Modified Newtonian Dynamics (MOND).

Newtonian uniform circular motion equates the gravitational acceleration to the centripetal acceleration so that

$$\frac{GMm}{r^2} = \frac{mv^2}{r}. \tag{10.65}$$

A time-dependent, non-singular gravitational field, such as the Dynamic Theory predicts, alters Equation (10.65) to

$$\frac{GMm\left(1 - H_0\tau\right)}{r^2}\left(1 - \frac{\lambda}{r}\right)e^{-\frac{\lambda}{r}} = \frac{mv^2}{r}, \tag{10.66}$$

where H_0 is Hubble's constant and

$$\lambda \equiv \frac{GM}{c^2} \tag{10.67}$$

as determined by planetary orbits. For this time dependent gravitational field the gravitational acceleration acting on an arm of a galaxy is due to the gravitational field of the mass M at a previous time. This previous time is given by the time that it takes for the field to travel from the site of the gravitational field to the point on the arm under consideration. This means that when all the mass is considered to be at the center of the galaxy the time that enters into Equation (10.66) is $\tau=-r/c$ so that, when $r>>\lambda$ the velocity of the arm of the galaxy would be given by

$$v = \sqrt{\frac{GM(1-H_o\tau)}{r}} - \sqrt{GM\left(\frac{1}{r}+\frac{H_0}{c}\right)}. \tag{10.68}$$

Equation (10.68) shows a very different character than the expression for the velocity for the time independent gravitational field. This expression shows that the velocity of the galaxy arms should not be expected to drop off as the time independent Newtonian gravitational field does.

We may look at the 5-dimensional approach of the Dynamic Theory by looking at the Lagrangian

$$L = \frac{1}{2}mc^2\left(\dot{\tau}\right)^2 + \frac{1}{2}m\dot{r}^2 + \frac{1}{2}m\left(r\dot{\theta}\right)^2$$

$$+ GMm\left(1-H_o\tau\right)\frac{e^{-\frac{\lambda}{r}}}{r} \tag{10.69}$$

where the universe time, τ, is treated as another variable and t is the local time. The universe time, τ, becomes a geometrical coordinate that makes the problem local-time independent in five dimensions.

The time Lagrange equation may then be written as

$$\frac{d}{ds}\left[\frac{\partial L}{\partial\dot{\tau}}\right] - \frac{\partial L}{\partial\tau} = 0 = \frac{d}{dt}\left[m\dot{\tau}\right] + H_o\lambda m\frac{e^{-\frac{\lambda}{r}}}{r}. \tag{10.70}$$

For a spherically symmetric field the radial equation is

$$\frac{d}{dt}\left[\frac{\partial L}{\partial \dot{r}}\right] - \frac{\partial L}{\partial r} = 0$$

$$= \frac{d}{dt}\left[m\dot{r}\right] - mr\dot{\theta}^2 \qquad . \qquad (10.71)$$

$$+ GMm\left(1 - H_o\tau\right)\left(1 - \frac{\lambda}{r}\right)\frac{e^{-\frac{\lambda}{r}}}{r^2}$$

The third Lagrange equation becomes

$$\frac{d}{dt}\left[\frac{\partial E}{\partial \dot{\theta}}\right] - \frac{\partial E}{\partial \theta} = 0 = \frac{d}{dt}\left[mr^2\dot{\theta}\right]. \qquad (10.72)$$

For the problem of spiral galaxy behaviour we may assume the $\lambda \ll r$ and write the equations of motion as

$$\ddot{\tau} = \frac{-H_o\lambda}{r}, \qquad (10.73)$$

$$\ddot{r} - r\dot{\theta}^2 = -\left(1 - H_o\tau\right)\frac{GM}{r^2}\left(1 - \frac{\lambda}{r}\right) \qquad (10.74)$$

and

$$\ddot{\theta} + \frac{2}{r}\dot{r}\dot{\theta} = 0. \qquad (10.75)$$

If we now look at uniform circular motion we find that Equation (10.73) becomes

$$\ddot{\tau} = \frac{-H_o\lambda}{r} = \text{constant} \Rightarrow \frac{d\dot{\tau}}{dt} = \frac{-H_o\lambda}{r} \qquad (10.76)$$

so that this may be integrated to get

$$\dot{\tau} = \dot{\tau}_o - \frac{H_o\lambda}{r}\left(t - t_o\right) \qquad (10.77)$$

which may be integrated again to get

$$\tau = \tau_o - \frac{H_o\lambda}{2r}t^2 + \left(\dot{\tau}_o + \frac{H_o\lambda}{r}t_o\right)t. \qquad (10.78)$$

Also for the assumed uniform circular motion Equation (10.74) may be written as

$$v^2 = \left(1 - H_o\tau\right)\frac{GM}{r}\left(1 - \frac{\lambda}{r}\right)$$
(10.79)

where v is the tangential velocity of the uniform circular motion. Putting Equation (10.78) into Equation (10.79) obtains

$$v^2 = \frac{GM}{r}\left\{\begin{array}{c} 1 - H_o\tau_o + \dfrac{H_o^2\lambda}{2r}t^2 \\ -\left(H_o\dot{\tau}_o + \dfrac{H_o^2\lambda}{r}t_o\right)t \end{array}\right\}\left(1 - \frac{\lambda}{r}\right).$$
(10.80)

We must keep in mind there are two times to be considered. First there is the time it takes for the gravitational change to travel from the center of the galaxy to the point of measurement in the galaxy arm. The second time is for the light signal to travel from the galaxy to the Earth.

Let us set $\tau_o = 0$ and $t_o = 0$ at the point in time when the light left the star on its way toward Earth. Now our Equation (10.80) becomes

$$v^2 \cong \frac{GM}{r}\left\{1 + \frac{H_o^2\lambda}{2r}t^2 - H_o\dot{\tau}_ot\right\}\left(1 - \frac{\lambda}{r}\right).$$
(10.81)

Time runs from the time the gravitational signal left the center of the galaxy at

$$t = \frac{-r}{c},$$
(10.82)

where r is the distance from the center of the galaxy.

Using Equation (10.82) in Equation (10.81) we find

$$v^2 \cong \frac{GM}{r}\left\{1 + \frac{H_o\dot{\tau}_or}{c}\left(1 - \frac{H_o^2\lambda}{2c^2}\right)\right\}.$$
(10.83)
$$\cong \frac{GM}{r}\left\{1 + \frac{H_o\dot{\tau}_or}{c}\right\}$$

Now we need to establish a value for $\dot{\tau}_o$. Look at the energy at time $t=0$ and $\tau=0$ with $r>>\lambda$, or

Gauge Gravitational Fields

$$E_o = \frac{1}{2}mc^2 \left(\dot{\tau}_o\right)^2 + \frac{1}{2}mv^2 - \frac{GMm}{r_o} \qquad (10.84)$$

This may be rewritten as

$$\frac{2E_o}{mc^2} = \left(\dot{\tau}_o\right)^2 + \frac{v^2}{c^2} - \frac{\lambda}{r_o} . \qquad (10.85)$$

Since the tangential velocities are non-relativistic this requires that

$$\dot{\tau}_o = \sqrt{\frac{2E_o}{mc^2} + \frac{\lambda}{r_o}} . \qquad (10.86)$$

Equation (10.86) shows that the initial conditions establish the point at which the tangential velocities begin to differ from those predicted by Newtonian gravity. In the absence of a m eans of evaluating the initial conditions we may turn to experimental results. First, suppose we write the acceleration in the arms of the galaxy as

$$a = a_N \left\{ 1 - \frac{H_o^2 \lambda}{2r} t^2 - H_o \dot{\tau}_o t \right\}$$
$$\cong a_N \left\{ 1 + H_o \dot{\tau}_o \frac{r}{c} \right\} \qquad (10.87)$$

We now use the data that shows the acceleration begins to deviate from Newtonian when the acceleration drops to a value of 1.2×10^{-10} m/sec^2 so that

$$a_N = \frac{GM}{r_c^2} \cong 1.2 \times 10^{-10} \Rightarrow r_c \cong \sqrt{\frac{GM}{1.2 \times 10^{-10}}} . \qquad (10.88)$$

Then requiring

$$\dot{\tau}_o = \frac{c}{r_c H_o} \qquad (10.89)$$

sets a value of $\dot{\tau}_o$ in keeping with the data. Equation (10.87) becomes

$$a \cong a_N \left\{ 1 + \frac{r}{r_c} \right\} \tag{10.90}$$

where we see the short range Newtonian acceleration and the long range acceleration predicted by MOND.

It should be noted that the approximate linearity of the tangential velocity with respect to time of Equations (10.81) and (10.83) displays an independence of the time it takes for light to travel from the galaxy to Earth. This apparent independence of time masks the fact that the gravitational strength of the galaxy, relative to the current epoch, depends upon the time of light travel to Earth.

Dark Energy

Data displaying evidence that provided the beginning of the hypothesized dark energy was first presented in 1998. To date no fundamental theory has had success in explaining these data.

The universe expansion factor may be taken from general relativity and is

$$\frac{\ddot{a}}{a} = -\frac{4}{3}\pi G \left(\rho + 3\frac{p}{c^2} \right). \tag{10.91}$$

The mean density and pressure are currently taken to include dark energy and are taken to obey the local conservation of energy relation

$$\dot{\rho} = -3\frac{\dot{a}}{a}\left(\rho + \frac{p}{c^2} \right). \tag{10.92}$$

The first integral of Equations (10.91) and (10.92) is the Friedman equation

$$\dot{a}^2 = \frac{8}{3}\pi G \rho a^2 + \text{constant}. \tag{10.93}$$

But consider what happens if one wishes to compare this with the cosmology produced by the non-singular, time dependent, gravitational gauge potential. Then Equation (10.91) becomes

$$m_g \frac{d^2x}{dt^2} = \frac{4\pi}{3} \frac{x^3 \rho(t) G m_g}{x^2} \left(1 - \frac{\lambda}{x}\right)\left(1 - H_o\tau\right)e^{-\frac{\lambda}{x}}$$

$$= \frac{4\pi}{3} G\rho(t)x \left(1 - \frac{\lambda}{x}\right)\left(1 - H_o\tau\right)e^{-\frac{\lambda}{x}} \tag{10.94}$$

where τ is the universe time.

Now let us replace x with the co-moving coordinate $x=R(t)r$ where $R(t)$ is the scale factor of the universe and r is the co-moving distance coordinate as is done in the standard model. When we also normalize the density to its value at the present epoch, ρ_o, by $\rho(t)=\rho_o R^{-3}(t)$ we obtain

$$\frac{d^2R}{dt^2} = \frac{4\pi G\rho R}{3}\left(1 - \frac{\lambda/r}{R}\right)\left(1 - H_o\tau\right)e^{-\frac{\lambda/r}{R}}. \tag{10.95}$$

If we multiply Equation (10.41) by dR/dt and integrate with respect to time we find

$$\int \dot{R}\frac{d^2R}{dt^2}dt = \frac{4\pi G\rho}{3}\int\left(1 - H_o\tau\right)R\dot{R}dt$$

$$\frac{\dot{R}^2}{2} - \frac{\dot{R}_o^2}{2} = \frac{4\pi G\rho}{3}\int\left(1 - H_o\tau\right)RdR \tag{10.96}$$

We now need to know how to integrate the right hand side of Equation (10.42). Suppose we consider the time it takes for light to travel from the distant star to Earth, or $t=-a/c$, where a is the distance from the star to Earth and the minus sign comes from looking backwards in time. The radius of the universe now has two parts. The first part is the radius of the universe when the light left the star on its journey to the Earth. Let this radius be R_o. Thus we see that

$$R = R_o + a \tag{10.97}$$

and

$$\dot{R} = \dot{a}. \tag{10.98}$$

Further, from considerations of the dark matter it was determined that the world time was given by

The Dynamic Theory

$$\tau = \tau_o - \frac{H_o GM}{2c^2 R}t^2 + \left(\dot{\tau}_o + \frac{H_o GM}{c^2 R}t_o \right)t . \qquad (10.99)$$

When we set both initial times to zero and use the value of

$$\lambda_U \equiv \frac{GM}{c^2}, \qquad (10.100)$$

Equation (10.99) becomes

$$\tau = -\frac{H_o \lambda_U}{2R}t^2 + \dot{\tau}_o t . \qquad (10.101)$$

Now we find that Equation (10.42) may be written as

$$\dot{a}^2 = \frac{8\pi G\rho}{6c^2} \left\{ \begin{array}{l} 2R_o c^2 a + \left(c^2 + H_o \dot{\tau}_o cR_o \right)a^2 \\ +\frac{1}{3}\left(H_o^2 \lambda_U + H_o \dot{\tau}_o 2c \right)a^3 \end{array} \right\} + K . \qquad (10.102)$$

If we set the constant of integration, K, to zero, then Equation (10.102) becomes

$$\dot{a}^2 = H_o^2 \Omega_M' \left\{ \begin{array}{l} R_o a + \frac{1}{2}\left(1 + \frac{H_o \dot{\tau}_o R_o}{c} \right)a^2 \\ +\frac{1}{6}\left(\frac{H_o^2 \lambda_U}{c^2} + \frac{2H_o \dot{\tau}_o}{c} \right)a^3 \end{array} \right\} . \qquad (10.103)$$

where we have used the definitions

$$\rho_c \equiv \frac{3H_o^2}{8\pi G}, \quad and \quad \Omega_M' \equiv \frac{\rho}{\rho_c} . \qquad (10.104)$$

In Equation (10.103) we find that the mass density term splits into three terms for a time-dependent gravitational field. For a time-independent gravitational field there was only one term.

An interesting aspect of Equation (10.103) is that the two new mass terms both involve the same time dependence factor as the one that causes the tangential velocity of the arms of spiral galaxies to differ from Newtonian behaviour. That is to say that should the two new terms provide a basis for the current experimental evidence for dark energy it comes from the same source as the basis for dark matter. The time

327

dependence of the gravitational field explains both phenomena.

Consider Equation (10.103) again and add the usual term for radiation so that we find

$$\left(\frac{\dot{a}}{a}\right)^2 = H_o^2 \left\{ \Omega'_M \left[\begin{array}{c} \dfrac{R_o}{a} + \dfrac{1}{2}\left(1 + \dfrac{H_o \dot{\tau}_o R_o}{c}\right) \\ + \dfrac{1}{6}\left(\dfrac{H_o^2 \lambda_U}{c^2} + \dfrac{2H_o \dot{\tau}_o}{c}\right)a \end{array} \right] + \Omega_{RO} \right\} \tag{10.105}$$

where we did not add a term for the cosmological constant. If this is to compare with the usual expression we could write

$$\left(\frac{\dot{a}}{a}\right)^2 = H_o^2 \left\{ \begin{array}{c} \Omega'_M \left(1+z\right)^3 + \Omega'_{DM}\left(1+z\right)^3 \\ + \Omega'_{DE}\left(1+z\right)^3 \\ + \Omega_{RO}\left(1+z\right)^4 \end{array} \right\} \tag{10.106}$$

wherein the sum of the terms are taken to be unity at $z=0$ and the integration constant has been taken to be zero. Equation (10.105) and (10.106) would require

$$\Omega'_M \left[\begin{array}{c} \dfrac{R_o}{a} + \dfrac{1}{2}\left(1 + \dfrac{H_o \dot{\tau}_o R_o}{c}\right) \\ + \dfrac{1}{6}\left(\dfrac{H_o^2 \lambda_U}{c^2} + \dfrac{2H_o \dot{\tau}_o}{c}\right)a \end{array} \right] + \Omega_{RO} = 1. \tag{10.107}$$

The relation between the red shift and a is

$$1 + z = \left[\frac{a\left(t_{obs}\right)}{a\left(t_{em}\right)}\right] = \frac{R_o + a}{R_o}. \tag{10.108}$$

By putting Equation (10.108) into (10.105) we find

$$\left(\frac{\dot{a}}{a}\right)^2 = H_o^2 \left\{ \Omega'_M \left[\begin{array}{c} z + \dfrac{1}{2}\left(1 + \dfrac{H_o \dot{\tau}_o c^3 a}{zc^4}\right) \\[2ex] + \dfrac{1}{6}\left(\dfrac{H_o^2 GM}{c^4} + \dfrac{2H_o \dot{\tau}_o}{c}\right)a \end{array}\right] + \Omega_{RO} \right\}. \quad (10.109)$$

The fact that these terms have expressions relating them argues that their relative values may be determined.

For example, if Ω_{RO} is taken to be small compared with the mass terms and Ω_M is set at the typical value of 0.25, then we would require

$$\dot{\tau}_o = \left(\frac{zc\left(21 - 6z - \dfrac{H_o^2 \lambda_U a}{c^2}\right)}{H_o a(3 + 2z)}\right). \quad (10.110)$$

Since the source of the light being measured left its origin some time after the universe completed the exponential inflationary expansion early in universe time, we would have

$$\dot{\tau}_o = \left(\frac{zc3(7 - 2z)}{H_o a(3 + 2z)}\right). \quad (10.111)$$

Putting this back into Equation (10.109) we find

$$\left(\frac{\dot{a}}{a}\right)^2 = H_o^2 \left\{ \Omega'_M \left[z + \frac{12 - 2z}{(3 + 2z)} + \frac{z(7 - 2z)}{(3 + 2z)} \right] + \Omega_{RO} \right\}. \quad (10.112)$$

There are three terms for the mass with different functions of z.

Now we would have

$$\Omega'_M = \frac{(3 + 2z)}{4(3 + 2z)} = 0.25 \quad (10.113)$$

as set above and we can then evaluate each term when $z=0$. Let us associate the middle term with Ω_M, the first term with Ω_{DMO} and the remaining term with Ω_{DEO}. We would then have the values

$$\left(\Omega_M\right]_{z=0} = \left(\Omega'_M\frac{12-2z}{(3+2z)}\right]_{z=0} = 1$$

$$\left(\Omega_{DM}\right]_{z=0} = \left(\Omega'_M z\right]_{z=0} = 0 \qquad (10.114)$$

$$\left(\Omega_{DE}\right]_{z=0} = \left(\Omega'_M\frac{z(7-2z)}{(3+2z)}\right]_{z=0} = 0$$

for $z=0$.

Our overall equation would then be

$$\left(\frac{\dot{a}}{a}\right)^2 = H_o^2\left\{\left[\begin{matrix}\Omega_M z + \Omega_M\dfrac{12-2z}{(3+2z)}\\[12pt] +\Omega_M\dfrac{z(7-2z)}{(3+2z)}\end{matrix}\right]+\Omega_{RO}\right\}, \qquad (10.115)$$

where Ω_M varies as $(1+z)^3$.

Comparing with Experiment

The expansion of the universe means the distance between two distant galaxies varies with time as

$$L(t) \propto a(t). \qquad (10.116)$$

The rate of change of the distance is the speed

$$v = \frac{dl}{dt} = Hl, \quad H = \frac{\dot{a}}{a} \qquad (10.117)$$

where H is the time dependent Hubble parameter.

A method of measuring of the expansion of the universe comes from measuring the shift of frequencies of light, the red shift, coming from distant stars. The observed wave length, λ_r, of a feature in the spectrum that had wavelength λ_e at emission is given by the relation

The Dynamic Theory

$$1 + z = \frac{\lambda_r}{\lambda_e} = \frac{a(t_r)}{a(t_e)}. \qquad (10.118)$$

When the velocity is given by cz then Hubble's law is written as

$$cz = Hl \qquad (10.119)$$

from which we see that

$$z = \frac{HL}{c}, \text{ or } H = \frac{cz}{L}. \qquad (10.120)$$

An additional feature of the new theory presented here is the expression for the red shift of light from distant stars. This has been shown to be

$$z_{\exp} = \frac{\Delta\lambda}{\lambda_e}$$

$$= \exp\left\{\frac{\left(\frac{-G}{c^2}\right)\left[\frac{M_r e^{-\frac{\lambda_r}{R_r}}}{R_r} - \frac{M_e e^{-\frac{\lambda_e}{R_e}}}{R_e}\right]}{+\left(\frac{HL}{c}\right)\frac{\left(\frac{M_r}{R_r}\right)}{\left(\frac{M_E}{R_E}\right)}}\right\} - 1, \qquad (10.121)$$

where the subscript r designates values at the time and point of reception, the subscript e represents values at the time and point of emission and the quantities, M_E and R_E, represent the mass and mean radius of the Earth. We have also used the subscript 'exp' on the red shift to indicate that it is the experimental value of red shift measured at the receiving location.

There are two parts to the red shift. One part is due to the gravitational fields at the points and times of emission and reception and the other part is due to the travel time between emission and reception. This is the part that involves the expansion of the universe. Therefore let us rewrite Equation (10.121) as

$$z_{\exp} = \exp\left\{\left(\frac{-G}{c^2}\right)\left[\frac{M_r e^{-\frac{\lambda_r}{R_r}}}{R_r} - \frac{M_e e^{-\frac{\lambda_e}{R_e}}}{R_e}\right]\right\}\exp\left\{\left(\frac{HL}{c}\right)\frac{\left(\frac{M_r}{R_r}\right)}{\left(\frac{M_E}{R_E}\right)}\right\} - 1, \quad (10.122)$$

and then rearrange it to get

$$z \equiv \frac{HL}{c} = \left\{\left(\frac{G}{c^2}\right)\left[\frac{M_r e^{-\frac{\lambda_r}{R_r}}}{R_r} - \frac{M_e e^{-\frac{\lambda_e}{R_e}}}{R_e}\right] + \log\left(1 + z_{\exp}\right)\right\}\frac{\left(\frac{M_E}{R_E}\right)}{\left(\frac{M_r}{R_r}\right)} \quad (10.123)$$

which is the red shift of the universe expansion.

Two simplifications may now be made. First, in many cases the gravitational component of the experimental red shift may be ignored. Secondly, if we are only using the red shift data measured at the Earth's surface then Equation (10.123) reduces to

$$z \equiv \frac{HL}{c} = \log\left(1 + z_{\exp}\right). \quad (10.124)$$

This is the red shift value to be used in the expansion velocity of the universe, Equation (10.115). so that we may write

$$\left(\frac{\dot{a}}{a}\right)^2 = H_o^2\left\{\begin{matrix}0.25\left(1 + \log\left(1 + z_{\exp}\right)\right)^3 \log\left(1 + z_{\exp}\right) \\[2ex] +0.25\left(1 + \log\left(1 + z_{\exp}\right)\right)^3 \frac{12 - 2\log\left(1 + z_{\exp}\right)}{\left(3 + 2\log\left(1 + z_{\exp}\right)\right)} \\[2ex] +0.25\left(1 + \log\left(1 + z_{\exp}\right)\right)^3 \frac{z\left(7 - 2\log\left(1 + z_{\exp}\right)\right)}{\left(3 + 2\log\left(1 + z_{\exp}\right)\right)}\end{matrix}\right\} + \Omega_{RO} \quad (10.125)$$

If one can simultaneously measure the red shift and the distance to the object then Equation (10.120) gives a value of Hubble's parameter that may be used in Equation (10.117) to get the universe expansion velocity.

332

The Dynamic Theory

Standard Candles

One reason for choosing the Type Ia supernova in the universe expansion research is the assumption that the mass of this type supernova are all the same; roughly the Chandrasekhar Limit mass of 1.39 solar masses. However, a time dependent gravitational field causes this limit to change with time. This may be seen by considering the Newtonian equation of hydrostatic equilibrium known as the Tolman-Oppenheimer-Volkov [TOV] equation, or

$$\frac{dp}{dr} = \frac{-GM(r)\left(1 - H_o\tau\right)\rho}{r^2}. \tag{10.126}$$

The gravitational field that is holding the star together against the internal pressure is diminishing in time. This means that the limiting mass increases in time. Supernova found closer to Earth will have more mass and therefore greater luminosity, than more distant supernova. A reduction in luminosity from the assumed constancy would show up in an analysis by making the more distant supernova appear further away than it really is. The natural conclusion, based on the time-independent gravitational field that produces the constant Chandrasekhar limiting mass, would be that the expansion of the universe is accelerating.

Using the Virial Theorem development by Collins who arrives at the Chandrasekhar limiting mass with the equation

$$\frac{R_o}{\left(\dfrac{2GM}{c^2}\right)} > 228\left(\frac{M_{Sun}}{M}\right)^{\frac{4}{9}} \approx 200 \tag{10.127}$$

the time dependent gravitational field requires that this relation become

$$\frac{R_o}{\left(\dfrac{2GM}{c^2}\right)\left(1 - H_o\tau\right)} > 228\left(\frac{M_{Sun}}{M}\right)^{\frac{4}{9}} \approx 200. \tag{10.128}$$

This gives the limiting mass as

$$M_L = M_{Ch}\left(1 - H_o\tau\right)^{-\frac{9}{4}},\qquad(10.129)$$

where M_{Ch} is the Chandrasekhar limiting mass. By differentiating Equation (10.129) with respect to universe time we find the limiting mass for the type Ia supernovae to change according to

$$\frac{dM_L}{d\tau} = \left(\frac{9H_o}{4}\right)M_{Ch}\left(1 - H_o\tau\right)^{-\frac{13}{4}}.\qquad(10.130)$$

Conclusions

A time-dependent gravitational field, that gets weaker in time, shows the physical effects of this past, stronger field in the dynamics of spiral galaxies. This weakening gravitational field also shows up in the analysis of the distances to, and red shift of light from, supernovas. Here it adds terms to the universe expansion velocity relations that are not present in the analysis of time-independent fields. It also changes the luminosity of the supernovas that were assumed to have constant luminosity. These effects of the time dependent gravitational field remove the need for hypothesizing new matter or energy to explain these effects.

There have been many attempts in the past to find different solutions to Einstein's field equations and to show how an expanding universe may be viewed in different ways. Portions of the above may be reminders of prior approaches. Therefore, it may prove useful to point out what is new in this section on cosmology.

Fundamentally there are three things that are new in this section. First, the fifth dimension is considered to be a real physical entity. All five dimensional theories that I know of in the past, whether by Kalusa-Klein, Einstein with his many collaborators, and others, did not consider the fifth dimension to be real and, therefore, required several terms in the resulting gauge field equations to be zero. Here these terms are non-zero and require that the gravitational potential and field be time-dependent. Second, it uses the Weyl Quantum Principle as its basis for quantum theory. This and the five dimensionality require that the gravitational potential be a non-singular potential. These two things also require the gravitational field to be a time-dependent gauge field not seen previously. The third aspect of the section is that the Weyl Gauge Principle requires that the unit of action be dependent upon the gauge function. This requires the red-shift from distant objects to have an exponential dependence upon both the time and distance between emission and reception. The new red-shift relation becomes important in both dark matter and dark energy predictions because both phenomena are witnessed by red-shifted light. The time dependence, or weakening, of the gravitational field is the major factor in predicting effects interpreted as dark matter. The time dependence of the gravitational field also provides the major factor in predictions with respect to dark energy as it is responsible for the diminishing of the luminosity of the distant supernovas used as standard candles and the expression for the expansion of the universe.

Chapter 11 Five Dimensional Fields

In this chapter we shall investigate properties of the five dimensional gauge fields. The five dimensional fields are derived in precisely the same manner as Weyl derived the four dimensional electromagnetic fields. The only difference is that now the indices run from 0 to 4 rather than from 0 to 3.

The potential five-vector was defined as

$$\phi_i \equiv \pm \frac{\partial \ln f^{\frac{1}{2}}}{\partial x^i} \qquad (11.1)$$

and the field tensor is given by

$$F_{ij} \equiv \phi_{i,j} - \phi_{j,i}. \qquad (11.2)$$

Now the field tensor given by Equation (11.2) has 25 components. We would like to determine the field equations for these components. The quickest, though not the only, way is to consider the five dimensions to be $x^0 = ict$, $x^\alpha = x^\alpha$, $\alpha = 1,2,3,4$. The field tensor is then defined to be

$$F_{ij} = \begin{vmatrix} 0 & iE_1 & iE_2 & iE_3 & iV_4 \\ -iE_1 & 0 & B_3 & -B_2 & V_1 \\ -iE_2 & -B_3 & 0 & B_1 & V_2 \\ -iE_3 & B_2 & -B_1 & 0 & V_3 \\ -iV_4 & -V_1 & -V_2 & -V_3 & 0 \end{vmatrix}. \qquad (11.3)$$

Using Bianchi's identities

$$\frac{\partial F_{ij}}{\partial x^k} + \frac{\partial F_{jk}}{\partial x^i} + \frac{\partial F_{ki}}{\partial x^j} = 0$$

and the various combinations of the indices $0, 1, 2, 3, 4$ we obtain the field equations

Five Dimensional Fields

$$\overline{\nabla} \bullet \overline{B} = 0$$

$$\overline{\nabla} \times \overline{E} + \frac{1}{c}\frac{\partial \overline{B}}{\partial t} = 0$$

$$\overline{\nabla} \times \overline{V} + a_0 \frac{\partial \overline{B}}{\partial \gamma} = 0 \tag{11.4}$$

$$\overline{\nabla} V_4 + \frac{1}{c}\frac{\partial \overline{V}}{\partial t} + a_0 \frac{\partial \overline{E}}{\partial \gamma} = 0.$$

The definition of the five-vector current density

$$\frac{\partial F_{ij}}{\partial x^i} \equiv \frac{4\pi}{c} J_i \tag{11.5}$$

yields the equations

$$\overline{\nabla} \bullet \overline{E} + a_0 \frac{\partial V_4}{\partial \gamma} = 4\pi\rho$$

$$\overline{\nabla} \times \overline{B} - \frac{1}{c}\frac{\partial \overline{E}}{\partial t} + a_0 \frac{\partial \overline{V}}{\partial \gamma} = \frac{4\pi \overline{J}}{c} \tag{11.6}$$

$$\overline{\nabla} \bullet \overline{V} + \frac{1}{c}\frac{\partial V_4}{\partial t} = -\frac{4\pi J_4}{c}.$$

In addition to these field equations there is the statement of conservation of charge where

$$\frac{\partial J_i}{\partial x^i} = 0, \quad i = 0,1,2,3,4,$$

so that

$$\frac{\partial \rho}{\partial t} + \overline{\nabla} \bullet \overline{J} + a_0 \frac{\partial J_4}{\partial \gamma} = 0. \tag{11.7}$$

For ease in future reference to these eight field equations they may be rewritten as

The Dynamic Theory

$$\overline{\nabla}\bullet\overline{B} = 0 \qquad [a]$$

$$\frac{1}{c}\frac{\partial\overline{B}}{\partial t}+\overline{\nabla}\times\overline{E} = \overline{0} \qquad [b]$$

$$\overline{\nabla}\times\overline{B}-\frac{1}{c}\frac{\partial\overline{E}}{\partial t}+a_0\frac{\partial\overline{V}}{\partial\gamma} = \frac{4\pi\overline{J}}{c} \qquad [c]$$

$$\overline{\nabla}\bullet\overline{E}+a_0\frac{\partial V_4}{\partial\gamma} = 4\pi\rho \qquad [d]$$

$$\frac{\partial\rho}{\partial t}+\overline{\nabla}\bullet\overline{J}+a_0\frac{\partial J_4}{\partial\gamma} = 0 \qquad [e]$$

$$\overline{\nabla}\times\overline{V}+a_0\frac{\partial\overline{B}}{\partial\gamma} = \overline{0} \qquad [f]$$

$$\overline{\nabla}V_0+\frac{1}{c}\frac{\partial\overline{V}}{\partial t} = a_0\frac{\partial\overline{E}}{\partial\gamma} \qquad [g]$$

$$\overline{\nabla}\bullet\overline{V}+\frac{1}{c}\frac{\partial V_4}{\partial t} = -\frac{4\pi J_4}{c} \qquad [h] \tag{11.8}$$

Energy-Momentum Tensor

If we follow the approach of relativistic electrodynamics, we may define the tensor $\{T\}$ in terms of the field tensor $\{F\}$ according to

$$T_{jk} \equiv \left(\frac{1}{4\pi}\right)\left[F_{j\ell}F^{\ell k}+\frac{1}{4}\delta_{jk}F_{st}F_{st}\right] \tag{11.9}$$

Using the field tensor to calculate the components of the energy-momentum tensor we find that the components are given by

$$T_{0\alpha} = \frac{-1}{4\pi}[(\overline{E}\times\overline{B})_{\alpha}+V_4V_{\alpha}] \ , \ \alpha = 1,2,3 \ ,$$

(11.10)

$$T_{00} = \frac{1}{8\pi}[E^2+B^2+V_4^2+V^2] \ , \tag{11.11}$$

$$T_{04} = \frac{i}{4\pi} [\,\overline{E} \bullet \overline{V}\,] \ , \tag{11.12}$$

$$T_{4\alpha} = \frac{1}{4\pi} [V_4 E_\alpha + (\overline{V} \bullet \overline{B})_\alpha] \ , \quad \alpha = 1, 2, 3 \ ,$$

(11.13)

$$T_{44} = \frac{1}{8\pi} [V_4^2 + B^2 - E^2 - V^2] \ , \tag{11.14}$$

and

$$T_{\alpha\beta} = \frac{1}{4\pi} \left\{ \begin{array}{l} E_\alpha E_\beta + B_\alpha B_\beta - V_\alpha V_\beta \\ -\dfrac{1}{2} \delta_{\alpha\beta} [\,E^2 + B^2 + V_4^2 - V^2\,] \end{array} \right\} \ , \tag{11.15}$$

where $\alpha, \beta = 1, 2, 3.$

Equations (11.8) form a set of eight Maxwell type equations which obviously reduce to Maxwell's four equations.

Five Dimensional Waves

The wave equations for the new gravitational field quantities may be derived using standard assumptions. First we have

$$\frac{\partial}{\partial t}(\overline{\nabla} \bullet \overline{V}) + \frac{1}{c}\frac{\partial^2 V_4}{\partial t^2} = \frac{4\pi}{c}\frac{\partial J_4}{\partial t}$$

$$= \overline{\nabla} \bullet \frac{\partial \overline{V}}{\partial t} + \frac{1}{c}\frac{\partial^2 V_4}{\partial t^2} \tag{11.16}$$

and

$$\overline{\nabla} \bullet (\overline{\nabla} V_4) + \frac{1}{c}\overline{\nabla} \bullet \frac{\partial \overline{V}}{\partial t} = -a_0 \overline{\nabla} \bullet \frac{\partial \overline{E}}{\partial \gamma} = \frac{1}{c}\overline{\nabla} \bullet \frac{\partial \overline{V}}{\partial t} + \overline{\nabla} \bullet \overline{\nabla} V \ .$$

Therefore, the gravitational scalar wave equation becomes

$$\nabla^2 V_4 - \frac{1}{c^2}\frac{\partial^2 V_4}{\partial t^2} = \frac{4\pi}{c^2}\frac{\partial J_4}{\partial t} - a_0 \overline{\nabla} \bullet \frac{\partial \overline{E}}{\partial \gamma} \ . \tag{11.17}$$

The Dynamic Theory

For the vector field we have:

$$\overline{\nabla}(\overline{\nabla}\bullet\overline{V}) + \frac{\overline{\nabla}}{c}\frac{\partial V_4}{\partial t} = -\overline{\nabla}(\frac{4\pi}{c}J_4)$$

and

$$\overline{\nabla}\times(\overline{\nabla}\times\overline{V}) + \nabla^2\overline{V} + \frac{1}{c}\frac{\partial}{\partial t}\overline{\nabla}V_4 = \frac{4\pi}{c}\overline{\nabla}J_4 \quad ;$$

therefore

$$\nabla^2\overline{V} - \frac{1}{c^2}\frac{\partial^2\overline{V}}{\partial t^2} = \frac{4\pi}{c}\overline{\nabla}J_4 + \frac{a_0}{c}\frac{\partial^2\overline{E}}{\partial t\partial\gamma}$$
$$+ a_0(\overline{\nabla}\times\frac{\partial\overline{B}}{\partial\gamma}) \quad . \tag{11.18}$$

But using [d] of Equations (11.8) in Equation (11.17) yields

$$\nabla^2 V_4 - \frac{1}{c^2}\frac{\partial^2 V_4}{\partial t^2} = \frac{4\pi}{c^2}\frac{\partial J_4}{\partial t}$$
$$- a_0\frac{\partial}{\partial\gamma}4\pi\rho - a_0\frac{\partial V_4}{\partial t} \tag{11.19}$$

and using [c] of Equations (11.8) in Equation (11.18) gives us

$$\nabla^2\overline{V} - \frac{1}{c^2}\frac{\partial^2\overline{V}}{\partial t^2} = \frac{4\pi}{c}\overline{\nabla}J_4 + a_0\frac{\partial}{\partial\gamma}\frac{4\pi}{c}\overline{J}$$
$$+ 2\frac{\partial\overline{E}}{\partial t} - a_0\frac{\partial\overline{V}}{\partial\gamma}. \tag{11.20}$$

Now the wave equations for the usual vector and scalar potentials are

$$\nabla^2\overline{A} - \frac{1}{c^2}\frac{\partial^2\overline{A}}{\partial t^2} = -\frac{4\pi}{c}\overline{J} \tag{11.21}$$

and

$$\Delta^2\phi - \frac{1}{c^2}\frac{\partial^2\phi}{\partial t^2} = -4\pi\rho \quad . \tag{11.22}$$

We may differentiate these with respect to the mass density and substitute them into our wave equations and get

Five Dimensional Fields

$$\nabla^2 V_4 - \frac{1}{c^2}\frac{\partial^2}{\partial t^2}V_4 = \frac{4\pi}{c^2}\frac{\partial J_4}{\partial t} + a_0^2\frac{\partial^2 V_4}{\partial \gamma^2} \qquad (11.23)$$

and

$$\nabla^2 \overline{V} - \frac{1}{c^2}\frac{\partial^2 \overline{V}}{\partial t^2} = \frac{4\pi}{c}\overline{\nabla} J_4$$

$$+ a_0\frac{\partial}{\partial \gamma}\left[2\frac{\partial \overline{E}}{\partial t} - a_0\frac{\partial \overline{V}}{\partial \gamma}\right] , \qquad (11.24)$$

where $V_4 \equiv V_4 + a_0\dfrac{\partial \phi}{\partial \gamma}$ and $\overline{V} = \overline{V} - a_0\dfrac{\partial \overline{A}}{\partial \gamma}$.

In this manner the eight field equations lead to wave equations involving the ten field components. When put into the usual 3-dimensional space and time wave equation form, the components are grouped into two groups. One group consists of field components directed perpendicular to the direction of wave propagation. This group may be called the "transverse" components and it consists of electric, magnetic and gravitational vector field components. The other group has no component perpendicular to the direction of wave propagation. This group of components may be called the "non-transverse" group. The non-transverse components include electric and gravitational vector fields and the gravitational scalar field. The appearance of the scalar component makes it difficult to term this group as "longitudinal" waves because no all of the components in the group are vector components in the direction of propagation. This is the basis of the name of "non-transverse" for this group.

The wave equations given by the Dynamic Theory may be summarized as:

$$\nabla^2 \overline{E} - \left(\frac{4\pi\mu\sigma}{c^2} \right) \frac{\partial \overline{E}}{\partial t} - \left(\frac{\mu\sigma}{c^2} \right) \frac{\partial^2 \overline{E}}{\partial t^2} + a_0^2 \frac{\partial^2 \overline{E}}{\partial \gamma^2} = 0,$$

$$\nabla^2 \overline{B} - \left(\frac{4\pi\mu\sigma}{c^2} \right) \frac{\partial \overline{B}}{\partial t} - \left(\frac{\mu\varepsilon}{c^2} \right) \frac{\partial^2 \overline{B}}{\partial t^2} + a_0^2 \frac{\partial^2 \overline{B}}{\partial \gamma^2} = 0,$$

$$\nabla^2 \overline{V} - \left(\frac{4\pi\mu\sigma_4}{c^2} \right) \frac{\partial \overline{V}}{\partial t} - \left(\frac{\mu\varepsilon}{c^2} \right) \frac{\partial^2 \overline{V}}{\partial t^2} + a_0^2 \frac{\partial^2 \overline{V}}{\partial \gamma^2} = \frac{a_0 4\pi\mu}{c} \left[\frac{\partial(\sigma \overline{E})}{\partial \gamma} - \sigma_4 \frac{\partial \overline{E}}{\partial \gamma} \right], \quad (11.25)$$

and

$$\nabla^2 V_4 - \left(\frac{4\pi\mu\sigma_4}{c^2} \right) \frac{\partial V_4}{\partial t} - \left(\frac{\mu\varepsilon}{c^2} \right) \frac{\partial^2 V_4}{\partial t^2} + a_0^2 \frac{\partial^2 V_4}{\partial \gamma^2} = 0.$$

In these wave equations the E and the B represent the normal electric and magnetic fields. The vector V field is the gravitational field while the scalar, V_4, is related to the gravitational potential. When the usual assumption of sinusoidal waves is made and a trial solution of the form $E_y = E_{yo}exp[\ -i(ct-k_yx-k_4\gamma)]$ transverse components are given by

$$B_z = \left(\frac{k_y c}{\omega} \right) E_y$$

$$V_y = \left(\frac{-a_0 c k_4}{\omega} \right) E_y \qquad (11.26)$$

$$J_y = \sigma_y E_y$$

the remaining non-transverse components are

$$V_x = \left[\frac{-\left(\mu\varepsilon\omega^2 + i4\pi\mu\omega\sigma_4\right)}{a_0 c\left(k_4 - i\dfrac{\partial\ln\varepsilon}{\partial\gamma}\right)} \right] E_x$$

$$V_4 = \left[\frac{-k_x c}{a_0 c\left(k_4 - i\dfrac{\partial\ln\varepsilon}{\partial\gamma}\right)} \right] E_x \qquad (11.27)$$

$$J_x = \sigma_x E_x$$

$$J_4 = -\sigma_4 V_4$$

with three indicial relations

$$k_y^2 c^2 = \mu\varepsilon\omega^2 + i4\pi\mu\omega\sigma_4 - a_0^2 c^2 k_4\left(k_4 - i\frac{\partial\ln\varepsilon}{\partial\gamma}\right)$$

$$k_x^2 c^2 \left[\frac{k_4 + i\dfrac{\partial\ln\mu}{\partial\gamma}}{k_4 - i\dfrac{\partial\ln\varepsilon}{\partial\gamma}} \right] = \mu\varepsilon\omega^2 + i4\pi\mu\omega\sigma_x$$

$$\qquad (11.28)$$

$$-a_0^2 c^2 k_4\left(k_4 + i\frac{\partial\ln\mu}{\partial\gamma}\right)$$

$$k_x\sigma_x = \frac{\partial}{\partial\gamma}\left[\frac{i\sigma_4 k_x}{k_4 - i\dfrac{\partial\ln\varepsilon}{\partial\gamma}} \right] + \left[\frac{k_4\sigma_4 k_x}{k_4 - i\dfrac{\partial\ln\varepsilon}{\partial\gamma}} \right]$$

where it may be seen that the non-transverse components are independent of the transverse components.

Energy Density

The field energy density may be defined by

$$\xi \equiv \frac{1}{8\pi}\left[\overline{E}\bullet\overline{E} + \overline{B}\bullet\overline{B} + \overline{V}\bullet\overline{V} + V_0^2 \right] \ , \quad (11.29)$$

and the electrical Poynting vector may be defined by

The Dynamic Theory

$$\overline{S}_E \equiv \frac{c}{4\pi}(\overline{E} \times \overline{B}) \ . \qquad (11.30)$$

Now the electrical Poynting vector represents the outward flow of the electromagnetic field energy through a surface. If we take the total vector, whose components are $T_{0\alpha}$, to be the total flow of energy, then the vector with components $(c/4\pi)V_4 V_\alpha$ must be the outward flow of energy due to changes of the mass density within the surface. Let us designate the mass energy vector as

$$\overline{S}_m \equiv \frac{c}{4\pi}(V_4 \overline{V}) \ , \qquad (11.31)$$

so that the total energy vector is

$$\overline{S} = \overline{S}_E + \overline{S}_m \qquad (11.32)$$

whose components are

$$S_\alpha = \frac{c}{4\pi}[\,(\overline{E} \times \overline{B})_\alpha + V_4 V_\alpha\,] = -(\frac{c}{i})T_{0\alpha} \ .$$

(11.33)

Neutrinos

What sort of particle would be associated with the non-transverse waves? The unity scale factor would certainly quantize these waves through the gauge potentials. However, this time the quantization will come through the gravitational potential. This may be seen by writing the scale factor for the five dimensional gauge potentials, or

$$\int \tilde{\phi}_j dx^j = 2\pi i N \qquad (11.34)$$

where now the $j=0,1,2,3,4$. Again we may allow all gauge potentials to vanish except the one of interest. In this case we set all gauge potentials to zero except $\tilde{\phi}_4$. Then the quantization required by Weyl's Quantum Principle now quantizes the gravitational potential. This quantization then quantizes the non-transverse waves in Equation (11.27).

Five Dimensional Fields

We now have a non-transverse particle. This non-transverse particle does not readily interact with electrons in the atoms that it passes for it cannot accelerate them in the direction the wave travels and because this particle must travel at the speed of light. This particle has all of the attributes we currently assign to a neutrino.

Recalling the determination of the phat photon energy and our neutrino, we see that the mass energy vector determines the energy of the neutrino just as the electric Poynting vector determines the energy of the photon.

Weyl's Quantum Principle requires that the gauge potentials be quantized since

$$\int \tilde{\phi}_j dx^j = 2\pi i N \qquad (11.35)$$

where $i^2 = -1$ and there is no summation over the j's and this must be true for any path whatsoever. If the quantum principle must hold for any path, then by choosing all dx^j to be zero except one, it may be seen that each gauge potential must be quantized and, therefore, the principle was written as

$$\int N_j \phi_j dx^j = 2\pi i N \qquad (11.36)$$

The gravitational potential may be investigated by considering N_4 to be nonzero and $N_0 = N_x = N_y = N_z = 0$. The concept of a neutrino, as a particle, is one that is electrically neutral. The wave solution above shows that the gravitational gauge potential gives rise to the non-transverse waves. The quantum principle requires that the gravitational potential component of the gauge potential to be quantized, as

$$\phi_4 = N_4 B \cos 2\pi \left(\frac{x}{\lambda} - vt + k_4 \gamma \right) \qquad (11.37)$$

where the dependence upon x, t and y was chosen to ensure that the electric field, given by

$$E_x(x,t,\gamma) = \frac{-a_o \partial \phi_4(x,t,\lambda)}{\partial(\gamma)}$$

$$= -2\pi a_o N_4 k_4 B \sin 2\pi\left(\frac{x}{\lambda} - vt + k_4\gamma\right)$$

(11.38)

satisfies the wave equations. This expression may be used to find the average value of the mass Poynting vector, or

$$I_n = \left(\frac{c}{4\pi}\right)\langle V_4 V_x\rangle = \left(\frac{c}{4\pi}\right)\left\langle\left[\frac{-k_x c}{a_o c k_4}\right]\left[\frac{\mu\varepsilon\omega^2}{a_o c k_4}\right]E_x^2\right\rangle$$

$$= \frac{-k_x c}{4\pi a_o k_4}\left(\frac{\mu\varepsilon\omega^2}{a_o c k_4}\right)\langle E_x^2\rangle$$

(11.39)

The average value of the square of non-transverse electric field is given by

$$\langle E_x^2(x,t)\rangle = \left(2\pi a_o N_4 k_4 B\right)^2 \int_{t=0}^{t=\frac{1}{v}} \sin^2\left(\frac{x}{\lambda} - vt + k_4\gamma\right)dt$$

(11.40)

$$= \frac{\left(2\pi a_o N_4 k_4 B\right)^2}{v}$$

Therefore, neutrino energy flow is given by

$$I = \frac{-k_x c}{4\pi a_o k_4}\left(\frac{\mu\varepsilon\omega^2}{a_o c k_4}\right)\frac{\left(2\pi a_o N_4 k_4 B\right)^2}{v}$$

(11.41)

$$= -k_x c^2 4\pi^3 B^2 N_4^2 v$$

Now a neutrino for which $N_4=1$ have an would energy flow of $I=h_n v$ when

$$B = \sqrt{\frac{h_n}{4\pi^3 k_x c^2}}$$

(11.42)

In general the neutrino energy would be given by

$$\varepsilon_n = N_4^2 h_n v$$

(11.43)

which is identical in form to the expression for the energy of the phat photon. However, the value remains unspecified because the non-transverse propagation constant, k_x, is, as yet, unspecified.

Having obtained the prediction of neutrino energy one may predict a neutrino mass using the Einstein energy expression and see that the neutrino mass would be

$$m_n = \frac{\varepsilon_n}{c^2} = \frac{N_4^2 h_n v}{c^2}.$$ (11.44)

One may see by this prediction of neutrino energy that while neutrino emission is not required during neutron breakup neutrinos are required for the non-transverse waves. It is a rather interesting turn of events to see that neutrinos are the result of gravitational phenomena and not the result of nuclear phenomena. This argues that the neutrino experiments have been measuring gravitational waves. More precisely we perhaps should call these waves electrogravitic waves due to their electric and gravitational components and in analogy to the electromagnetic waves that have both electric and magnetic components.

Thus, in the five dimensional gauge fields there are two types of waves predicted. First, there are the transverse electromagnetic waves that get an additional gravitational component carrying a very small amount of energy. When the electromagnetic waves are quantized by the unity scale factor requirement the particles may be called phat photons. Secondly, there are the non-transverse electrogravitic waves that are accompanied by a scalar gravitational potential wave. When the electrogravitic waves are quantized the particle may be called a neutrino. Though the five dimensional gauge fields produce Einstein's general theory of relativity when mass is conserved, they do not produce a purely gravitational wave. The inductive coupling with the electromagnetic fields precludes a p ure gravitational wave made up of only gravitational components.

Stress tensor

The Dynamic stress tensor may be defined as the three-dimensional tensor whose elements are

$$T^{D}_{\alpha\beta} = \frac{1}{4\pi}\left\{\begin{array}{c} E_\alpha E_\beta + B_\alpha B_\beta - V_\alpha V_\beta \\ -\frac{1}{2}\delta_{\alpha\beta}\,[\,\xi - 2V^2\,] \end{array}\right\}. \qquad (11.45)$$

The Maxwell stress tensor is defined in electrodynamics as the three-dimensional tensor with elements

$$T^{M}_{\alpha\beta} = \frac{1}{4\pi}\left\{ E_\alpha E_\beta + B_\alpha B_\beta - \frac{1}{2}\delta_{\alpha\beta}\,[\,E^2 + B^2\,]\right\}.$$

(11.46)

In terms of the Maxwell stress tensor, the Dynamic stress tensor may be written as

$$T^{D}_{\alpha\beta} = T^{M}_{\alpha\beta} - \{V_\alpha V_\beta - \frac{1}{2}\delta_{\alpha\beta}\,[\,V^2 - V_4^2\,]\} . \qquad (11.47)$$

Then in terms of the above defined quantities

$$\{T\} = \begin{pmatrix} \xi & -\frac{1}{c}\overline{S} & \frac{i}{4\pi}(\overline{E}\bullet\overline{V}) \\ -\frac{i}{c}\overline{S} & \{T^D\} & \frac{i}{4\pi}V_4\overline{E}+(\overline{V}\times\overline{B})] \\ \frac{i}{4\pi}(\overline{E}\bullet\overline{V}) & [V_4\overline{E}+(\overline{V}\times\overline{B})] & \frac{1}{8\pi}[V_4^2 + B^2 - E^2 - V^2] \end{pmatrix}.$$

(11.48)

Suppose we calculate the trace of the energy-momentum tensor:

Five Dimensional Fields

$$t_R\{T\} = T_{jj} = T^D_{jj} + \xi + \frac{1}{8\pi}\left(V_4^2 + B^2 - E^2 - V^2\right)$$

$$= \frac{1}{4\pi}\left(B^2 + V_4^2\right) + T^4_{\alpha\alpha}$$

$$= \frac{1}{4\pi}\left(B^2 + V_4^2\right) + \frac{1}{4\pi}\left[\begin{array}{c} E^2 + B^2 - V^2 \\ -\frac{3}{2}(E^2 + B^2 + V_4^2 - V^2) \end{array}\right] \qquad (11.49)$$

$$= \frac{1}{4\pi}\left(\frac{1}{2}B^2 - \frac{1}{2}V^2 - \frac{1}{2}V_4^2\right)$$

$$= \frac{1}{8\pi}\left(B^2 + V^2 - E^2 - V_4^2\right) \ .$$

Force Density Vector

The force density vector may be defined in terms of the divergence of the energy-momentum tensor. Therefore, suppose we calculate the five dimensional divergence of the tensor $\{T\}$, or

$$\frac{\partial T_{jk}}{\partial x^k} = \frac{1}{4\pi}\frac{\partial}{\partial x^k}\left[F_{j\ell}F_{\ell k} + \frac{1}{4}\delta_{jk}F_{st}F_{st}\right] \ . \qquad (11.50)$$

Because of the anti-symmetry of F_{jk}, the first term may be written as

$$\frac{\partial F_{j\ell}}{\partial x^k}F_{\ell k} = \frac{\partial F_{\ell j}}{\partial x^k}F_{k\ell} \ . \qquad (11.51)$$

By interchanging the indices k and l

$$\frac{\partial F_{j\ell}}{\partial x^k}F_{\ell k} = \frac{\partial F_{kj}}{\partial x^\ell}F_{\ell k} = \frac{1}{2}\left(\frac{\partial F_{j\ell}}{\partial x^k} + \frac{\partial F_{j\ell}}{\partial x^\ell}\right)F_{\ell k} \ . \qquad (11.52)$$

Using the Bianchi identity

$$\frac{\partial F_{j\ell}}{\partial x^k} + \frac{\partial F_{kj}}{\partial x^\ell} + \frac{\partial F_{\ell k}}{\partial x^j} = 0 \ ,$$

the terms contained within the parentheses of Equation (11.52) may be written as

The Dynamic Theory

$$\frac{\partial F_{j\ell}}{\partial x^k}F_{\ell k}=\frac{1}{2}\frac{\partial F_{\ell k}}{\partial x^j}F_{\ell k}=-\frac{1}{4}\frac{\partial(F_{\ell k}F_{\ell k})}{\partial x^i} \ .$$

Substituting this back into the expression for the divergence, the last term will be canceled because l, k, s, and t are dummy indices. Then the divergence becomes

$$\frac{\partial T_{jk}}{\partial x^k}=\frac{1}{4\pi}F_{j\ell}\frac{\partial F_{\ell k}}{\partial x^k} \ . \tag{11.53}$$

By interchanging the indices k and l on the right-hand side we obtain

$$\frac{\partial T_{jk}}{\partial x_k}=\frac{1}{4\pi}F_{jk}\frac{\partial F_{k\ell}}{\partial x^\ell} \ . \tag{11.54}$$

The Dynamic force density five-vector may now be defined as

$$K \equiv Div_{\ 5}\{T\} \ .$$

Therefore, the components of K are given by

$$K_j=\frac{1}{4\pi}F_{jk}\frac{\partial F_{k\ell}}{\partial x^\ell} \ . \tag{11.55}$$

But the five-vector current density is given by

$$\frac{\partial F_{k\ell}}{\partial x^\ell}=\frac{-4\pi}{c}J_k \ . \tag{11.56}$$

The components of the five-vector force density become

$$K_j=\frac{-1}{4\pi}F_{kj}\left(\frac{4\pi}{c}J_k\right)=\frac{-1}{c}J_kF_{kj} \ . \tag{11.57}$$

Now, since $J_k = (ic\rho, J, J_4)$, then

$$K_0=\frac{i}{c}[\overline{J}\bullet\overline{E}+J_4V_4] \ ,$$

$$\overline{K}=\rho[\overline{E}+\frac{1}{c}(\overline{u}\times\overline{B})]+\frac{J_4}{c}\overline{V} \tag{11.58}$$

and

$$K_4=\rho V_4-\frac{\overline{J}\bullet\overline{V}}{c} \ , \tag{11.59}$$

where $J=\rho u$. These then are the components of the force density five-vector resulting from a gauge field in the

Five Dimensional Fields

Dynamic Theory. These components reduce to the four components of the Lorentz force density should $V_4=V=0$.

With the interpretation that the four force density components with subscript 1 through 4 are the force density vectors which appear in the First Law as F_α , then the force density vector provides the connection between the First Law and the geometry of the sigma manifold discussed previously. Thus, the existence of the vector field φ_i is also demanded by the Dynamic Theory and need not exist as a separate assumption.

Energy Flow

Consider the zeroth component of the Dynamic force density five-vector

$$K_0 = \frac{\partial T_{0k}}{\partial x^k} = \frac{\partial T_{00}}{\partial x^0} + \frac{\partial T_{0\alpha}}{\partial x^\alpha} + \frac{\partial T_{04}}{\partial x^4} \ .$$

Then

$$\frac{i}{c}[\overline{J} \bullet \overline{E} + J_4 V_4] = \frac{\partial \xi}{\partial(ict)} - \frac{i}{c}\frac{\partial S_\alpha}{\partial x^\alpha} + \frac{i}{4\pi}\frac{\partial(\overline{E} \bullet \overline{V})}{\partial x^4}$$

or

$$\frac{1}{c}[\overline{J} \bullet \overline{E} + J_4 V_4] = \frac{1}{c}\frac{\partial \xi}{\partial t} + \frac{1}{c}\frac{\partial S_\alpha}{\partial x^\alpha} - \frac{1}{4\pi}\frac{\partial(\overline{E} \bullet \overline{V})}{\partial x^4} \ ,$$

or, since $x^4 = \gamma/a_0$,

$$-\overline{J} \bullet \overline{E} - J_4 V_4 = \frac{\partial \xi}{\partial t} + \frac{\partial S_\alpha}{\partial x^\alpha} - \frac{a_0 c}{4\pi}\frac{\partial(\overline{E} \bullet \overline{V})}{\partial \gamma} \ .$$

Rearranging the terms

$$div\ \overline{S} + \frac{\partial \xi}{\partial t} = -\overline{J} \bullet \overline{E} - J_4 V_4 + \frac{a_0 c}{4\pi}\frac{\partial(\overline{E} \bullet \overline{V})}{\partial \gamma}$$

and separating out the electrical Poynting vector leads to

$$div\ \overline{S}_E + \frac{\partial \xi}{\partial t} = -\overline{J} \bullet \overline{E} - J_4 V_4 - div\ \overline{S}_m$$

$$+ \frac{a_0 c}{4\pi}\frac{\partial(\overline{E} \bullet \overline{V})}{\partial \gamma} \ . \tag{11.60}$$

352

This then is the five-dimensional energy flow equation.

Momentum Conservation

The expression for the conservation of momentum may be obtained from the space portion of the force density five-vector

$$K = \frac{\partial T_{\alpha k}}{\partial x^k} = \partial \frac{T_{\alpha 0}}{\partial x^0} + \frac{\partial T_{\alpha \beta}}{\partial x^\beta} + \frac{\partial T_{\alpha 4}}{\partial x^4}, \quad \alpha, \beta = 1, 2, 3.$$

But $\dfrac{\partial T_{\alpha \beta}}{\partial x^\beta}$ is the three-dimensional divergence of the Dynamic stress tensor $\{T^D\}$, therefore,

$$\overline{K} = -\frac{1}{c^2} \frac{\partial \overline{S}}{\partial t} + div \, \{T^D\} + \frac{a_0}{4\pi} \frac{\partial}{\partial \gamma} [V_4 \overline{E} + (\overline{V} \times \overline{B})] \quad . (11.61)$$

If we consider a volume in which all the material is contained and outside of which the field vanishes, then integrating over this volume yields

$$\int_v \left\{ \begin{array}{c} \overline{K} + \dfrac{1}{c^2} \dfrac{\partial \overline{S}}{\partial t} \\[2ex] -\dfrac{a_0}{4\pi} \dfrac{\partial}{\partial \gamma} [V_4 \overline{E} + (\overline{V} \times \overline{B})] \end{array} \right\} dv = \int_v div\{T^D\} \, dv \quad . (11.62)$$

The integral of K gives the total force (i.e., the time derivative of the mechanical momentum p less the vector $\left(\dot{\gamma} v^\alpha / \sqrt{1 - \beta^2} \right)$. Now define the vector

$$\overline{g} \equiv \frac{\overline{S}}{c^2} - \frac{a_0}{4\pi} \int \{ \frac{\partial}{\partial \gamma} [V_0 \overline{E} + (\overline{V} \times \overline{B}) + \frac{\dot{\gamma} v^\alpha}{\sqrt{1 - \beta^2}} \} dt \quad . (11.63)$$

Then define $\int_v \overline{g} \, d\sigma \equiv \overline{G}$, so that

$$\frac{d}{dt} (\overline{p} + \overline{G}) = \int_v div\{T^D\} \, dv \quad .$$

Using the divergence theorem the volume integral may be converted to a surface integral so that

$$\frac{d}{dt}(\overline{p}+\overline{G}) = \int_s \{T^D\} \bullet \overline{n}\, da \quad . \qquad (11.64)$$

If the field vanishes outside of V, it must do so also on the boundary surface s, hence

$$\frac{d}{dt}(\overline{p}+\overline{G}) = 0 \quad . \qquad (11.65)$$

Therefore, it is not the mechanical momentum p but the quantity $p+G$ which is conserved. Therefore, we must interpret G as the momentum of the field and

$$\overline{g} = \frac{\overline{S}}{c^2} - \frac{a_0}{4\pi}\int\left(\frac{\partial}{\partial\gamma}\left[\overline{V}\bullet\overline{E}+(\overline{V}\times\overline{B})\right]+\frac{4\pi\dot\gamma\overline{v}}{\sqrt{1-\beta^2}}\right)dt \qquad (11.66)$$

as the momentum density of the field.

Gauge Field Pressure

Now the gauge field pressure may be derived. The Dynamic stress tensor is given by

$$T^D_{\alpha\beta} = \frac{1}{4\pi}\left\{ \begin{array}{l} E_\alpha E_\beta + B_\alpha B_\beta - V_\alpha V_\beta \\[4pt] -\frac{1}{2}\delta_{\alpha\beta}\left[E^2+B^2+V_4^2-V^2\right] \end{array} \right\} \qquad (11.67)$$

$$= -\frac{1}{8\pi}\left[E^2+B^2-V^2\right]-\frac{3}{8\pi}V_4^2 \quad .$$

Now separate the three-dimensional dynamic stress tensor into a traceless and an isotropic tensor, or

$$T_{\alpha\beta}^{D} = \frac{1}{4\pi}\left\{ \begin{array}{l} E_\alpha E_\beta + B_\alpha B_\beta - V_\alpha V_\beta \\ -\frac{1}{2}\gamma_{\alpha\beta}\left[E^2 + B^2 + V_4^2 - V^2 \right] \end{array} \right\}$$

$$= \frac{1}{4\pi}\left\{ \begin{array}{l} E_\alpha E_\beta + B_\alpha B_\beta - V_\alpha V_\beta \\ -\frac{1}{8\pi}\delta_{\alpha\beta}\left[E^2 + B^2 + V_4^2 - V^2 \right] \end{array} \right\}$$

$$= \frac{1}{4\pi}\left[E_\alpha E_\beta + B_\alpha B_\beta - V_\alpha V_\beta \right]$$

$$-\left(\frac{2}{3}\right)\left(\frac{1}{8\pi}\right)\delta_{\alpha\beta}\left[E^2 + B^2 - V^2 \right]$$

$$-\left(\frac{1}{3}\right)\left(\frac{1}{8\pi}\right)\delta_{\alpha\beta}\left[E^2 + B^2 + 3V_4^2 - V^2 \right]$$

$$\equiv t_{\alpha\beta} + \tau'_{\alpha\beta}$$

(11.68)

where

$$t_{\alpha\beta} \equiv \frac{1}{4\pi}\left\{ E_\alpha E_\beta + B_\alpha B_\beta + V_\alpha V_\beta - \left(\frac{1}{3\pi}\right)\delta_{\alpha\beta}\left[E^2 + B^2 - V^2 \right] \right.$$

and

$$\tau'_{\alpha\beta} \equiv \left(\frac{1}{24\pi}\right)\delta_{\alpha\beta}\left[E^2 + B^2 + 3V_4^2 - V^2 \right] \; .$$

Now

$$t_r\{t_{\alpha\beta}\} = \left(\frac{1}{4\pi}\right)\left[E^2 + B^2 - V^2 - \left(E^2 + B^2 - V^2 \right) \right] \equiv 0$$

and

$$t_r\{\tau'_{\alpha\beta}\} = \left(\frac{1}{8\pi}\right)\left[E^2 + B^2 + 3V_4^2 - V^2 \right] \; . \quad (11.69)$$

Consider the definition

$$\frac{1}{3} t \delta_{\alpha\beta} \equiv \tau'$$

$$= \left(\frac{1}{24\pi}\right) \delta_{\alpha\beta} \left[E^2 + B^2 + 3V_4^2 - V^2 \right] \ .$$

(11.70)

Then

$$t = \left(\frac{1}{8\pi}\right) [E^2 + B^2 + 3V_4^2 - V^2] \qquad (11.71)$$

and

$$\tau'_{\alpha\beta} = \begin{pmatrix} t & 0 & 0 \\ 0 & t & 0 \\ 0 & 0 & t \end{pmatrix}$$

The isotropic part of the stress tensor is usually called the "pressure."

Therefore, define $3p \equiv t$ in accordance with customary notation, so that

$$p = \left(\frac{1}{24\pi}\right)\left[E^2 + B^2 + 3V_4^2 - V^2 \right] \ . \qquad (11.72)$$

With the exception of the factor of 3 this reduces to the "radiation pressure" for an electromagnetic field when $V=V_4=0$.

Zero pressure radiation

One may note that the pressure given by Equation (11.72) may be zero since it is the sum and difference of squares, or $p = 0$, when

$$V^2 = E^2 + B^2 + 3V_4^2 \ . \qquad (11.73)$$

This may prove to be an important point when considering boundary conditions in cosmology or the study of elementary particles.

For example, the five dimensional gauge wave equations lead to an energy density for transverse waves

The Dynamic Theory

given by Equation (11.29) that is strictly the sum of squares of the ten wave components, or

$$\xi \equiv \frac{1}{8\pi}\left[\overline{E}\bullet\overline{E}+\overline{B}\bullet\overline{B}+\overline{V}\bullet\overline{V}+V_4^2\right].$$

On the other hand, the radiation pressure is given by Equation (11.72)

$$p=\left(\frac{1}{24\pi}\right)\left[E^2+B^2+3V_4^2-V^2\right]$$

which displays the fact that there may be non-zero wave components at zero pressure.

For wave solutions of the form

$$\overline{E}(\overline{x},t,\gamma)=\overline{E}_o\sin\left(\overline{k}\bullet\overline{x}-\omega t+k_4\gamma\right)$$

$$=\overline{E}_o\sin 2\pi\left(\frac{x}{\lambda}-vt+\frac{\gamma}{\eta}\right)$$

It may be shown that $\overline{V}=-\dfrac{a_o c}{\eta 2\pi v}\overline{E}$ and $\overline{B}=\dfrac{\overline{k}c\times\overline{E}}{2\pi v}$ and

$$k^2 c^2=\mu\varepsilon\omega^2-a_o^2 c^2 k_4\left(k_4+\frac{\partial\ln\mu}{\partial\gamma}\right).$$

This leads to the situation that for vanishing radiation pressure, $p=0$, the radiation energy density does not vanish, but is given by

$$\xi=\frac{1}{8\pi}2\left(\overline{V}\bullet\overline{V}\right).$$

The wave solutions may be used to determine the ratio of the radiation pressure to the energy density, which is

$$\frac{p}{\xi}=1-\frac{a_o^2 c^2 k_4^2}{\omega^2}. \tag{11.74}$$

We might get a sense of what this means by taking the data from the Nichols and Hull experiments, assuming no experimental error, as accounting for the reduction of pressure compared with the energy density. We find the value of $a_ock_4\approx4.1x10^{13}$ sec^{-1} when $\lambda=550$ nanometers. The expression for the ratio at zero radiation pressure requires that $a_ock_4=\omega$.

In writing about cosmology no one has offered a reasonable explanation for a means for support for a non-zero background radiation pressure. By using Equation (11.29) and Equation (11.72) one may see that a non-zero radiation pressure is not required. Rather, one sees that in a zero radiation pressure boundary condition a non-zero energy density is required. It would appear that herein lays an explanation for the cosmological background radiation without associating such radiation with a hot big bang theory. This provides an alternative explanation for experimental data currently taken to be primary supporting evidence for the big bang. An alternative explanation for this experimental data weakens its support for the big bang.

Alternate Communication Concept

The prediction of non-transverse components independent of the transverse components gives rise to the question as to whether or not a communication system might be devised based upon t he non-transverse components alone.

Such a communication system was investigated by the author under contract to Lockheed as a covert communication system for the then developing stealth fighter. The author failed to design an antenna for such waves. This failure might have been expected of someone trained in the design of electric field antennae as I had been so trained.

Consider, for a moment, the differences between the transverse waves and the non-transverse waves. While they both have an electric field component only the transverse waves may cause electrons to move as the transverse electric field may accelerate the electrons in a direction 90° from the direction of wave propagation. This is the basis the common electromagnetic wave antennae. It might be called a wave-particle antennae for a wave moves a particle and this particle motion may be detected. This would be the operation of a receiving antenna. The reciprocal, or sending, antenna works on the ability of a moving particle to create a wave. This also might be called a wave-particle antenna. On the other hand, the non-transverse waves cannot cause a charged particle to move since the electric field would try to cause the particle to move in the direction of wave propagation. Such motion is impossible for the wave is itself travelling at the speed of light and cannot interact with a charged particle so as to accelerate it in the direction of wave propagation. Therefore, a wave-particle type of antenna cannot detect non-transverse waves.

After several years of intermittent investigations, the author came to the conclusion that the inertia of the charged particle was the main stumbling block in the way towards a working antenna design. A field-field antenna seemed necessary as fields did not have inertia. Further, intermittent research revealed that an appropriately designed interface between two different materials with carefully selected indices of refraction could be placed at an certain angle with respect to a polarized transverse wave could cause some of the transverse wave energy to emerge as non-transverse wave energy. This then became the model for a field-field antenna design. A sending antenna would convert transverse wave components into non-transverse wave components in a manner similar to filtering light. The receiving antenna would reverse the operation. Incoming non-transverse waves would strike the appropriately designed interface and have some of the non-transverse wave energy converted into transverse wave energy that could then be used to cause charged particles to move and be detected in the usual fashion.

A time varying electric field can create a magnetic field without involving inertia. This is, of course, the concept used in an electric motor. The electric generator uses a magnetic field to produce an electric field. In both of these cases one field has been used to create another field directly even though particles are used to create the electric field in the motor and the electric field in the generator then moves a p article. These might be called field-field antennas.

A communication system using non-transverse waves needs a field-field antenna since fields do not have inertia. Further, an appropriately designed interface between two different materials with carefully selected indices of refraction could be placed at a certain angle with respect to a polarized transverse wave could cause some of the transverse wave energy to emerge as non-transverse wave energy. This then became the model for a field-field antenna design. A sending antenna would convert transverse wave components into non-transverse wave components in a manner similar to filtering light. The receiving antenna would reverse the operation. Incoming non-transverse waves would strike the appropriately designed interface and have some of the non-transverse wave energy converted into transverse wave energy that could then be used to cause charged particles to move and be detected in the usual fashion.

The mechanical means for making a receiving field-field antenna is one in which causes a non-transverse wave to create a transverse wave. Placing an interface, such as a special glass with a specific index of refraction, at an angle to the incoming non-transverse wave, may do this. The boundary conditions discussed below will involve the indices of refraction of the different media involved as well as the angle between the incoming non-transverse wave and the plane of the interface to cause the incoming wave to be split into reflected and refracted waves. Though the predominate energy in the refracted wave will be in another non-transverse wave there is an angle of incidence that will maximize the refracted transverse wave for a given set of indices of refraction. This refracted transverse wave may then be detected in the usual fashion. A field-field antenna for the transmission of non-transverse waves would essentially be the reverse of the receiving antenna. Current means of creating transverse waves may be employed to create the input to the sending field-field antenna where, again, a special media, chosen for its index of refraction, is placed at an angle to the polarized transverse input wave. The boundary conditions then require a small amount of the refracted energy to appear in a non-transverse wave.

A communication system based upon only the non-transverse components would represent an alternate to the currently used EM radio antennas. One advantage to such as alternate communication system would be a lack of interference from current EM systems. A second advantage may be extended ranges. Further, almost all of the current communication systems' components could be used in the alternate system with the exception of the antenna.

Boundary Conditions

The alternate communication concept requiring the conversion of transverse components into non-transverse components and back into transverse components uses the reflection and refraction of waves at a boundary between two materials with different indexes of refraction. In order to establish the boundary conditions for the combination of transverse and non-transverse components at such a boundary, new field and material concepts arise. For example, where the electric displacement is related to the electric field through the concept of electric polarization for Maxwellian waves the Electromagnetogravtic waves requires the addition of the concept of gravitational polarization such that:

$$\overline{D} = \overline{E} + 4\pi\overline{P} + 4\pi\overline{P}_g . \tag{11.75}$$

Five Dimensional Fields

Additionally, the magnetic intensity is related to the magnetic induction field by both a magnetic susceptibility and a gravitational susceptibility as:

$$\bar{H} = \bar{B} - 4\pi\left(\bar{M} + \bar{M}_g\right) = \bar{B} - 4\pi\left(\chi_m\bar{H} + \chi_g\bar{H}\right). \qquad (11.76)$$

Further, the gravitational intensity is related to the gravitational induction field in a similar fashion so that:

$$\frac{\bar{V}}{\beta} = \mu_g\bar{G} = \bar{G} + 4\pi\chi_g\bar{G} = \bar{G} + 4\pi\bar{N} \qquad (11.77)$$

where N has not yet been given a name.

These concepts may be used in determining the boundary conditions at a material interface. The derivations of the boundary conditions are contained in Appendix B. the boundary conditions may be listed here as:

1. the normal component $)$ is continuous,
2. the tangential component of $\bar{E}$ must be continuous,
3. the normal component of $\bar{B}$ is continuous across the boundary,
4. the tangential component of $\bar{H}$ must be continuous across the boundary
5. the normal component of the gravitational field $\bar{G}$ must be continuous across the boundary,
6. the tangential component of $\bar{V}$, the gravitational induction, must be continuous across the boundary, and
7. the scalar, V_4, must be continuous across the boundary because it is a scalar.

It perhaps should be noted that the new physical notions that appear in the foregoing could prove extremely important should one consider going into materials development.

Reflection and Refraction

In order to establish whether transverse wane energy may be converted into non-transverse wave energy and back again into transverse wave energy we must develop the equations governing the reflection and refraction of these waves. First, we may start with a pure transverse wave impinging upon an interface between two different mediums to see if there exists a polarization of the transverse wave that will convert at least some of the transverse wave energy into non-transverse wave energy. Then we will need to determine if another interface will convert some of the non-transverse wave energy into transverse wave energy.

We shall now consider the case where the incident wave impinges upon the boundary interface at an oblique angle θ_o. The wave is polarized so that the electric component is parallel with the interface (perpendicular to the plane of incidence). See Figure 11.1.

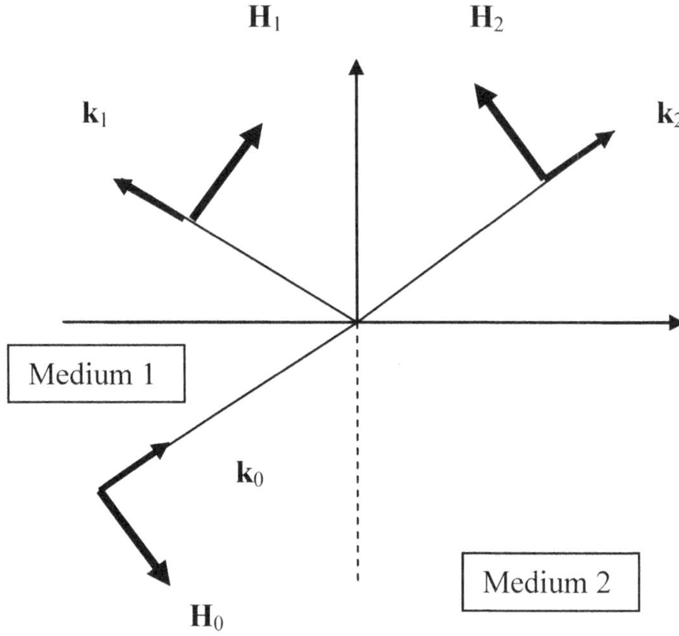

Figure 11.1. Reflection and refraction – oblique incidence – parallel magnetic field. The xz-plane is the plane of incidence. The vectors E_0 and E_1 are directed out of the paper, and E_2 is directed in.

For the incident wave we have

$$\left.\begin{aligned}
\overline{E}_0 &= \overline{E}_0^0 \, e^{-i(wt - \overline{k}_0 \cdot \overline{r} - k_{40}\gamma_0)} \\
\overline{H}_0 &= \overline{H}_0^0 \, e^{-i(wt - \overline{k}_0 \cdot \overline{r} - k_{40}\gamma_0)} \\
&= \left(\frac{c}{\omega}\right)\left(\overline{k}_0 \times \overline{E}_0^0\right) = n_0 \hat{k} \times \overline{E}_0^0 \\
and \quad \overline{V}_0 &= \overline{V}_0^0 \, e^{-i\left(wt - \overline{k}_0 \cdot \overline{r} - k_{40}\gamma_0\right)} \\
&= \left(\frac{-a_o c k_{40}}{\omega}\right) \overline{E}_0 = -m_0 \overline{E}_0 \quad .
\end{aligned}\right\}$$

For the reflected waves

$$\left.\begin{array}{l} \overline{E}_1 = \overline{E}_1^{\,0}\, e^{-i(wt - \overline{k}_{e1}\bullet \overline{r} - k_{4e1}\gamma_1)} \\[2ex] \overline{H}_1 = \left(\dfrac{c}{\omega}\right)\left(\overline{k}_{e1}\times \overline{E}_1\right) = n_1\overline{E}_1 \\[2ex] and\ \ \overline{V}_1 = \left(\dfrac{-A_{e1}}{\omega}\right)\overline{E}_1 = -m_1\overline{E}_1 \end{array}\right\} \qquad (11.78)$$

The refracted waves are given by

$$\left.\begin{array}{l} \overline{E}_2 = \overline{E}_2^{\,0}\, e^{-i(wt - \overline{k}_{e2}\cdot \overline{r} - k_{4e2}\gamma_2)} \\[2ex] \overline{H}_2 = \left(\dfrac{c}{\omega}\right)\left(\overline{k}_{e2}\times \overline{E}_2\right) = n_2\overline{E}_2 \\[2ex] and\ \ \overline{V}_2 = \left(\dfrac{-A_{e2}}{\omega}\right)\overline{E}_2 = -m_2\overline{E}_2 \end{array}\right\} \qquad (11.79)$$

In order to calculate the energy conversion efficiency knowledge of the value of A_e's and k_e's is necessary.

In Chapter 7 of the 1993 book titled, "The Dynamic Theory – A New View of Space-Time-Matter" the viscous effect of the geometry of a four dimensional space-time hyper surface embedded into a five dimensional space-time-mass manifold by the conservation of mass allows the evaluation of the constant a_o by comparing the prediction with shock wave experiments. This evaluation shows a value of approximately 3.65×10^7 kg/m^4. Since the value of the speed of light is known, it remains to get an estimate for the k_4.

The five dimensional gauge fields predicts an energy density that is given by

$$\xi = \frac{1}{8\pi}\left[E^2 + B^2 + V^2 + V_4^2\right]$$

and a radiation pressure given by

$$p = \frac{1}{8\pi}\left[E^2 + B^2 - V^2 + V_4^2 \right].$$

When only transverse waves are considered where V_4 is zero, this gives us a ratio of

$$\frac{p}{\xi} = \frac{\frac{1}{8\pi}\left[E^2 + B^2 - V^2 \right]}{\frac{1}{8\pi}\left[E^2 + B^2 + V^2 \right]} = \frac{\left[1 + n^2 - \frac{\left(a_o c k_4 \right)^2}{\omega^2} \right]}{\left[1 + n^2 + \frac{\left(a_o c k_4 \right)^2}{\omega^2} \right]}$$

for vacuum propagation. This ratio is slightly less than unity and can be compared with the Nichols and Hull experiments where the ratio was found to be 0.99433 (see Dennis Suhre's 2000 report). Taking the Nichols and Hull value the product and the frequency of white light as 540 THz and $n=1$ for air requires

$$a_o c k_4 \equiv A_e = 40.72 x 10^{12} \text{ sec}^{-1}.$$

Now we have an estimate for both a_o and k_4 for the vacuum. They are

$$a_o = 3.65 x 10^7 \text{ kg/m}^4$$

$$k_4 = 0.0037186 \text{ m}^3 / kg$$

These values may now be used in estimating the conversion of transverse wave energy into longitudinal wave energy and then back into transverse wave energy in the experiment. However, for these calculations it may be an advantage to use the analogy of the index of refraction such that

$$m \equiv \frac{a_o c k_4}{\omega} = \frac{A_e}{\omega}.$$

The tangential components of E, H, and V, can be continuous across the boundary only if the phases of the field vectors are all equal at the interface. That is

$$\overline{k}_{eo}\bullet\overline{r}+k_{4eo}\gamma_1=\overline{k}_{e1}\bullet\overline{r}+k_{4e1}\gamma_1=\overline{k}_{e2}\bullet\overline{r}+k_{4e2}\gamma_2$$

$$\overline{k}_{bo}\bullet\overline{r}+k_{4bo}\gamma_1=\overline{k}_{b1}\bullet\overline{r}+k_{4b1}\gamma_1=\overline{k}_{b2}\bullet\overline{r}+k_{4b2}\gamma_2.$$

$$\overline{k}_{vo}\bullet\overline{r}+k_{4vo}\gamma_1=\overline{k}_{v1}\bullet\overline{r}+k_{4v1}\gamma_1=\overline{k}_{v2}\bullet\overline{r}+k_{4v2}\gamma_2$$

$$(11.80)$$

For each component the propagation vectors $\overline{k}_{eo}$, $\overline{k}_{e1}$, and $\overline{k}_{e2}$ are coplanar, so if r is chosen to lie in the interface and in the plane of the propagation vectors, then we have,

$$k_{eo}\sin\theta_{eo}+k_{4eo}\gamma_1=k_{e1}\sin\theta_{e1}+k_{4e1}\gamma_1=k_{e2}\sin\theta_{e2}+k_{4e2}\gamma_2.$$

$$k_{bo}\sin\theta_{bo}+k_{4bo}\gamma_1=k_{b1}\sin\theta_{b1}+k_{4b1}\gamma_1=k_{b2}\sin\theta_{b2}+k_{4b2}\gamma_2$$

$$k_{vo}\sin\theta_{vo}+k_{4vo}\gamma_1=k_{v1}\sin\theta_{v1}+k_{4v1}\gamma_1=k_{v2}\sin\theta_{v2}+k_{4v2}\gamma_2.$$

For $\overline{k}_{eo}=\overline{k}_{e1}$ we find

$$\sin\theta_{e1}=\sin\theta_{eo}-\left(\frac{\gamma_1}{k_{eo}}\right)(k_{4e1}-k_{4eo})=\sin\theta_{eo}-\left(\frac{\gamma_1}{n_0 a_o}\right)(m_1-m_0).$$

From this we find that $sin\theta_{eo}=sin\theta_{e1}$ if $\gamma_1=0$ or if $m_1=m_0$. Not wanting to restrict ourselves unnecessarily by assumptions, let's continue.

For other components we have

$$\sin\theta_{b1}=\sin\theta_{bo}-\left(\frac{\gamma_1}{k_{bo}}\right)(k_{4b1}-k_{4bo})$$

and

$$\sin\theta_v=\sin\theta_{V1}-\left(\frac{\gamma_1}{k_{vo}}\right)(k_{4v1}-k_{4v2}).$$

Again using Equation (11.80) we find

$$k_{e1}\sin\theta_{e1}-k_{e2}\sin\theta_{e2}=k_{4e2\gamma_2}-k_{4e1\gamma_1}\qquad(11.81)$$

and

$$k_{eo}\sin\theta_{eo}-k_{e2}\sin\theta_{e2}=k_{4e2\gamma_2}-k_{4eo\gamma_1}.\qquad(11.82)$$

However, for $k_{eo}=k_{e1}$, subtracting Equation (11.82) from Equation (11.81) yields

$$\gamma_1(k_{4eo}-k_{4e1})=0$$

so that we have as a required result

Five Dimensional Fields

$$k_{4eo} = k_{4el} \Rightarrow m_0 = m_1. \tag{11.83}$$

In a similar fashion we have

$$k_{4bo} = k_{4bl}$$

and

$$k_{4vo} = k_{4vl}.$$

With this result Equation (11.81) becomes

$$\sin\theta_{e2} = \left(\frac{k_{el}}{k_{e2}}\right)\sin\theta_{el} - \frac{(k_{4e2}\gamma_2 - k_{4el}\gamma_1)}{k_{e2}} = \left(\frac{n_0}{n_2}\right)\sin\theta_{el} - \frac{(m_2\gamma_2 - m_0\gamma_1)}{n_2 a_o} \tag{11.84}$$

Similarly, for the other components we have

$$\sin\theta_{b2} = \left(\frac{k_{bl}}{k_{b2}}\right)\sin\theta_{bl} - \frac{(k_{4b2}\gamma_2 - k_{4bl}\gamma_1)}{k_{b2}} = \left(\frac{n_0}{n_2}\right)\sin\theta_{bl} - \frac{(m_2\gamma_2 - m_0\gamma_1)}{n_2 a_o} \tag{11.85}$$

and

$$\sin\theta_{v2} = \left(\frac{k_{vl}}{k_{v2}}\right)\sin\theta_{vl} - \frac{(k_{4v2}\gamma_2 - k_{4vl}\gamma_1)}{k_{V2}} = \left(\frac{n_0}{n_2}\right)\sin\theta_{vl} - \frac{(m_2\gamma_2 - m_1\gamma_1)}{n_2}.$$

$$(11.86)$$

Because of Equation (11.79), we must have

$$\theta_{eo} = \theta_{el} \ ; \ \theta_{b0} = \theta_{bl} \ ; \ \theta_{vo} = \theta_{vl} \ .$$

Now the tangential components of $\overline{E}$, $\overline{H}$, and V must be continuous at the interface. Therefore,

$$(\overline{E}_0 + \overline{E}_1) \times \hat{n} = \overline{E}_2 \times \hat{n} \ ,$$
$$(\overline{H}_0 + \overline{H}_1) \times \hat{n} = \overline{H}_2 \times \hat{n} \ , \tag{11.87}$$

and

$$(\overline{V}_0 + \overline{V}_1) \times \hat{n} = \overline{V}_2 \times \hat{n} \ .$$

Equation (11.87) may be written in terms of the electric component then we would have

$$(\overline{E}_0 + \overline{E}_1) \times \hat{n} = \overline{E}_2 \times \hat{n},$$

$$\left[(\overline{k}_{eo} \times \overline{E}_0) + (\overline{k}_{el} \times \overline{E}_1)\right] x\hat{n} = (\overline{k}_{e2} \times \overline{E}_2) \times \hat{n}, \text{ and} \tag{11.88}$$

$$(A_{eo}\overline{E}_0 + A_{el}\overline{E}_1) \times \hat{n} = A_{e2}(\overline{E}_2 \times \hat{n}).$$

Now suppose we expand the triple cross products in the middle equation of Equations (11.88), then

368

The Dynamic Theory

$$\left[\left(\hat{n}\bullet\overline{k}_{eo}\right)\overline{E}_0-\left(\hat{n}\bullet\overline{E}_0\right)\overline{k}_{eo}\right]+\left[\left(\hat{n}\bullet\overline{k}_{e1}\right)\overline{E}_1-\left(\hat{n}\bullet\overline{E}_1\right)\overline{k}_{e1}\right]=$$

$$\left[\left(\hat{n}\bullet\overline{k}_{e2}\right)\overline{E}_2-\left(\hat{n}\bullet\overline{E}_2\right)\overline{k}_{e2}\right].$$

All the products $(\hat{n}\bullet\overline{E};)$ vanish, and

$$\hat{n}\bullet\overline{k}_{ej}=(-1)^j\,k_{ej}\cos\theta_{ej};\ j=0,\ 1,\ 2,\ \text{so that}$$

$$k_{eo}\cos\theta_{eo}\overline{E}_0-k_{e1}\cos\theta_{e1}\overline{E}_1=k_{e2}\overline{E}_2\cos\theta_{e2}.$$

If we use the fact that $\theta_{eo}=\theta_{e1}$, and $k_{eo}=k_{e1}$ and rearrange the terms we arrive at

$$\left(E_0^0-E_1^0\right)\cos\theta_{eo}=\left(\frac{k_{e2}}{k_{e1}}\right)\cos\theta_{e2}E_2^0e^{i\left(k_{4e2}\gamma_2-k_{4e1}\gamma_1\right)}$$

$$=\left(\frac{n_2}{n_0}\right)\cos\theta_{e2}E_2^0e^{i\left(k_{4e2}\gamma_2-k_{4e1}\gamma_1\right)} \qquad (11.89)$$

Since the electric vectors are all parallel to the boundary surface, we must have

$$E_O^O+E_1^O=E_2^Oe^{i\left(k_{4e2}\gamma_2-k_{4e1}\gamma_1\right)} \qquad (11.90)$$

We may combine Equation (11.89) and (11.90) by subtraction to eliminate E_{20} and obtain

$$E_1^0=\frac{\left[\cos\theta_{eo}-\left(\dfrac{n_2}{n_0}\right)\cos\theta_{e2}\right]}{\left[\cos\theta_{eo}+\left(\dfrac{n_2}{n_0}\right)\cos\theta_{e2}\right]}E_O^O \qquad (11.91)$$

which is identical to the classical relationship between the incident and reflected waves.

Eliminating E_0^0 we have

$$E_2^0=\frac{2\cos\theta_{eo}}{\left[\cos\theta_{eo}+\left(\dfrac{n_2}{n_0}\right)\cos\theta_{e2}\right]}E_O^Oe^{i\left(k_{4e1}\gamma_1-k_{4e2}\gamma_2\right)} \qquad (11.92)$$

Equation (11.92) reduces to the classical expression for vanishing mass densities.

Equations (11.84) and (11.85) may be used to determine the refracted angles for each component while Equations (11.91) and (11.92) determine the magnitude of the reflected and refracted electric field components. The magnitudes of the reflected and refracted magnetic and gravitational components may be found using Equations (11.78) and (11.79).

Converting Pure Transverse Energy into Longitudinal Energy

To complete the analysis of reflection and refraction we now need to look at the case where the magnetic intensity vector is parallel with the interface as shown in Figure 11.2. However, in this case a purely transverse wave will be used as the incident wave while both transverse and longitudinal waves will be expected in both the reflected and the refracted waves.

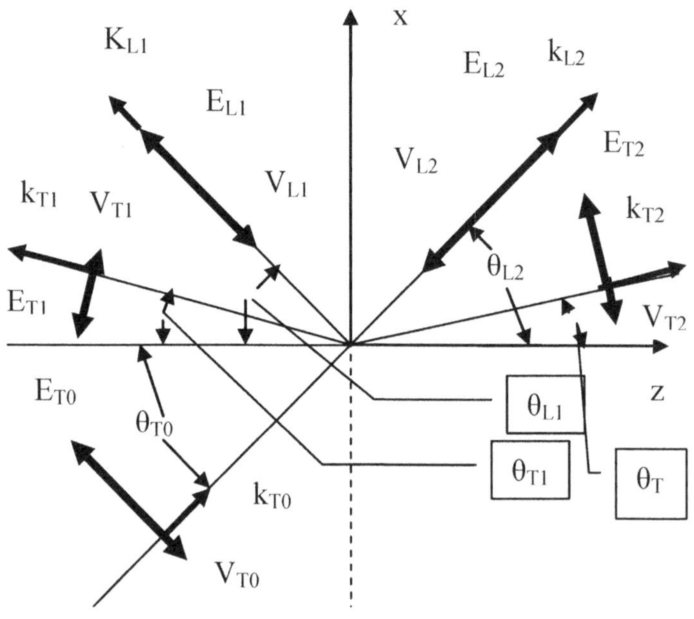

Figure 11.2. Reflection and refraction – oblique incidence. The xz-plane is the plane of incidence. The magnetic vectors are all parallel with the plane of incidence.

For the incident wave we have

$$
\left.
\begin{aligned}
\overline{E}_0 &= \overline{E}_0^0 \, e^{-i\left(wt - \overline{k}_{eo} \bullet \overline{r} - k_{4eo}\gamma_1\right)} \\[4pt]
\overline{H}_0 &= \overline{H}_0^0 \, e^{-i\left(wt - \overline{k}_{bo} \bullet \overline{r} - k_{4bo}\gamma_1\right)} \\[4pt]
&= \left(\frac{c}{\omega}\right)\left(\overline{k}_{eo} \times \overline{E}_0\right) = n_0\left(\hat{k}_{eo} \times \overline{E}_0\right) \\[4pt]
and \quad \overline{V}_0 &= \overline{V}_o^0 \, e^{-i\left(wt - \overline{k}_{vo} \bullet \overline{r} - k_{4vo}\gamma_1\right)} \\[4pt]
&= \left(\frac{-A_{eo}}{\omega}\right)\overline{E}_0 = -m_0\,\overline{E}_0 \ .
\end{aligned}
\right\}
$$

For the reflected transverse waves

$$
\left.
\begin{aligned}
\overline{E}_{t1} &= \overline{E}_{t1}^0 \, e^{-i\left(wt - \overline{k}_{et1} \bullet \overline{r} - k_{4et1}\gamma_1\right)} \\[4pt]
\overline{H} &= \left(\frac{c}{\omega}\right)\left(\overline{k}_{et1} \times \overline{E}_{t1}\right) = n_1\left(\hat{k} \times \overline{E}_{t1}\right) \\[4pt]
and \quad \overline{V}_{t1} &= \left(\frac{-A_{et1}}{\omega}\right)\overline{E}_{t1} = -m_1\overline{E}_{t1}
\end{aligned}
\right\} \qquad (11.93)
$$

The refracted transverse waves are given by

$$
\left.
\begin{aligned}
\overline{E}_{t2} &= \overline{E}_{t1}^0 \, e^{-i\left(wt - \overline{k}_{et2} \cdot \overline{r} - k_{4et2}\gamma_2\right)} \\[4pt]
\overline{H}_2 &= \left(\frac{c}{\omega}\right)\left(\overline{k}_{et2} \times \overline{E}_{t2}\right) = n_2\left(\hat{k}_{et2} \times \overline{E}_{t2}\right) \\[4pt]
and \quad \overline{V}_{t2} &= \left(\frac{-A_{et}}{\omega}\right)\overline{E}_{t2} = -m_2\overline{E}_{t2}
\end{aligned}
\right\} \qquad (11.94)
$$

For the reflected longitudinal waves, in a non-conducting medium,

$$\overline{E}_{l1} = \overline{E}^{\,0}_{l1}\, e^{-i\left(wt - \overline{k}_{el1}\bullet\overline{r} - k_{4el1}\gamma_1\right)}$$

$$\overline{V}_{l1} = \left(\frac{-k_{l1}}{a_o c k_{41}}\right)\overline{E}_{l1} = \left(\frac{-n_{l1}}{m_1}\right)\overline{E}_{l1}$$

$$and \;\; V_{41} = \left(\frac{-k_{l1}}{a_o c k_{41}}\right)E_{l1} = \left(\frac{-n_{l1}}{m_1}\right)E_{l1} \qquad (11.95)$$

The refracted longitudinal waves are given by

$$\overline{E}_{l2} = \overline{E}^{\,0}_{l2}\, e^{-i\left(wt - \overline{k}_{el2}\bullet\overline{r} - k_{4el2}\gamma_2\right)}$$

$$\overline{V}_{l2} = \left(\frac{-k_{l2}}{a_o c k_{42}}\right)\overline{E}_{l2} = \left(\frac{-n_{l2}}{m_2}\right)\overline{E}_{l2}$$

$$and \;\; V_{42} = \left(\frac{-k_{l2}}{a_o c k_{42}}\right)E_{l2} = \left(\frac{-n_{l2}}{m_2}\right)E_{l2} \qquad (11.96)$$

Along with these solutions there are three indicial relations that must be satisfied. For a non-conducting medium these relations reduce to two, which are

$$k_t^2 c^2 = \mu\varepsilon\omega^2 - a_0^2 c^2 k_4 \left(k_4 - i\frac{\partial \ln\varepsilon}{\partial\gamma}\right)$$

$$k_l^2 c^2 \left[\frac{k_4 + i\dfrac{\partial \ln\mu}{\partial\gamma}}{k_4 - i\dfrac{\partial \ln\varepsilon}{\partial\gamma}}\right] = \mu\varepsilon\omega^2 - a_0^2 c^2 k_4 \left(k_4 + i\frac{\partial \ln\mu}{\partial\gamma}\right) \qquad .(11.97)$$

The indicial relations of Equations (11.97) indicate that propagation constants of the transverse and the longitudinal waves are different and this must be considered. During these calculations this consideration may be met with the definitions

372

$$n \equiv \frac{k_i c}{\omega} = \sqrt{\mu\varepsilon}\sqrt{1 - \frac{a_0^2 c^2 k_4}{\mu\varepsilon\omega^2}\left(k_4 - i\frac{\partial \ln \varepsilon}{\partial \gamma}\right)}$$

$$n_l \equiv \frac{k_l c}{\omega} = \sqrt{\mu\varepsilon}\sqrt{\frac{k_4 - i\dfrac{\partial \ln \varepsilon}{\partial \gamma}}{k_4 + i\dfrac{\partial \ln \mu}{\partial \gamma}}}\sqrt{1 - \frac{a_0^2 c^2 k_4}{\mu\varepsilon\omega^2}\left(k_4 + i\frac{\partial \ln \mu}{\partial \gamma}\right)} \qquad .(11.98)$$

Obviously the last of Equations (11.98) reduces to the transverse index of refraction for a vacuum.

The tangential components of $\bar{E}$, $\bar{H}$, and $\bar{V}$, can be continuous across the boundary only if the phases of the field vectors are all equal at the interface. That is

$$\bar{k}_{eto} \bullet \bar{r} + k_{4eto}\gamma_1 = \bar{k}_{et1} \bullet \bar{r} + k_{4et1}\gamma_1 = \bar{k}_{et2} \bullet \bar{r} + k_{4et2}\gamma_2$$

$$= \bar{k}_{el1} \bullet \bar{r} + k_{4el1}\gamma_1 = \bar{k}_{el2} \bullet \bar{r} + k_{4el2}\gamma_2$$

$$\bar{k}_{bo} \bullet \bar{r} + k_{4bo}\gamma_1 = \bar{k}_{b1} \bullet \bar{r} + k_{4b1}\gamma_1 = \bar{k}_{b2} \bullet \bar{r} + k_{4b2}\gamma_2 \qquad (11.99)$$

$$\bar{k}_{vto} \bullet \bar{r} + k_{4vto}\gamma_1 = \bar{k}_{vt1} \bullet \bar{r} + k_{4vt1}\gamma_1 = \bar{k}_{vt2} \bullet \bar{r} + k_{4vt2}\gamma_2$$

$$= \bar{k}_{vl1} \bullet \bar{r} + k_{4vl1}\gamma_2 = \bar{k}_{vl2} \bullet \bar{r} + k_{4vl2}\gamma_2.$$

For each component the propagation vectors are coplanar, so if r is chosen to lie in the interface and in the plane of the propagation vectors, then we have,

$$k_{eto}\sin\theta_{t0} + k_{4eto}\gamma_1 = k_{et1}\sin\theta_{t1} + k_{4et1}\gamma_1 = k_{et2}\sin\theta_{t2} + k_{4et2}\gamma_2$$

$$= k_{el1}\sin\theta_{l1} + k_{4el1}\gamma_1 = k_{el2}\sin\theta_{l2} + k_{4el2}\gamma_2$$

$$k_{bo}\sin\theta_{t0} + k_{4bo}\gamma_1 = k_{b1}\sin\theta_{t1} + k_{4b1}\gamma_1 = k_{b2}\sin\theta_{t2} + k_{4b2}\gamma_2 \qquad (11.100)$$

$$k_{vto}\sin\theta_{t0} + k_{4vto}\gamma_1 = k_{vt1}\sin\theta_{t1} + k_{4vt1}\gamma_1 = k_{vt2}\sin\theta_{t2} + k_{4vt2}\gamma_2$$

$$= k_{vl1}\sin\theta_{l1} + k_{4vl1}\gamma_2 = k_{vl2}\sin\theta_{l2} + k_{4vl2}\gamma_2.$$

Rewriting the various combinations in the first of Equations (11.100) we get

$$n_0 \sin\theta_{t0} = n_1 \sin\theta_{t1} + \frac{m_1}{a_o}\gamma_1 - \frac{m_0}{a_o}\gamma_0$$

$$n_1 \sin\theta_{t1} = n_2 \sin\theta_{t2} + \frac{m_2}{a_o}\gamma_2 - \frac{m_1}{a_o}\gamma_1$$

$$n_2 \sin\theta_{t2} = n_{t1} \sin\theta_{t1} + \frac{m_{t1}}{a_o}\gamma_1 - \frac{m_2}{a_o}\gamma_2$$

$$n_{t1} \sin\theta_{t1} = n_{t2} \sin\theta_{t2} + \frac{m_{t2}}{a_o}\gamma_2 - \frac{m_{t1}}{a_o}\gamma_1$$

$$n_0 \sin\theta_{t0} = n_2 \sin\theta_{t2} + \frac{m_2}{a_o}\gamma_2 - \frac{m_0}{a_o}\gamma_1$$

$$n_0 \sin\theta_{t0} = n_{t1} \sin\theta_{t1} + \frac{m_{t1}}{a_o}\gamma_1 - \frac{m_0}{a_o}\gamma_0$$

$$n_0 \sin\theta_{t0} = n_{t2} \sin\theta_{t2} + \frac{m_{t2}}{a_o}\gamma_2 - \frac{m_0}{a_o}\gamma_1$$

$$n_1 \sin\theta_{t1} = n_{t1} \sin\theta_{t1} + \frac{m_{t1}}{a_o}\gamma_1 - \frac{m_1}{a_o}\gamma_1$$

$$n_1 \sin\theta_{t1} = n_{t2} \sin\theta_{t2} + \frac{m_{t2}}{a_o}\gamma_2 - \frac{m_1}{a_o}\gamma_1 \; . \qquad (11.101)$$

For $k_{et0} = k_{et1}$ or $n_0 = n_1$, we find

$$\sin\theta_{t1} = \sin\theta_{t0} - \left(\frac{\gamma_1}{k_{t1}}\right)\left(k_{4et1} - k_{4et0}\right)$$

$$= \sin\theta_{t0} + \left(m_0 - m_1\right)\frac{\gamma_1}{n_1 a_o} . \qquad (11.102)$$

From this we find that $\sin\theta_{t0} = \sin\theta_{t1}$ if $\gamma_1 = 0$ or if $k_{4et0} = k_{4et1}$. Notice that for air where $\gamma_1 \approx 0$ the reflected angle is approximately equal to the incident angle as in the classical case.

For other components we have

$$\sin\theta_{l1} = \left(\frac{k_0}{k_{l1}}\right)\sin\theta_{t0} - \left(\frac{\gamma_1}{k_{l1}}\right)\left(k_{4el1} - k_{4et0}\right)$$

$$\equiv \left(\frac{n_0}{n_{l1}}\right)\sin\theta_{t0} - \left(\frac{\gamma_1}{n_{l1}a_o}\right)\left(m_{l1} - m_o\right) \quad ,(11.103)$$

where we the relationship between k_{l1} and k_{t1} is given by Equation (11.98) above but we don't yet know the difference between the longitudinal and transverse k_4's. However, we can see that for air where $\gamma_1 \approx 0$ the angle of the reflected longitudinal component is the same as the incident angle if $n_{l1} = n_1 = n_0$.

$$\sin\theta_{t2} = \frac{k_{et0}}{k_{et2}}\sin\theta_{t0} - \left(\frac{1}{k_{et2}}\right)\left(k_{4et2}\gamma_2 - k_{4et0}\gamma_1\right)$$

$$= \frac{n_0}{n_2}\sin\theta_{t0} - \left(\frac{1}{n_2 a_o}\right)\left(m_2\gamma_2 - m_0\gamma_1\right)$$

$$(11.104)$$

and

$$\sin\theta_{l2} = \left(\frac{k_{et0}}{k_{el2}}\right)\sin\theta_{t0} + \left(\frac{1}{k_{el2}}\right)\left(k_{4et0}\gamma_1 - k_{4el2}\gamma_2\right)$$

$$\equiv \left(\frac{n_0}{n_{l2}}\right)\sin\theta_{t0} + \left(\frac{1}{n_{l2}a_o}\right)\left(m_0\gamma_1 - m_{l2}\gamma_2\right) \quad . \quad (11.105)$$

Equations (11.102), (11.103), (11.104), and (11.105) specify the angles θ_{t1}, θ_{t2}, θ_{l1}, and θ_{l2} in terms of the incident angle.

Again using Equation (11.100) we find

$$k_{et1}\sin\theta_{t1} - k_{et2}\sin\theta_{t2} = k_{4et2}\gamma_2 - k_{4et1}\gamma_1 \quad (11.106)$$

$$k_{el1}\sin\theta_{l1} - k_{et2}\sin\theta_{t2} = k_{4et2}\gamma_2 - k_{4el1}\gamma_1 \quad (11.107)$$

$$k_{el1}\sin\theta_{l1} - k_{el2}\sin\theta_{l2} = k_{4el2}\gamma_2 - k_{4el1}\gamma_1 \quad (11.108)$$

$$k_{et2}\sin\theta_{l2} - k_{et2}\sin\theta_{t2} = k_{4et2}\gamma_2 - k_{4el2}\gamma_2$$

$$= \gamma_2\left(k_{4et2} - k_{4el2}\right) \quad (11.109)$$

and

Five Dimensional Fields

$$k_{ell} \sin \theta_{ll} - k_{etl} \sin \theta_{tl} = k_{4etl}\gamma_1 - k_{4ell}\gamma_1$$
$$= \gamma_1 \left(k_{4etl} - k_{4ell} \right). \quad (11.110)$$

Subtracting Equation (11.107) from Equation (11.106) then using Equation (11.103) yields

$$\sin \theta_{tl} = \left(\frac{n_0}{n_1} \right) \sin \theta_{to} + \frac{\gamma_1}{n_1 a_o} \left[m_{l1} + m_0 - 2m_1 \right]. \quad (11.111)$$

Equations (11.111) and (11.103) require that $m_{l1}=m_1$.

Subtracting Equation (11.108) from (11.107) yields

$$\sin \theta_{l2} = \frac{n_0}{n_{l2}} \sin \theta_{t0} + \frac{1}{n_{l2} a_o} \left(m_0 \gamma_1 - m_{l2}\gamma_2 \right). \quad (11.112)$$

Equations (11.105) and (11.112) are identical if $m_{l2}=m_2$.

Substituting (11.104) and (11.105) into (11.112) obtains an identity.

Adding (11.108) and (11.109) we get

$$n_{l1} \sin \theta_{tl} = n_2 \sin \theta_{t2} + m_2 \gamma_2 - m_1 \gamma_1 \quad (11.113)$$

so that by subtracting (11.113) from (11.112) we get

$$n_{l2} \sin \theta_{l2} = n_{l1} \sin \theta_{l1} + \gamma_1 \frac{m_{l1}}{a_o} - \gamma_2 \frac{m_{l2}}{a_o}. \quad (11.114)$$

Substituting (11.111) into (11.114) produces

$$\sin \theta_{l2} = \frac{n_0}{n_{l2}} \sin \theta_{to} + \frac{1}{n_{l2} a_o} \left(m_{l1}\gamma_1 - m_{l2}\gamma_2 \right). \quad (11.115)$$

Equating (11.115) and (11.105) we get

$$m_{l1} = m_0.$$

Turning now to the middle equation of (11.100) we find three more relations as

$$\sin \theta_{t1} = \frac{n_0}{n_1} \sin \theta_{t0} + \frac{1}{n_1 a_o} \left(m_0 \gamma_0 - m_1 \gamma_1 \right)$$

$$\sin \theta_{t2} = \frac{n_1}{n_2} \sin \theta_{t1} + \frac{1}{n_2 a_o} \left(m_1 \gamma_1 - m_2 \gamma_2 \right) \quad (11.116)$$

$$\sin \theta_{t2} = \frac{n_0}{n_2} \sin \theta_{t0} + \frac{1}{n_2 a_o} \left(m_0 \gamma_0 - m_2 \gamma_2 \right)$$

376

which repeat what we learned before in (11.102) and (11.104).

Now considering the third equation of (11.100) we find that for gravitational propagation constants equal to the electric propagation constants these equations duplicate the first of Equations (11.100).

A summary of where we stand now may be contained in the equations:

$$\left. \begin{aligned} \sin\theta_{t1} &= \frac{n_0}{n_1}\sin\theta_{t0}, \\[2ex] \sin\theta_{l1} &= \left(\frac{n_0}{n_{l1}}\right)\sin\theta_{t0} \\[2ex] \sin\theta_{t2} &= \frac{n_0}{n_2}\sin\theta_{t0} - \left(\frac{1}{n_2 a_o}\right)(m_2\gamma_2 - m_0\gamma_1) \\[2ex] \sin\theta_{l2} &= \left(\frac{n_0}{n_{l2}}\right)\sin\theta_{t0} - \left(\frac{1}{n_{l2} a_o}\right)(m_2\gamma_2 - m_0\gamma_1) \end{aligned} \right\} \quad (11.117)$$

From these equations it ma y be seen that the longitudinal reflected and refracted waves will have different angles as n_{l1} is different from n_1 and n_{l2} is different from n_2.

Now the tangential components of $\bar{E}$, $\bar{H}$, and $\bar{V}$ must be continuous at the interface. Therefore, for the tangential electric components,

$$E_{t0}^{0}\cos\theta_{t0}\,e^{-i(wt - k_{et0}x\sin\theta_{t0} - k_{4et0}\gamma_0)} - E_{t1}^{0}\cos\theta_{t1}\,e^{-i(wt - k_{et1}x\sin\theta_{t1} - k_{4et1}\gamma_1)}$$

$$+ E_{l1}^{0}\sin\theta_{l1}\,e^{-i(wt - k_{el1}x\sin\theta_{l1} - k_{4el1}\gamma_1)} \qquad .(11.118)$$

$$= E_{t2}^{0}\cos\theta_{t2}\,e^{-i(wt - k_{et2}x\sin\theta_{t2} - k_{4et2}\gamma_2)}$$

$$+ E_{l2}^{0}\sin\theta_{l2}\,e^{-i(wt - k_{el2}x\sin\theta_{l2} - k_{4el2}\gamma_2)}$$

The common factor $e^{-j\omega t}$ may be cancelled from all terms leaving

$$E_{t0}^{0}\cos\theta_{t0}\,e^{i(k_{et0}x\sin\theta_{t0}+k_{4et0}\gamma_{0})}-E_{t1}^{0}\cos\theta_{t1}\,e^{i(k_{et1}x\sin\theta_{t1}+k_{4et1}\gamma_{1})}$$

$$+E_{t1}^{0}\sin\theta_{t1}\,e^{i(k_{et1}x\sin\theta_{t1}+k_{4et1}\gamma_{1})}$$

$$=E_{t2}^{0}\cos\theta_{t2}\,e^{i(k_{et2}x\sin\theta_{t2}+k_{4et2}\gamma_{2})}$$

$$+E_{t2}^{0}\sin\theta_{t2}\,e^{i(k_{et2}x\sin\theta_{t2}+k_{4et2}\gamma_{2})}$$

. (11.119)

Each term in Equation (11.119) depends upon x through an exponential factor. The only way in which Equation (11.119) can be satisfied for all values of x is if all exponential factors are all the same, that is if

$$k_{eto}\sin\theta_{t0}+k_{4eto}\gamma_{0}=k_{et1}\sin\theta_{et1}+k_{4et1}\gamma_{1}$$
$$=k_{et2}\sin\theta_{et2}+k_{4et2}\gamma_{2}$$
$$=k_{el1}\sin\theta_{el1}+k_{4el1}\gamma_{1}$$
$$=k_{el2}\sin\theta_{el2}+k_{4el2}\gamma_{2}$$

. (11.120)

Equation (11.120) reproduces Equation (11.100) above and determines the reflected and refracted angles in terms of the incident angle.

Returning to Equation (11.119) and canceling the common factors, since the exponents are all equal at the interface, leaves

$$\cos\theta_{t0}E_{t0}^{0}-\cos\theta_{t0}E_{t1}^{0}+\sin\theta_{t1}E_{t1}^{0}=\cos\theta_{t2}E_{t2}^{0}+\sin\theta_{t2}E_{t2}^{0}. \quad (11.121)$$

Equation (11.121) is an equation which must be satisfied by the electric components. There are four unknowns in this equation. We need an additional three equations in order to have a solvable system of equations. In addition there are four other conditions, obtained from continuity of the normal component of the electric displacement, the normal component of the gravitational field ($\bar{G}$), the tangential component of the gravitational intensity ($\bar{V}$), and of the tangential component of the magnetic intensity.

Continuity of the tangential component of the magnetic intensity gives

$$n_{0}E_{t0}^{0}+n_{1}E_{t1}^{0}=n_{2}E_{t2}^{0}. \quad (11.122)$$

The continuity of the tangential component of the gravitational intensity gives

$$m_0 \cos\theta_{t0} E_{t0}^0 - m_1 \cos\theta_{t1} E_{t1}^0 + \frac{n_{t1}}{m_1}\sin\theta_{t1} E_{l1}^0$$

$$= m_2 \cos\theta_{t2} E_{t2}^0 + \frac{n_{t2}}{m_2}\sin\theta_{t2} E_{l2}^0$$

$$.(11.123)$$

Continuity of the electric displacement gives

$$-\varepsilon_0 \sin\theta_{t0} E_{t0}^0 - \varepsilon_1 \sin\theta_{t1} E_{t1}^0$$

$$-\varepsilon_1 \cos\theta_{t1} E_{l1}^0 = -\varepsilon_2 \sin\theta_{t2} E_{t2}^0 + \varepsilon_2 \cos\theta_{t2} E_{l2}^0$$

$$.(11.124)$$

For continuity of the gravitational potential we must have

$$E_{l1}^0 = \frac{n_{t2} m_1}{n_{t1} m_2} E_{l2}^0 \qquad (11.125)$$

From Equation (11.122) we find

$$E_{t1}^0 = \frac{n_2}{n_1} E_{t2}^0 - E_{t0}^0 . \qquad (11.126)$$

Equations (11.125) and (11.126) allow us to eliminate E_{t1}^0 and E_{l1}^0 from Equations (11.121) and (11.122). Equation (11.121) becomes

$$\left(\cos\theta_{t2} + \frac{n_2}{n_1}\cos\theta_{t1}\right) E_{t2}^0 = 2\cos\theta_{t0} E_{t0}^0$$

$$+ \left(\frac{n_{t2} m_1}{n_{t1} m_2}\sin\theta_{t1} - \sin\theta_{t2}\right) E_{l2}^0$$

$$.(11.127)$$

Using Equations (11.117), (11.127) may be written as

$$\left(\cos\theta_{t2} + \frac{n_2}{n_1}\cos\theta_{t1}\right)E_{t2}^0 = 2\cos\theta_{t0}\,E_{t0}^0$$

$$+ \left[\begin{array}{c}\left(\dfrac{n_2 m_1}{n_{t1}m_2} - \dfrac{n_0}{n_{t2}}\right)\sin\theta_{t0} \\[2ex] + \left(\dfrac{m_2\gamma_2 - m_0\gamma_1}{n_{t2}a_o}\right)\end{array}\right]E_{t2}^0 \qquad . \quad (11.128)$$

From (11.128) it may be seen that, when $\gamma_l \approx 0$, the small angle deviation from the classical solution depends primarily upon

$$\frac{m_2\gamma_2}{n_{t2}a_o}\,E_{t2}^0 \qquad . \qquad (11.129)$$

Equation (11.127) collapses to the classical equation when there is no longitudinal component.

Using (11.122) we find

$$\left(\cos\theta_{t1} + \frac{n_1}{n_2}\cos\theta_{t2}\right)E_{t1}^0 = \left(\cos\theta_{t0} - \frac{n_0}{n_2}\cos\theta_{t2}\right)E_{t0}^0$$

$$+ \left(\frac{n_{t2}m_1}{n_{t1}m_2}\sin\theta_{t1} - \sin\theta_{t2}\right)E_{t2}^0 \quad (11.130)$$

Equation (11.130) collapses to the classical expression for the reflected transverse component when the longitudinal component vanishes.

Equation (11.124) becomes

$$\left(\begin{array}{c}\varepsilon_2\cos\theta_{t2} \\[1ex] +\varepsilon_1\dfrac{n_{t2}m_1}{n_{t1}m_2}\cos\theta_{t1}\end{array}\right)E_{t2}^0 = \left(\varepsilon_2\sin\theta_{t2} - \frac{n_2}{n_1}\varepsilon_1\sin\theta_{t1}\right)E_{t2}^0 \qquad . \,(11.131)$$

$$+\left(\varepsilon_1\sin\theta_{t1} - \varepsilon_0\sin\theta_{t0}\right)E_{t0}^0$$

this reduces to

$$\left(\begin{array}{c} \varepsilon_2 \cos\theta_{l2} \\ +\varepsilon_1 \dfrac{n_{l2}m_1}{n_{l1}m_2}\cos\theta_{l1} \end{array} \right) E_{l2}^0 = \left(\varepsilon_2 \sin\theta_{l2} - \dfrac{n_2}{n_1}\varepsilon_1 \sin\theta_{l1} \right) E_{l2}^0 \quad (11.132)$$

because medium 1 is the same as medium 0.

Substituting (11.132) into (11.127) we get

$$E_{t2}^0 = \frac{2\cos\theta_{t0}\, E_{t0}^0}{\left[\left(\cos\theta_{t2} + \dfrac{n_2}{n_1}\cos\theta_{t1} \right) - \dfrac{A1\,B1}{C1} \right]}$$

$$A1 \equiv \left(\frac{n_{l2}m_1}{n_{l1}m_2}\sin\theta_{l1} - \sin\theta_{l2} \right)$$

$$B1 \equiv \left(\varepsilon_2 \sin\theta_{l2} - \varepsilon_1 \frac{n_2}{n_1}\sin\theta_{l1} \right) \qquad (11.133)$$

$$C1 \equiv \left(\varepsilon_2 \cos\theta_{l2} + \varepsilon_1 \cos\theta_{l1}\frac{n_{l2}m_1}{n_{l1}m_2} \right)$$

Converting Pure Longitudinal Energy into Transverse Energy

Now we need to show how a purely longitudinal wave can have some of its energy converted into transverse wave energy by an interface between two media. Consider a pure longitudinal wave striking an interface at an angle as shown in Figure 11.3.

For the incident wave we have

$$\left. \begin{array}{c} \overline{E}_{l0} = \overline{E}_{l0}^{\,0}\, e^{-i(wt - \overline{k}_{el0}\bullet\overline{r} - k_{4el0}\gamma_0)} \\[2mm] \overline{V}_{i0} = \left(\dfrac{-\mu\varepsilon\omega}{a_o c k_{40}} \right)\overline{E}_{l0} = \left(\dfrac{-n_{l0}}{m_o} \right)\overline{E}_{l0} \\[2mm] and \ \ V_{40} = \left(\dfrac{-k_{l0}}{a_o c k_{40}} \right)E_{l0} = \left(\dfrac{-n_{l0}}{m_o} \right)E_{l0} \end{array} \right\}.$$

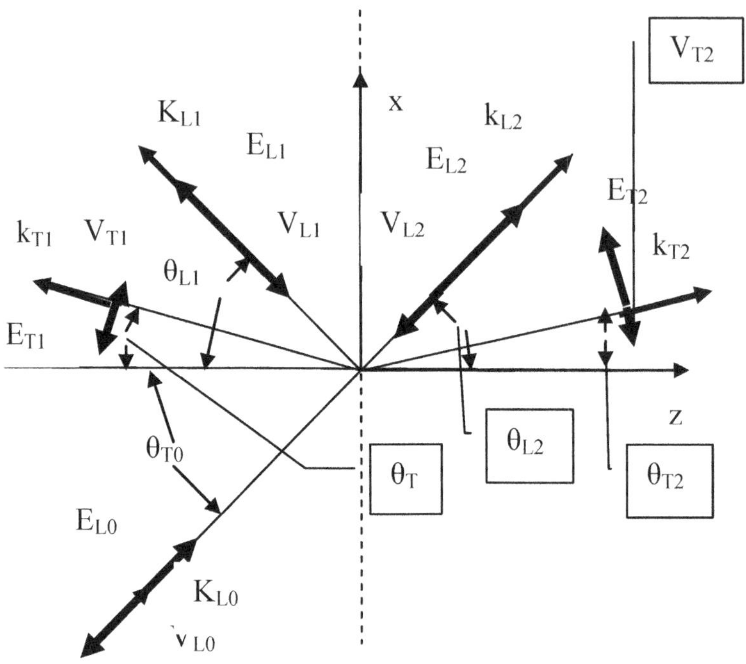

Figure 11.3. Reflection and refraction of a pure longitudinal wave at an oblique incidence. The xz-plane is the plane of incidence.

For the reflected transverse waves

$$\overline{E}_{t1} = \overline{E}_{t1}^{0}\, e^{-i(wt - \overline{k}_{et1}\bullet\overline{r} - k_{4et1}\gamma_1)}$$

$$\overline{H} = \left(\frac{c}{\omega}\right)\left(\overline{k}_{et1}\times\overline{E}_{t1}\right) = n_1\left(\hat{k}_{et1}\times\overline{E}_{t1}\right)$$

$$and \quad \overline{V}_{t1} = \left(\frac{-A_{et1}}{\omega}\right)\overline{E}_{t1} = -m_1\overline{E}_{t1} \qquad (11.134)$$

The refracted transverse waves are given by

$$\left.\begin{array}{c} \overline{E}_{t2} = \overline{E}_{t1}^{0}\, e^{-i(wt-\overline{k}_{et2}\cdot\overline{r}\,-\,k_{4et2}\gamma_2)} \\[2mm] \overline{H}_2 = \left(\dfrac{c}{\omega}\right)\left(\overline{k}_{et2}\times\overline{E}_{t2}\right) = n_2\left(\hat{k}_{et2}\times\overline{E}_{t2}\right) \\[2mm] and \quad \overline{V}_{t2} = \left(\dfrac{-A_{et}}{\omega}\right)\overline{E}_{t2} = -m_2\overline{E}_{t2} \end{array}\right\} . \quad (11.135)$$

For the reflected longitudinal waves, in a non-conducting medium,

$$\left.\begin{array}{c} \overline{E}_{l1} = \overline{E}_{l1}^{0}\, e^{-i(wt-\overline{k}_{el1}\bullet\overline{r}\,-\,k_{4el1}\gamma_1)} \\[2mm] \overline{V}_{l1} = \left(\dfrac{-\mu\varepsilon\omega}{a_o c k_{41}}\right)\overline{E}_{l1} = \left(\dfrac{-n_{l1}}{m_1}\right)\overline{E}_{l1} \\[2mm] and \;\; V_{41} = \left(\dfrac{-k_{l1}}{a_o c k_{41}}\right)E_{l1} = \left(\dfrac{-n_{l1}}{m_1}\right)E_{l1} \end{array}\right\} . \quad (11.136)$$

The refracted longitudinal waves are given by

$$\left.\begin{array}{c} \overline{E}_{l2} = \overline{E}_{l2}^{0}\, e^{-i(wt-\overline{k}_{el2}\bullet\overline{r}\,-\,k_{4el2}\gamma_2)} \\[2mm] \overline{V}_{l2} = \left(\dfrac{-\mu\varepsilon\omega}{a_o c k_{42}}\right)\overline{E}_{l2} = \left(\dfrac{-n_{l2}}{m_2}\right)\overline{E}_{l2} \\[2mm] and \;\; V_{42} = \left(\dfrac{-k_{l2}}{a_o c k_{42}}\right)E_{l2} = \left(\dfrac{-n_{l2}}{m_2}\right)E_{l2} \end{array}\right\} . \quad (11.137)$$

The tangential components of $\overline{E}$, $\overline{H}$, and $\overline{V}$, can be continuous across the boundary only if the phases of the field vectors are all equal at the interface. That is

$$\overline{k}_{elo}\bullet\overline{r}+k_{4elo}\gamma_0=\overline{k}_{et1}\bullet\overline{r}+k_{4et1}\gamma_1=\overline{k}_{et2}\bullet\overline{r}+k_{4et2}\gamma_2$$

$$=\overline{k}_{el1}\bullet\overline{r}+k_{4el1}\gamma_1=\overline{k}_{el2}\bullet\overline{r}+k_{4el2}\gamma_2$$

$$\overline{k}_{b1}\bullet\overline{r}+k_{4b1}\gamma_1=\overline{k}_{b2}\bullet\overline{r}+k_{4b2}\gamma_2 \qquad (11.138)$$

$$\overline{k}_{vlo}\bullet\overline{r}+k_{4vlo}\gamma_0=\overline{k}_{vt1}\bullet\overline{r}+k_{4vt1}\gamma_1=\overline{k}_{vt2}\bullet\overline{r}+k_{4vt2}\gamma_2$$

$$=\overline{k}_{vl1}\bullet\overline{r}+k_{4vl1}\gamma_1=\overline{k}_{vl2}\bullet\overline{r}+k_{4vl2}\gamma_2.$$

For each component the propagation vectors are coplanar, so if r is chosen to lie in the interface and in the plane of the propagation vectors, then we have,

$$k_{el0}\sin\theta_{l0}+k_{4elo}\gamma_0=k_{et1}\sin\theta_{t1}+k_{4et1}\gamma_1$$

$$=k_{et2}\sin\theta_{t2}+k_{4et2}\gamma_2$$
$$=k_{el1}\sin\theta_{l1}+k_{4el1}\gamma_1 \qquad (11.139)$$
$$=k_{el2}\sin\theta_{l2}+k_{4el2}\gamma_2$$

For these relations we find

$$\left[\begin{array}{l}\sin\theta_{l1}=\dfrac{k_{el0}}{k_{el1}}\sin\theta_{lo}\\[2em]-\left(\dfrac{1}{k_{el1}}\right)\left(\gamma_1 k_{4el1}-\gamma_0 k_{4el0}\right)\end{array}\right]=\dfrac{n_{l0}}{n_{l1}}\sin\theta_{lo}-\dfrac{1}{n_{l1}a_o}\left(\gamma_1 m_1-\gamma_0 m_0\right). \quad (11.140)$$

From this we find that $\sin\theta_{lo}=\sin\theta_{l1}$ if $\gamma_0=0$ or if $k_{4el0}=k_{4el1}$.

For other components we have

$$\left[\begin{array}{l}\sin\theta_{t1}=\dfrac{k_{el0}}{k_{et1}}\sin\theta_{lo}\\[2em]-\left(\dfrac{1}{k_{et1}}\right)\left(\gamma_1 k_{4et1}-\gamma_0 k_{4elo}\right)\end{array}\right]=\dfrac{n_{l0}}{n_1}\sin\theta_{lo}-\left(\dfrac{1}{n_1 a_o}\right)\left(\gamma_1 m_1-\gamma_0 m_0\right), \quad (11.141)$$

Five Dimensional Fields

384

The Dynamic Theory

$$\sin\theta_{t2} = \frac{k_{el0}}{k_{et2}}\sin\theta_{t0} - \left(\frac{1}{k_{et2}}\right)\left(k_{4et2}\gamma_2 - k_{4el0}\gamma_0\right)$$

$$= \frac{n_{t0}}{n_2}\sin\theta_{t0} - \frac{1}{n_2 a_o}\left(m_2\gamma_2 - m_0\gamma_0\right) \tag{11.142}$$

and

$$\sin\theta_{l2} = \left(\frac{k_{el0}}{k_{el2}}\right)\sin\theta_{l0} - \left(\frac{1}{k_{el2}}\right)\left(k_{4el2}\gamma_2 - k_{4el0}\gamma_0\right)$$

$$= \frac{n_{l0}}{n_{l2}}\sin\theta_{l0} - \frac{1}{n_{l2}a_o}\left(m_2\gamma_2 - m_0\gamma_0\right) \tag{11.143}$$

Equations (11.140), (11.141), (11.142), and (11.143) specify the angles θ_{t1}, θ_{t2}, θ_{l1}, and θ_{l2} in terms of the incident angle, θ_{t0}. To summarize

$$\sin\theta_{l1} = \frac{n_{t0}}{n_{l1}}\sin\theta_{t0} - \frac{1}{n_{l1}a_o}\left(\gamma_1 m_1 - \gamma_0 m_0\right)$$

$$\sin\theta_{t1} = \frac{n_{t0}}{n_1}\sin\theta_{lo} - \frac{1}{n_1 a_o}\left(\gamma_1 m_1 - \gamma_0 m_0\right)$$

$$\sin\theta_{t2} = \frac{n_{t0}}{n_2}\sin\theta_{t0} - \frac{1}{n_2 a_o}\left(m_2\gamma_2 - m_0\gamma_0\right)$$

$$\sin\theta_{l2} = \frac{n_{t0}}{n_{l2}}\sin\theta_{t0} - \frac{1}{n_{l2}a_o}\left(m_2\gamma_2 - m_0\gamma_0\right)$$

Now the tangential components of $\overline{E}$, $\overline{H}$, and $\overline{V}$ must be continuous at the interface. Therefore,

$$\begin{bmatrix} E_{l0}^0\sin\theta_{l0}e^{-i(wt-k_{el0}x\sin\theta_{l0}-k_{4el0}\gamma_0)} \\ -E_{t1}^0\cos\theta_{t1}e^{-i(wt-k_{et1}x\sin\theta_{t1}-k_{4et1}\gamma_1)} \\ +E_{l1}^0\sin\theta_{l1}e^{-i(wt-k_{el1}x\sin\theta_{l1}-k_{4el1}\gamma_1)} \end{bmatrix}$$

$$= \begin{bmatrix} E_{t2}^0\cos\theta_{t2}e^{-i(wt-k_{et2}x\sin\theta_{t2}-k_{4et2}\gamma_2)} \\ +E_{l2}^0\sin\theta_{l2}e^{-i(wt-k_{el2}x\sin\theta_{l2}-k_{4el2}\gamma_2)} \end{bmatrix} \tag{11.144}$$

The common factor $e^{-j\omega t}$ may be cancelled from all terms leaving

$$\begin{bmatrix} E_{I0}^{0} \sin\theta_{I0}\, e^{i(k_{eI0}x\sin\theta_{I0}+k_{4eI0}\gamma_{0})} \\ -E_{I1}^{0}\cos\theta_{I1}\, e^{i(k_{eI1}x\sin\theta_{I1}+k_{4eI1}\gamma_{1})} \\ +E_{I1}^{0}\sin\theta_{I1}\, e^{i(k_{eI1}x\sin\theta_{I1}-+k_{4eI1}\gamma_{1})} \end{bmatrix} \qquad .(11.145)$$

$$= \begin{bmatrix} E_{I2}^{0}\cos\theta_{I2}\, e^{i(k_{eI2}x\sin\theta_{I2}+k_{4eI2}\gamma_{2})} \\ +E_{I2}^{0}\sin\theta_{I2}\, e^{i(k_{eI2}x\sin\theta_{I2}+k_{4eI2}\gamma_{2})} \end{bmatrix}$$

Each term in Equation (11.145) depends upon x through an exponential factor. The only way in which Equation (11.145) can be satisfied for all values of x is if all exponential factors are all the same, that is if

$$k_{eI0}\sin\theta_{I0} + k_{4eI0}\gamma_{0} = k_{eI1}\sin\theta_{eI1} + k_{4eI1}\gamma_{1}$$

$$= k_{eI2}\sin\theta_{eI2} + k_{4eI2}\gamma_{2}$$

$$= k_{eI1}\sin\theta_{eI1} + k_{4eI1}\gamma_{1} \qquad .(11.146)$$

$$= k_{eI2}\sin\theta_{eI2} + k_{4eI2}\gamma_{2}$$

Equation (11.146) reproduces Equation (11.139) above and determines the reflected and refracted angles in terms of the incident angle.

Returning to Equation (11.145) and canceling the common factors leaves

$$\begin{bmatrix} \sin\theta_{I0}E_{I0}^{0} - \cos\theta_{I1}E_{I1}^{0} \\ +\sin\theta_{I1}E_{I1}^{0} \end{bmatrix} = \cos\theta_{I2}E_{I2}^{0} + \sin\theta_{I2}E_{I2}^{0}. \quad (11.147)$$

Equation (11.147) is one equation which must be satisfied by the electric components; in addition there are five other conditions, obtained from continuity of the normal component of the electric displacement, the normal component of the gravitational field ($\overline{G}$), the tangential component of the gravitational intensity ($\overline{V}$), the tangential component of the magnetic intensity and the gravitational potential.

Continuity of the electric displacement gives

$$\begin{bmatrix} \varepsilon_1 \cos\theta_{l0} E_{l0}^0 \\ -\varepsilon_1 \sin\theta_{l1} E_{l1}^0 \\ -\varepsilon_1 \cos\theta_{l1} E_{l1}^0 \end{bmatrix} = -\varepsilon_2 \sin\theta_{l2} E_{l2}^0 + \varepsilon_2 \cos\theta_{l2} E_{l2}^0 \qquad (11.148)$$

while continuity of the tangential component of the magnetic intensity gives

$$n_1 E_{l1}^0 = n_2 E_{l2}^0 . \qquad (11.149)$$

The continuity of the tangential component of the gravitational intensity gives

$$\begin{bmatrix} \dfrac{n_{l0}}{m_0} \sin\theta_{l0} E_{l0}^0 \\ -m_1 \cos\theta_{l1} E_{l1}^0 \\ +\dfrac{n_{l1}}{m_1} \sin\theta_{l1} E_{l1}^0 \end{bmatrix} = m_2 \cos\theta_{l2} E_{l2}^0 + \dfrac{n_{l2}}{m_2} \sin\theta_{l2} E_{l2}^0 . \qquad (11.150)$$

Continuity of the gravitational potential requires that

$$\frac{n_{l0}}{m_0} E_{l0} + \frac{n_{l1}}{m_1} E_{l1} = \frac{n_{l2}}{m_2} E_{l2} . \qquad (11.151)$$

Since $n_{l0}=n_{l1}$ and $m_0=m_1$, Equation (11.151) may be written as

$$E_{l2} = \frac{n_{l0} m_2}{n_{l2} m_0} \left(E_{l0} + E_{l1} \right) . \qquad (11.152)$$

Using (11.152) and (11.149) in (11.147) gives us

$$\left(\frac{n_2}{n_1} \cos\theta_{l1} + \cos\theta_{l2} \right) E_{l2}^0 = \begin{bmatrix} \left(\sin\theta_{l0} - \dfrac{n_{l0} m_2}{n_{l2} m_0} \sin\theta_{l2} \right) E_{l0}^0 \\ + \left(\sin\theta_{l1} - \dfrac{n_{l0} m_2}{n_{l2} m_0} \sin\theta_{l2} \right) E_{l1}^0 \end{bmatrix} . \qquad (11.153)$$

Because medium 1 is the same as medium 0 Equation (11.153) reduces to

$$\text{Five Dimensional Fields}$$

$$E_{t2}^0 = \frac{\left(\sin\theta_{t0} - \frac{n_{t0}m_2}{n_{t2}m_0}\sin\theta_{t2}\right)\left(E_{t0}^0 + E_{t1}^0\right)}{\left(\frac{n_2}{n_1}\cos\theta_{t1} + \cos\theta_{t2}\right)} \qquad .(11.154)$$

Using (11.152) and (11.149) in (11.148) produces

$$E_{t2}^0 = \frac{\left[\begin{array}{c}\left(\varepsilon_2\dfrac{n_{t0}m_2}{n_{t2}m_0}\cos\theta_{t2} - \varepsilon_1\cos\theta_{t0}\right)E_{t0}^0 \\[2em] +\left(\varepsilon_2\dfrac{n_{t0}m_2}{n_{t2}m_0}\cos\theta_{t2} + \varepsilon_1\cos\theta_{t1}\right)E_{t1}^0\end{array}\right]}{\left(\varepsilon_2\sin\theta_{t2} - \varepsilon_1\dfrac{n_2}{n_1}\sin\theta_{t1}\right)} .(11.155)$$

Equating (11.154) and (11.155)

$$E_{t1}^0 = -\frac{\left[A1B1 - \left(\varepsilon_2\dfrac{n_{t0}m_2}{n_{t2}m_0}\cos\theta_{t2} - \varepsilon_1\cos\theta_{t0}\right)C1\right]}{\left[A1B1 - \left(\varepsilon_2\dfrac{n_{t0}m_2}{n_{t2}m_0}\cos\theta_{t2} + \varepsilon_1\cos\theta_{t1}\right)C1\right]} E_{t0}^0$$

$$A1 \equiv \left(\sin\theta_{t0} - \frac{n_{t0}m_2}{n_{t2}m_0}\sin\theta_{t2}\right) \qquad . \quad (11.156)$$

$$B1 \equiv \left(\varepsilon_2\sin\theta_{t2} - \varepsilon_1\frac{n_2}{n_1}\sin\theta_{t1}\right)$$

$$C1 \equiv \left(\frac{n_2}{n_1}\cos\theta_{t1} + \cos\theta_{t2}\right)$$

Then using Equation (11.156) in (11.155) we find

$$E_{t2}^0 = A1\left\{\frac{-2\varepsilon_1\cos\theta_{t0}}{A1B1 - C1E1}\right\}E_{t0}^0$$

$$E1 \equiv \left(\varepsilon_2\frac{n_{t0}m_2}{n_{t2}m_0}\cos\theta_{t2} + \varepsilon_1\cos\theta_{t1}\right) \qquad (11.157)$$

Self-Energy of Charged Particles

One of the difficulties in Maxwellian electromagnetism is the infinite self-energy that is predicted for a charged particle. This "electron catastrophe," or singularity, does not exist with the non-singular electrostatic potential, and the self-energy of a charged particle may be found using the free energy concept from classical thermodynamics.

In classical electromagnetic theory the self-energy of a charged particle is discussed but its value has not been established. This is because the expression for the self-energy is a function of the radius associated with the physical extent of the charge distribution. Thus, the radius of the charged particle must be known before the self-energy can be determined.

Currently the self-energy of a ch arged particle is equated with the energy associated with its inertial mass by $E = mc^2$. Then the radius associated with its energy is taken as the "radius" of the particle. There is no intention that this radius be the physical radius of the particle though it compares favorably with experimental values.

The question arises here of whether or not the Dynamic Theory, with its thermodynamic energy concepts and the non-singular electrostatic potential, can theoretically predict the self-energy and/or the radius of the physical extent of the mass or charge distribution of the particle. One of the beneficial aspects of the generalization of physical theory as done in the Dynamic Theory is the possibility of using conceptualizations and procedures developed in one branch of physics in another branch. This aspect of the theory appears applicable here. The self-energy of a charged particle is the notion that a c ertain amount of energy be associated with the existence of the particle and its charge. This notion may be associated with the notion of free energy used in thermodynamics, for, if the self-energy of the charged particle is its free energy, then it represents the energy which may be "freed" upon converting the particle into energy. Conversely, this would represent the energy required to assemble the charged particle.

With the conceptualization of free energy the second law provides the condition for a stable equilibrium state, namely that a charged particle in an equilibrium state must exist at a minimum of its free energy. Thus, if the self-energy, or free energy, of a charged particle is sought, then minimizing its free energy will yield the desired result.

The free energy may be defined, in analogy with the thermodynamic case, as

$$G \equiv U - \phi S - x^\alpha F_\alpha, \tag{11.158}$$

where α depends upon the applicable work terms which here will be taken as the three spatial dimensions, so that $\alpha=1,2,3$. The First Law is given by

$$\text{d}E = dU - F_\alpha dx^\alpha$$

while the Second Law yields

$$\phi dS = dU - F_\alpha dx^\alpha$$

for a quasi-static, reversible process. Therefore, the differential change in the system energy is

$$dU = \phi dS + F_\alpha dx^\alpha. \tag{11.159}$$

Differentiating Equation (11.158) gives the differential change in the free energy as

$$dG = dU - \phi dS - S d\phi - F_\alpha dx^\alpha - x^\alpha dF_\alpha. \tag{11.160}$$

Substituting Equation (11.159) into (11.160) yields

$$dG = - S d\phi - x^\alpha dF_\alpha. \tag{11.161}$$

The force in Equation (11.161) is considered to be the Lorentz force

$$F_\alpha = q[\,\overline{E} + (\,\overline{v}x\overline{B}\,)\,]_\alpha$$

so that Equation (11.161) becomes

$$dG = - S d\phi - x^\alpha d\{\, q[\,\overline{E} + (\,\overline{v}x\overline{B}\,)\,]_\alpha\}.$$

The Dynamic Theory

If we wish to consider the change in free energy with respect to a ch ange in the charge at a constant velocity, we find that, since ϕ is a function of velocity only, $d\phi = 0$. The specification of constant velocity stems from the desire to obtain the self-energy of a charged particle; therefore, the particle should be considered as sitting still, so that it will have no kinetic energy. The differential change of free energy for a stationary particle is then

$$dG = -Sd\phi - x^\alpha d\{q[\,\overline{E} + (\,\overline{vx}\,\overline{B}\,)\,]_\alpha\},$$

$$= -Sd\phi - x^\alpha\{dq[\,\overline{E} + (\,\overline{vx}\,\overline{B}\,)\,]_\alpha + qd[\,\overline{E} + [(\,\overline{vx}\,\overline{B}\,)\,]_\alpha\}.$$

so that for ϕ = constant

$$\left(\frac{\partial G}{\partial q}\right)_\phi = -x^\alpha[\,\overline{E} + (\,\overline{vx}\,\overline{B}\,)\,]_\alpha - x^\alpha q\left(\frac{\partial[\,\overline{E} + (\,\overline{vx}\,\overline{B}\,)\,]_\alpha}{\partial q}\right)_\phi.$$

but $\overline{E} + (\,\overline{v} \times \overline{B}\,)$ is independent of the charge q and, therefore,

$$\left(\frac{\partial G}{\partial q}\right)_\phi = -x^\alpha[\,\overline{E} + (\,\overline{vx}\,\overline{B}\,)\,]_\alpha.$$

If the charge is not in motion, then

$$\left(\frac{\partial G}{\partial q}\right)_\phi = -x^\alpha E_\alpha \tag{11.162}$$

since v = 0.

If G is the self-energy of a charged particle, then by Equation (11.162)

$$\frac{\partial G}{\partial q} = -rE_r \ ,$$

the change in the self-energy is given with respect to a change in the charge q. If we assume a spherically symmetric charge density, ρ, then

$$dq = \rho dv = 4\pi r^2 \rho dr \ .$$

We may then find the free energy by the integration

Five Dimensional Fields

$$\int_{G_0}^{G} dG = -\int_{q_0}^{q} r\,E_r dq = -\int_{0}^{R} 4\pi r^3 E_r \rho\, dr \quad , \qquad (11.163)$$

where R represents the radius within which the charge density ρ is contained.

The field relating the electric field and the charge density is

$$\overline{\nabla} \bullet (\varepsilon \overline{E}) = 4\pi \rho \quad . \qquad (11.164)$$

In spherical coordinates Equation (11.164) gives us

$$\left(\frac{\varepsilon}{r^2}\right)\frac{\partial(r^2 E_r)}{\partial r} = 4\pi\rho \quad ,$$

when we consider only a radially symmetric field E_r. Thus,

$$\rho dr = \left(\frac{\varepsilon}{r\pi r^2}\right) d(r^2 E_r) \quad . \qquad (11.165)$$

Substituting Equation (11.165) into Equation (11.163) the self-energy is then found to be

$$G = -\varepsilon \int (r E_r) d(r^2 E_r) + G_0 \quad . \qquad (11.166)$$

If we now use the non-singular electric field given by

$$E_r = \frac{e}{4\pi\varepsilon r^2}\left(1 - \frac{\lambda}{r}\right)e^{-\left(\frac{\lambda}{r}\right)}$$

in the integral of Equation (11.166), we find

$$G = -e\int_{0}^{R} \frac{e}{4\pi\varepsilon r}\left(1 - \frac{\lambda}{r}\right)e^{-\left(\frac{\lambda}{r}\right)} d\left[\left(\frac{e}{4\pi\varepsilon}\right)\left(1 - \frac{\lambda}{r}\right)e^{-\left(\frac{\lambda}{r}\right)}\right] + G_0 \quad ,$$

which may be integrated so that

$$G = \left[\frac{-e^2}{2(4\pi)^2\varepsilon\lambda}\right]\left[\frac{\left(\frac{\lambda}{R}\right)^3 + \frac{3}{4}\left(\frac{\lambda}{R}\right)^2}{-\frac{1}{2}\left(\frac{\lambda}{R}\right) - \frac{1}{4}}\right]e^{-\left(\frac{2\lambda}{R}\right)} + G_0 \quad . \qquad (11.167)$$

To find the specific value of the self-energy, we must find the R that minimizes G. Therefore, set

$$\frac{\partial G}{\partial R} = 0 \quad .$$

392

After carrying out the required differentiation and simplifying, this is satisfied if

$$R^2 + \left(\frac{3}{5}\lambda\right)R - \left(\frac{2}{5}\lambda\right) = 0 \ . \tag{11.168}$$

Equation (11.168) only has one positive root, which is $R = 0.4\lambda$. Substituting this result into Equation (11.167), the self-energy becomes

$$G = \left[\frac{-e^2}{(4\pi)^2\varepsilon\lambda}\right](0.063379) + G_0 \ ,$$

or

$$G = \frac{-7.26235 \ x \ 10^{-3} \ MeV \ - \ fermi}{\lambda \ (fermi)} + G_0 \ ,$$

(11.169)
when λ is given in units of Fermi.

An example may be a proton for which λ is approximately 1 Fermi if the proton-proton scattering data is considered. Then if $\lambda \approx 1$ Fermi,

$$G = -7.26235 \ x \ 10^{-3} \ MeV + G_{0p},$$

so that

$$G_{op} = 938.263 \ \text{MeV} \tag{11.170}$$

is the part of the proton rest energy independent of its charge. The charge energy of the proton would then be

$$G_{ch} = -7.26235 \ x \ 10^{-3} \ MeV \ ,$$

which is negligibly small compared to the non-charge energy G_{op}.

What is the nature of the energy G_{op}? It is not energy caused by the presence of electric charge on the protons. Also the self-energy, G, was found for a resting particle. If we associate the resting self-energy, G, with the rest mass as

$$G_p = m_{op}c^2 = G_{ch} + G_{op},$$

and G_{ch} is the portion of the proton's rest energy that is due to its charge, then G_{op} must be that portion of the rest energy that is due to the proton mass above. In this case, the proton mass energy, G_{op}, is given by Equation (11.170).

Suppose we consider an electron and assume that $\lambda_e \approx 10^{-3}$ Fermi. Then

$$G = 0.511 MeV = -7.26235\ MeV + G_{0e-},$$

or the mass energy of the electron would then be $G_{oe-} = 7.773\ MeV$ whereas its charge energy is $G_{ce-} = -7.26235\ MeV$.

We find that the non-singular electrostatic potential can specify the self-energy of a charged particle when its mechanical free energy is minimized. However, for our efforts we have been handed additional questions concerning the particles mass energy that, for now, will remain unanswered.

Chapter 12 Hydrodynamic Systems

At first blush it may appear that since almost all of the preceding parts of this book are related to fields there seems no reason, or support, for turning now to the field of hydrodynamics. However, hydrodynamics may represent the field best able to provide an understanding of just how the fifth dimension of mass density fits into the picture of nature painted by the Dynamic Theory. Also, we will see that the restrictions placed upon the five dimensional manifold by the requirement of conservation of mass leads to the classical Navier-Stokes equations in which the fifth dimension leaves its foot print in the expression for viscosity that may be experimentally measured. The ability to measure a macroscopic effect of the fifth dimension was one of the first predictions made using the Dynamic Theory.

The equation of motion for the fifth dimension, mass density, appears as a generalization of the principle of the conservation of mass. Further, in classical hydrodynamic systems five equations in five unknowns are used. It seems logical then to expect the five equations of motion appearing in the five dimensional Dynamic Theory to be generalizations of the classical equations. An added incentive to investigate the possibilities of this generalization is gained when electromagnetically contained ionized plasmas with mass conversion are considered. For, if the five equations are generalizations of the classical hydrodynamic equations, then the use of the five dimensional fields allowing mass conversion should provide an entirely new viewpoint of electromagnetically contained plasma leading to a controlled fusion reactor.

Since it is suspected that the five equations of motion resulting from the application of the principle of increasing entropy to a thermo-mechanical system are generalizations of the classical equations, it then becomes necessary to show that this is indeed the case. This seems possible by restricting the system so that it corresponds to the usual system considered.

First, from the Dynamic Theory approach, the manifold required for a description of the system is the five dimensional manifold of space, time, and mass density. Within this manifold the continuity equation of mass no longer holds for the general system. We can, however, restrict our system by first requiring that the system remain on a hyper-surface within the five dimensional manifold. For a system so restricted, any of the five dimensions may be considered as functions of the other four. In particular, since by custom in hydrodynamics the mass density is considered to be a function of space and time, we may consider the mass density to be the variable chosen to be function of the others or

$$\gamma = \gamma\left(x^0, x^1, x^2, x^3\right)$$

so that

$$d\gamma = \left(\frac{\partial \gamma}{\partial x^\alpha}\right) dx^\alpha \quad .$$

Such a system will be constrained to be on a hyper-surface embedded within the five dimensional manifold of space, time, and mass density as shown and upon this hyper-surface will be described in a four dimensional manifold of space and time.

If we further restrict our system by requiring that the total derivative of the mass density to be zero or

$$d\gamma = 0 = \frac{d\gamma}{dx^\alpha} dx^\alpha$$

then

$$\frac{d\gamma}{dt} = 0 = \frac{\partial \gamma}{\partial t} + \frac{\partial \gamma}{\partial x^1} v^1 + \frac{\partial \gamma}{\partial x^2} v^2 + \frac{\partial \gamma}{dx^3} v^3$$

or

$$\frac{\partial \gamma}{\partial t} + grad\ \gamma \bullet \bar{v} = 0 \ , \tag{12.1}$$

which is the usual continuity equation. Thus, by restricting the system to this particular hyper-surface we have constrained the system to obey the continuity equation as does a usual hydrodynamic system.

Not only does this restriction place our system within the space-time manifold where we may compare the resulting four equations of motion with the equations of motion in relativistic theories but, since the seven gauge field equations plus the charge continuity equation must hold in the five dimensional manifold they must also hold on t he hyper surface. This allows the new field quantities to be expressed as functions of the $\bar{E}$ and $\bar{B}$ fields and the partial derivatives of the mass densities. Further, it appears that the additional $\bar{B}$ field equations may be used to determine a dependence of the $\bar{E}$ and $\bar{B}$ fields upon the mass density and/or its changes.

Then by comparing the equations of motion obtained here for the system restricted to the mass conservation hyper-surface with the relativistic Navier-Stokes equations it should be possible to identify the viscous coefficients with the field quantities and perhaps see how the viscosity depends upon these fields.

Previously we saw that conservation of mass embedded a four dimensional hyper-surface into the five dimensional manifold of space, time and mass. At that time we did not look into the properties of surfaces other than to look at the equations of curvature for the hyper-surface. We saw that these equations were Einstein's field equations of his general theory. Here, though, we will need to use a little more of the properties of the hyper-surface so we should spend a little time looking into these properties.

Since we have restricted the system to a h yper-surface where the mass density is a f unction of space and time, then the surface is defined by five equations of the type

Hydrodynamic Systems

$$x^i = x^i(u^0, u^1, u^2, u^3) \ . \tag{12.2}$$

Further, since $x^4 = y/a_0$ and $x^4 = x^4(x^0, x^1, x^2, x^3)$, then Equation (12.2) becomes

$$x^0 = u^0, x^1 = u^1, x^2 = u^2, x^3 = u^3$$

and

$$x^4 = f(u^0, u^1, u^2, u^3) \ .$$

Since u^0, u^1, u^2, and u^3 are independent variables, the locus defined by Equation (12.2) is four dimensional, and these equations give the coordinates x^i of a point on the hyper-surface when u^0, u^1, u^2, and u^3 are assigned particular values. This point of view leads one to consider the surface as a four dimensional manifold S embedded in a five dimensional enveloping space. We can also study surfaces without reference to the surrounding space, and consider parameters u^0, u^1, u^2, and u^3 as coordinates of points in the surface.

If we assign to u^0 in Equation (12.2) some fixed value u^0, we obtain a three dimensional manifold

$$x^i = x^i(u^0, u^1, u^2, u^3), (i = 0, 1, 2, 3, 4)$$

which is a three dimensional manifold lying on the hyper-surface S defined by Equation (12.2). By assigning fixed values for any three of the four hyper-surface variables we obtain a net of curves, on the hyper-surface, which may be called coordinate curves.

Obviously the parametric representation of a hyper-surface in the form of Equation (12.2) is not unique, and there are infinitely many curvilinear coordinate systems which can be used to locate points on a given hyper-surface S. Thus, if one introduces a transformation

$$u^0 = u^0(u^{-0}, u^{-1}, u^{-2}, u^{-3}) \ ,$$
$$u^1 = u^1(u^{-0}, u^{-1}, u^{-2}, u^{-3}) \ ,$$
$$u^2 = u^2(u^{-0}, u^{-1}, u^{-2}, u^{-3}) \ , \text{ and} \tag{12.3}$$
$$u_3 = u^3(u^{-0}, u^{-1}, u^{-2}, u^{-3}) \ ,$$

398

The Dynamic Theory

where the u^α (u^{-0}, u^{-1}, u^{-2}, u^{-3}) are of class C^1 and are such that the Jacobian

$$J = \frac{\partial(u^0, u^1, u^2, u^3)}{\partial(u^{-0}, u^{-1}, u^{-2}, u^{-3})}$$

does not vanish in some region of the variables u, then one can insert the values from Equation (12.3) in Equation (12.2) and obtain a different set of parametric equations

$$x^i = f^i(u^{-0}, u^{-1}, u^{-2}, u^{-3}) \tag{12.4}$$

defining the hyper-surface S. Equations (12.3) can be looked upon as representing a transformation of coordinates in the hyper-surface.

First Fundamental Quadratic Form

The properties of hyper-surfaces that can be described without reference to the space in which the hyper-surface is embedded are termed "intrinsic" properties. A study of intrinsic properties is made to depend on a certain quadratic differential form describing the metric character of the hyper-surface. We proceed to derive this quadratic form for our restricted system.

It will be convenient to adopt certain conventions concerning the meaning of indices to be used. We will be dealing with two distinct sets of variables: those referring to the five dimensional space into which the hyper-surface is embedded (these are five in number) and with four coordinates u^0, u^1, u^2, and u^3 referring to the four dimensional manifold S. In order not to confuse these sets of variables we shall use Latin letters for the indices referring to the space variables and Greek letters for the hyper-surface variables. Thus, Latin indices will assume values 0, 1, 2, 3, 4 and Greek indices will have the range of values 0, 1, 2, 3. A transformation T of space coordinates from one system X to another $\underline{X}$ will be written as

Hydrodynamic Systems

$$T = x^i = x^i \left(x^{-0}, x^{-1}, x^{-2}, x^{-3}, x^{-4} \right)$$

a transformation of Gaussian hyper-surface coordinates, such as described by Equation (12.3) will be denoted by

$$u^\alpha = u^\alpha \left(u^{-0}, u^{-1}, u^{-2}, u^{-3} \right).$$

A repeated Greek index in any term denotes the summation from 0 to 3; a repeated Latin index represents the sum from 0 to 4. Unless a statement to the contrary is made, we shall suppose that all functions appearing in the discussion are of class C^2 in the regions of their definitions.

Consider the hyper-surface S defined by

$$x^i = x^i (u^0, u^1, u^2, u^3) \ , \qquad (12.5)$$

where the x^i are coordinates covering the five dimensional space in which the hyper-surface S is embedded, and a curve C on S defined by

$$u_\alpha = u^\alpha (\tau) \ , \quad \tau_1 \le \tau \le \tau_2 \qquad (12.6)$$

where the u^α's are the Gaussian coordinates covering S. Viewed from the surrounding space, the curve defined by Equation (12.5) is a curve in a five dimensional manifold, which we shall assume, for the present, is the Riemannian entropy manifold of the Dynamic Theory, and its element of arc is given by the formula

$$(dq^0)^2 = g_{ij} \, dx^i dx^j \, . \qquad (12.7)$$

From Equation (12.5) we have

$$dx^i = \frac{\partial x^i}{\partial u^\alpha} \, du^\alpha \qquad (12.8)$$

where, as is clear from Equation (12.6)

$$du^\alpha = \frac{du^\alpha}{d\tau} \, d\tau \ .$$

Substituting from Equation (12.7) and Equation (12.8), we get

The Dynamic Theory

$$\left(dq^0\right)^2 = \hat{g}_{ij} \frac{\partial x^i}{\partial u^\alpha} \frac{\partial x^j}{\partial u^\beta} du^\alpha du^\beta$$

$$= A_{\alpha\beta} du^\alpha du^\beta \ ,$$

where

$$A_{\alpha\beta} \equiv \hat{g}_{ij} \frac{\partial x^i}{\partial u^\alpha} \frac{\partial x^j}{\partial u^\beta} \ . \tag{12.9}$$

The expression for $(dq^0)^2$, namely

$$\left(dq^0\right)^2 = A_{\alpha\beta} du^\alpha du^\beta \ ,$$

is the square of the linear element of C lying on the hyper-surface S, and the right hand member of Equation (12.9) can be called the First Fundamental Quadratic form of the hyper-surface. The length of arc of the curve is given by

$$q_2^0 - q_1^0 = \int_{\tau_1}^{\tau_2} \gamma \sqrt{A_{\alpha\beta} \dot{u}^\alpha \dot{u}^\beta} \ d\tau \ ,$$

where

$$\dot{u}^\alpha = \frac{du^\alpha}{d\tau}$$ and q^0 is the specific entropy. The total change in the entropy along the curve C would then be

$$\gamma \left(q_2^0 - q_1^0\right) = \int_{\tau_1}^{\tau_2} \gamma \sqrt{A_{\alpha\beta} u^\alpha u^\beta} \ d\tau \ . \tag{12.10}$$

Consider a transformation of surface coordinates

$$u^\alpha = u^\alpha \left(\bar{u}^0, \bar{u}^1, \bar{u}^2, \bar{u}^3\right) \tag{12.11}$$

with a non-vanishing Jacobian

$$J = \left| \frac{\partial u^\alpha}{\partial \bar{u}^\beta} \right| .$$

It follows from Equation (12.11) that

$$du^\alpha = \frac{\partial u^\alpha}{\partial \bar{u}^\beta} d\bar{u}^\beta \ ,$$

and hence

$$\left(dq^{0}\right)^{2} = A_{\alpha\beta}\frac{\partial u^{\alpha}}{\partial \overline{u}^{\gamma}}\frac{\partial u^{\beta}}{\partial \overline{u}^{\delta}}\,d\overline{u}^{\gamma}\,d\overline{u}^{\delta}\ .$$

If we set

$$A_{\gamma\delta} = A_{\alpha\beta}\frac{\partial u^{\alpha}}{\partial \overline{u}^{\gamma}}\frac{\partial u^{\beta}}{\partial \overline{u}^{\delta}}\ ,$$

we see that the set of quantities $A_{\alpha\beta}$ represents a symmetric covariant tensor of rank two with respect to the admissible transformations Equation (12.11) of hyper-surface coordinates. The fact that the $A_{\alpha\beta}$ are components of a tensor is also evident from Equation (12.9), since $(dq^{0})^{2}$ is an invariant and the quantities $A_{\alpha\beta}$ are symmetric. The tensor $A_{\alpha\beta}$ is called the covariant metric tensor of the hyper-surface.

Since the form, Equation (12.9), is positive definite, the determinant

$$A = \left| A_{\alpha\beta} \right| > 0$$

and we can define the reciprocal tensor $A^{\alpha\beta}$.

The properties of surfaces concerning the study of the first fundamental quadratic form

$$\left(dq^{0}\right)^{2} = A_{\alpha\beta}du^{\alpha}du^{\beta} \tag{12.12}$$

constitute a body of what is known as the 'intrinsic geometry of surfaces.' They take no account of the distinguishing characteristics of surfaces as they might appear to observer located in the surrounding space. Two surfaces, a cylinder and a cone, for example, appear to be entirely different when viewed from the enveloping space, and yet their intrinsic geometries are completely indistinguishable since the metric properties of cylinders and cones can be described by the identical expressions for square of the element of arc. If a coordinate system exists on each of the two surfaces such that the linear elements on them are characterized by the same metric coefficients $A_{\alpha\beta}$, the surfaces are called "isometric."

Thus, if our description of the restricted system is done only in terms of the intrinsic geometry of the hyper-surface we may lose sight of features which may characterize our system when viewed from the enveloping space. Therefore, in order to characterize the shape of the surface we must develop a view which involves the enveloping space.

Second Fundamental Quadratic Form

Sometimes it may be beneficial or desirable to describe the hyper surface from the surrounding space. An entity that provides a characteristic of the shape of the surface as it appears from the enveloping space is the normal line to the surface. The behavior of the normal line as its foot is displaced along the surface depends on the shape of the surface, and it occurred to Gauss to describe certain properties of surfaces with the aid of a quadratic form that depends in a fundamental way on the behavior of the normal line. Before we introduce this new quadratic form let us recall the definition Equation (12.9), or

$$A_{\alpha\beta} \equiv \hat{g}_{ij} \frac{\partial x^i}{\partial u^\alpha} \frac{\partial x^j}{\partial u^\beta} \, (i, j = 0, 1, 2, 3, 4)\,(\alpha, \beta = 0, 1, 2, 3) \ .$$

We note that the foregoing formulas depend on both the Latin and Greek indices, and we recall that the Latin indices run from 0 to 4 and refer to the surrounding space, whereas the Greek indices assume values 0, 1, 2, a nd 3 and are associated with the embedded hyper-surface. Furthermore, the dx^i and g_{ij}'s are tensors with respect to the transformations induced on the space variables x^i, whereas such quantities du^α and $A_{\alpha\beta}$ are tensors with respect to the transformation of Gaussian surface coordinates u^α. Equation (12.9) is a cu rious one since it contains partial derivatives $\partial x^i / \partial u^\alpha$ depending on both Latin and Greek indices. Since both $A_{\alpha\beta}$ and g_{ij} in Equation (12.9) are tensors, this formula suggests that $\partial x^i / \partial u^\alpha$ can be regarded either as a co ntravariant space vector or as a co variant surface vector. Let us investigate this set of quantities more closely.

Hydrodynamic Systems

Let us take a small displacement on the hyper-surface S, specified by the surface vector du^α. The same displacement, as is clear from Equation (12.8) is described by the space vector with components

$$dx^i = \frac{\partial x^i}{\partial u^\alpha} du^\alpha \quad . \tag{12.13}$$

The left hand member of this expression is independent of the Greek indices, and hence it is invariant relative to a change of the surface coordinates u^α. Since du^α is an arbitrary surface vector, we conclude that

$$\frac{\partial x^i}{\partial u^\alpha} \tag{12.14}$$

is a covariant surface vector. On the other hand, if we change the space coordinates, the du^α, being a surface vector, is invariant relative to this change, so the Equation (12.14) must be a contravariant space vector. Hence we can write Equation (12.14) as

$$x^i_\alpha \equiv \frac{\partial x^i}{\partial u^\alpha} \tag{12.15}$$

where the indices properly describe the tensor character of this set of quantities.

Let $\bar{A}$ and $\bar{B}$ be a pair of surface vectors drawn from one point P of S.

Then using Eqn. (7.14) they can be represented in the form

$$A^i = x^i_\alpha A^\alpha \ and \ B^i = x^i_\alpha B^\alpha \ . \tag{12.16}$$

The five dimensional vector product, defined by

$$N^k = \varepsilon^{kij} A_i B_j \quad , \tag{12.17}$$

is the vector normal to the tangent plane determined by the vectors $\bar{A}$ and $\bar{B}$, and the unit vector n perpendicular to the tangent plane, so oriented that $\bar{A}$, $\bar{B}$, and n form a right handed system, is

$$\bar{n} = \frac{\varepsilon^{kij} A_i B_j}{\left| \varepsilon^{\alpha\beta} A_\alpha B_\beta \right|}. \tag{12.18}$$

404

The Dynamic Theory

We call the vector n the unit normal vector to the hyper-surface S at P. Clearly, n is a function of coordinates $(u^0,\ u^1,\ u^2,\ u^3)$, and as the point $P(u^0,u^1,u^2,u^3)$ is displaced to a new position $P(u^0+du^0,u^1+du^1,u^2+du^2,u^3+du^3)$, the vector n undergoes a change

$$d\bar{n} = \frac{\partial \bar{n}}{\partial u^\alpha} du^\alpha \qquad (12.19)$$

whereas the position vector r is changed by the amount

$$d\bar{r} = \frac{\partial \bar{r}}{\partial u^\alpha} du^\alpha \quad . \qquad (12.20)$$

Let us form the scalar product

$$d\bar{n} \bullet d\bar{r} = \frac{\partial \bar{n}}{\partial u^\alpha} \bullet \frac{\partial \bar{r}}{\partial u^\beta} du^\alpha du^\beta \quad . \qquad (12.21)$$

If we define

$$b_{\alpha\beta} = \frac{1}{2}\left(\frac{\partial \bar{n}}{\partial u^\alpha} \bullet \frac{\partial \bar{r}}{\partial u^\beta} + \frac{\partial \bar{n}}{\partial u^\beta} \bullet \frac{\partial \bar{r}}{\partial u^\alpha} \right)$$

so that Equation (12.21) reads

$$d\bar{n} \bullet d\bar{r} = b_{\alpha\beta} du^\alpha du^\beta \quad , \qquad (12.22)$$

the left-hand member of Equation (12.22), being the scalar product of two vectors in a Riemannian space by being in the entropy manifold, is an invariant; moreover, from symmetry with respect to α and β, it is clear that the coefficients $du^\alpha\ du^\beta$ in the right-hand member of Equation (12.22) define a covariant tensor of rank two. The quadratic form

$$B \equiv b_{\alpha\beta} du^\alpha du^\beta \quad , \qquad (12.23)$$

called the second fundamental quadratic form of the hyper-surface, will be shown to play an essential part in the study of hyper-surfaces when they are viewed from the surrounding space, just as the first fundamental quadratic form $A \equiv d\bar{r} \bullet d\bar{r}$ or

$$A = A_{\alpha\beta} du^\alpha du^\beta \quad ,$$

did in the study of intrinsic properties of a hyper-surface.

Hydrodynamic Systems

We can rewrite the formula Equation (12.18) in terms of the components $x_\alpha{}^i$ of the base vectors a_α. We denote the covariant components of n by n^i and observe that its covariant components n_i are given by

$$n_i = \frac{\varepsilon_{ijk} A^j B^k}{A B \sin \theta} \qquad (12.24)$$

and

$$A B \sin \theta = \varepsilon_{\alpha\beta} A^\alpha A^\beta \ . \qquad (12.25)$$

Substituting in Equation (12.24) from Equation (12.16) and Equation (12.25), we get

$$\left(n_i \, \varepsilon_{\alpha\beta} - \varepsilon_{ijk} x_\alpha^j x_\beta^k \right) A^\alpha B^\beta = 0$$

and, since this relation is valid for all surface vectors, we conclude that

$$n_i \, \varepsilon_{\alpha\beta} = \varepsilon_{ijk} x_\alpha^j x_\beta^k \ . \qquad (12.26)$$

Multiplying Equation (12.26) by $\varepsilon^{\alpha\beta}$, and noting that $\varepsilon^{\alpha\beta}\varepsilon_{\alpha\beta}=2$, we get the desired result

$$n_i = \frac{1}{2} \varepsilon^{\alpha\beta} \varepsilon_{ijk} x_\alpha^j x_\beta^k \ . \qquad (12.27)$$

It is clear from the structure of this formula that n_i is a space vector which does not depend on the choice of surface coordinates. This fact is also obvious from purely geometric considerations. This displays that the second fundamental quadratic form of the hyper-surface, which is based upon the normal to the hyper-surface, is described by a space vector. It is this reason that the second fundamental quadratic form allows one to characterize the hyper-surface from the surrounding space.

Tensor Derivatives

We wish to reduce the second fundamental quadratic form Equation (12.23) analytically by the operation of tensor differentiation of tensor fields which are functions of both surface and space coordinates. To do this we shall first

The Dynamic Theory

present the concept of tensor differentiation introduced by A. J. McConnell.

Let us consider a curve C lying on a given hyper-surface S and a vector A^i defined along C. If τ is a parameter along C, we can compute the intrinsic derivative $\delta A^i / \delta \tau$ of A^i, namely,

$$\frac{\delta A^i}{\delta \tau} = \frac{dA^i}{dt} + \hat{g} \left\{ \begin{matrix} i \\ jk \end{matrix} \right\} A^j \frac{dx^k}{d\tau} \ , \qquad (12.28)$$

In formula Equation (12.28) the Christoffel symbols

$$\hat{g} \left\{ \begin{matrix} i \\ jk \end{matrix} \right\}$$ refer to the space coordinates x^i and are formed

from the metric coefficients g_{ij}. This is indicated by the prefix $\hat{g}$ on the symbol. On the other hand, if we consider a surface vector A defined along the same curve C, we can form the intrinsic derivative with respect to the surface variables, namely,

$$\frac{\delta S^\alpha}{\delta \tau} = \frac{dA^\alpha}{d\tau} + a \left\{ \begin{matrix} \alpha \\ \beta\gamma \end{matrix} \right\} A^\beta \frac{du^\gamma}{d\tau} \ . \qquad (12.29)$$

In this expression the Christoffel symbols

$$a \left\{ \begin{matrix} \alpha \\ \beta\gamma \end{matrix} \right\}$$ are formed from the metric coefficients $a_{\alpha\beta}$

associated with the Gaussian hyper-surface coordinates u_α. A geometric interpretation of these formulas is at hand when

the fields A^i and A^α are such that $\dfrac{\delta A^i}{\delta \tau} = 0$ and $\dfrac{\delta A^\alpha}{\delta \tau} = 0$. In

the first equation the vectors A^i form a parallel field with respect to C, considered as a space curve, whereas the equation

$\dfrac{\delta A^\alpha}{\delta \tau} = 0$ defines a parallel field with respect to C regarded

as a surface curve. The corresponding formulas for the intrinsic derivatives of the covariant vectors A_i and A_α are

$$\frac{\delta A_i}{\delta \tau} = \frac{d A_i}{d \tau} - \hat{g} \begin{Bmatrix} k \\ ij \end{Bmatrix} A_k \frac{dx^j}{d\tau} \qquad (12.30)$$

and

$$\frac{\delta A_\alpha}{\delta \tau} = \frac{d A_\alpha}{d \tau} - a \begin{Bmatrix} \gamma \\ \alpha\beta \end{Bmatrix} A_\gamma \frac{du^\beta}{d\tau} \ . \qquad (12.31)$$

Next consider a t ensor field T^i_α, which is a contravariant vector with respect to a transformation of space coordinate x^i and a covariant vector relative to a transformation of surface coordinates u^α. An example of a field of this type is the tensor $x^i_\alpha = \dfrac{\partial x^i}{\partial n^\alpha}$ introduced earlier. If T^i_α is defined over a s urface curve C, and the parameter along C is τ, then T^i_α is a function of τ. We introduce a parallel vector field A_i along C, regarded as a s pace curve, and a p arallel vector field B^α along C, viewed as a s urface curve, and form an invariant

$$\phi(\tau) = T^i_\alpha A_i B^\alpha \ .$$

The derivative of $\varphi(\tau)$ with respect to the parameter τ is given by the expression

$$\frac{d\phi}{d\tau} = \frac{dT^i_\alpha}{d\tau} A_i B^\alpha + T^i_\alpha \frac{dA^i}{d\tau} B^\alpha + T^i_\alpha A_i \frac{dB^\alpha}{d\tau} \ , \qquad (12.32)$$

which is obviously an invariant relative to both the space and surface coordinates. But, since the fields $A_i(\tau)$ and $B^\alpha(\tau)$ are parallel,

$$\frac{d A_i}{d \tau} = \hat{g} \begin{Bmatrix} k \\ ij \end{Bmatrix} A_k \frac{dx^j}{d\tau} \quad and \quad \frac{dB^\alpha}{d\tau} = \bar{a} \begin{Bmatrix} \alpha \\ \beta\gamma \end{Bmatrix} B^\beta \frac{du^\gamma}{d\tau} \ ,$$

and Equation (12.32) becomes

$$\frac{d\phi}{d\tau} = \left[\frac{dT^i_\alpha}{d\tau} + \hat{g} \begin{Bmatrix} i \\ jk \end{Bmatrix} T^j_\alpha \frac{dx^k}{d\tau} - \bar{a} \begin{Bmatrix} \delta \\ \beta\gamma \end{Bmatrix} T^i_\delta \frac{du^\gamma}{d\tau} \right] A_i B^\alpha \ . \qquad (12.33)$$

Since this is invariant for an arbitrary choice of parallel fields A_i and B^α, the quotient law guarantees that the expression in the brackets of Equation (12.33) is a t ensor of the same

408

character as T^i_α. We call this tensor the intrinsic tensor derivative of T^i_α with respect to the parameter τ, and write

$$\frac{\delta T^i_\alpha}{\delta t} = \frac{dT^i_\alpha}{d\tau} + \hat{g}\begin{Bmatrix} i \\ jk \end{Bmatrix} T^j_\alpha \frac{dx^k}{d\tau} = a\begin{Bmatrix} \delta \\ \alpha\gamma \end{Bmatrix} T^i_\delta \frac{du^\gamma}{d\tau} \quad .$$

If the field T^i_α is defined over the entire hyper-surface S, we can argue that, since

$$\frac{\delta T^i_\alpha}{\delta \tau} \equiv \left[\frac{\partial T^i_\alpha}{\partial u^\gamma} + g\begin{Bmatrix} i \\ jk \end{Bmatrix} T^j_\alpha x^k_\gamma - a\begin{Bmatrix} \delta \\ \gamma\alpha \end{Bmatrix} T^i_\delta \right] \frac{du^\gamma}{d\tau}$$

is a tensor field and $\dfrac{du^\gamma}{d\tau}$ is an arbitrary surface vector (for C is arbitrary), the expression in the bracket is a tensor of the type $T^i_{\alpha\gamma}$. We write

$$T^i_{\alpha\gamma} \equiv \frac{\partial T^i_\alpha}{\partial u^\gamma} + \hat{g}\begin{Bmatrix} i \\ jk \end{Bmatrix} T^j_\alpha x^k_\gamma - a\begin{Bmatrix} \delta \\ \alpha\gamma \end{Bmatrix} T^i_\delta \qquad (12.34)$$

and call $T^i_{\alpha,\gamma}$ the tensor derivative of T^i_α with respect to u^γ.

The extension of this definition to more complicated tensors is obvious from the structure of Equation (12.34). Thus the tensor derivative of $T^i_{\alpha\beta}$ with respect to u^γ is given by

$$T^i_{\alpha\beta,\gamma} = \frac{\partial T^i_{\alpha\beta}}{\partial u^\gamma} + g\begin{Bmatrix} i \\ jk \end{Bmatrix} T^i_{\alpha\beta} x^k_\gamma - a\begin{Bmatrix} \delta \\ \alpha\gamma \end{Bmatrix} T^i_{\delta\beta} - a\begin{Bmatrix} \delta \\ \beta\gamma \end{Bmatrix} T^i_{\alpha\delta} \quad . \quad (12.35)$$

If the surface coordinates at any point P or S are geodesic, and the space coordinates are orthogonal Cartesian, we see that at that point the tensor derivatives reduce to the ordinary derivatives. This leads us to conclude that the operations of tensor differentiations of products and sums follow the usual rules and that the tensor derivatives of g_{ij}, $A_{\alpha\beta}$, G_{ijk}, $\varepsilon_{\alpha\beta}$ and their associated tensors vanish. Accordingly, they behave as constants in the tensor differentiation.

The apparatus developed in this section permits us to obtain easily and in the most general form an important set of formulas due to Gauss. We will also deduce with its aid the second fundamental quadratic form of a surface already encountered.

We begin by calculating the tensor derivative of the tensor x_α^i, representing the components of the surface base vectors a_α. We have

$$x_{\alpha\beta}^i = \frac{\partial^2 x^i}{\partial u^\alpha \partial u^\beta} + \hat{g}\{^{\ i}_{jk}\}x_\alpha^j x_\beta^k - a\{^{\ \delta}_{\alpha\beta}\}x_\delta^i \ ,$$

from which we deduce that

$$x_{\alpha,\beta}^i = x_{\beta,\alpha}^i \ . \tag{12.36}$$

Since the tensor derivative of $a_{\alpha\beta}$ vanishes, we obtain, upon differentiating the relation

$$A_{\alpha\beta} = \hat{g}_{ij}\, x_\alpha^i x_\beta^j \ ,$$

$$\hat{g}_{ij}\, x_{\alpha,\gamma}^i x_\beta^j + \hat{g}_{ij}\, x_\alpha^i x_{\beta,\gamma}^j = 0 \ . \tag{12.37}$$

Interchanging α, β, γ cyclically leads to two formulas:

$$\hat{g}_{ij}\, x_{\beta,\alpha}^i x_\gamma^j + \hat{g}_{ij}\, x_\beta^i x_{\gamma,\alpha}^j = 0 \tag{12.38}$$

and

$$\hat{g}_{ij}\, x_{\gamma,\beta}^i x_\alpha^j + \hat{g}_{ij}\, x_\gamma^i x_{\alpha,\beta}^j = 0 \ . \tag{12.39}$$

If we add Equation (12.38) and Equation (12.39), subtract Equation (12.37), and take into account the symmetry relation Equation (12.36), we obtain

$$g_{ij}\, x_{\alpha,\beta}^i x_\gamma^j = 0 \ .$$

This is the orthogonality relation which states that $x_{\alpha,\beta}^i$ is a space vector normal to the surface, and hence it is directed along the unit normal n^i. Consequently, there exists a set of functions $b_{\alpha\beta}$ such that

$$x_{\alpha,\beta}^i = b_{\alpha\beta} n^i \ .$$

The quantities $b_{\alpha\beta}$ are the components of a symmetric surface tensor, and the differential quadratic form

$$B \equiv b_{\alpha\beta}\, du^\alpha\, du^\beta$$

is the desired second fundamental form.

The Dynamic Theory

Now since $n_j n^i = \delta^i_j$, and $n_i = \hat{g}_{ij} n^j$, then

$$b_{\alpha\beta} = \hat{g}_{ij}\, x^i_{\alpha,\beta} n^j \ ;$$

but since $n_i = \dfrac{1}{2}\varepsilon^{\alpha\beta}\, \varepsilon_{ijk}\, x^j_\alpha x^k_\beta$, then

$$b_{\alpha\beta} = \frac{1}{2}\varepsilon^{\gamma\delta}\, \varepsilon_{ijk}\, x^i_{\alpha,\beta} x^j_\gamma x^k_\delta \ . \tag{12.40}$$

We now have, in Equations (12.9) and (12.40), the formulas necessary to determine the first and second fundamental quadratic forms for our system constrained to a four dimensional hyper-surface within a five dimensional space.

Hydrodynamics

Our objective is to show that by appropriately constraining our system we arrive at the Navier-Stokes equations. Let us determine the first fundamental quadratic form.

First recall that our system was restricted so that $x^4 = x^4(x^0,\, x^1,\, x^2,\, x^3)$ or the mass density is a function of space and time; then we have the relations

$$x^0 = u^0,\ x^1 = u^1,\ x^2 = u^2,\ x^3 = u^3,$$

and

$$x^4 = f\,(x^0, x^1, x^2, x^3) = f\,(u^0, u^1, u^2, u^3) \ .$$

Since Equation (12.9) is

$$A_{\alpha\beta} = \hat{g}_{ij}\, \frac{\partial x^i}{\partial u^\alpha}\, \frac{\partial x^j}{\partial u^\beta} = \hat{g}_{ij}\, x^i_\alpha x^j_\beta \ ,$$

then

$$A_{00} = g_{00} = 2g_{04}\, f_0 + g_{44}(f_0)^2 \ ,$$

where $f_0 \equiv \dfrac{\partial f}{\partial u^0}$.

In a similar fashion we may determine the remaining coefficients and find that

$$A_{\alpha\beta} = \hat{g}_{\alpha\beta} + h_{\alpha\beta} \tag{12.41}$$

where

$$h_{\alpha\beta} = 2\hat{g}_{\alpha 4} f_{,\beta} + \hat{g}_{44} f_{,\alpha} f_{,\beta} \quad ; \quad \alpha, \beta = 0, 1, 2, 3 \quad , \qquad (12.42)$$

where the $h_{\alpha\beta}$ are functions of the partial derivatives of the mass density with respect to space and time in addition to space and time from the $\hat{g}_{i4}$ where i=0,1,2,3,4.

Though we may use Equation (12.40) to determine the metric coefficients for the second fundamental quadratic form, it is not necessary for the current presentation.

The hyper-surface which conservation of mass embeds in the five dimensional space is a four dimensional curvilinear space-time manifold. Thus the relativistic hydrodynamic equations are applicable here so long as the metric coefficients are determined as coefficients of the hyper-surface quadratic form.

The complete energy-momentum tensor for a fluid in a flat Riemannian space-time manifold is given by

$$T^{\alpha\beta} = \gamma \dot{u}^{\alpha} \dot{u}^{\beta} + \frac{P}{c^2} (\dot{u}^{\alpha} \dot{u}^{\beta} - g^{\alpha\beta}) \qquad (12.43)$$

where $\dot{u}^{\alpha} \equiv du^{\alpha}/ds$, s is the arc length. Then based upon this energy-momentum tensor the flow of a fluid under the effect of its own internal pressure force is given by setting the divergence of Equation (12.43) equal to zero, or

$$T^{\alpha\beta},_{\beta} = 0 \quad . \qquad (12.44)$$

If we reduce Equation (12.43) to the non-relativistic limit, the use of Equation (12.44) gives us

$$g^{\beta\delta} \tau_{\alpha\beta,\delta} = \gamma^{a}_{\alpha} \quad , \quad \alpha, \beta, \delta = 1, 2, 3 \quad ,$$

where $\tau^{\alpha\beta} = Pg^{\alpha\beta}$ is the three dimensional stress tensor of an ideal fluid.

If in Equation (12.43) we use the fact that the metric coefficients for the hyper-surface may be written as the sum in Equation (12.41), then we have

$$T^{\alpha\beta} = \gamma \, \dot{u}^{\alpha} \dot{u}^{\beta} + \frac{P}{c^2} \left(\dot{u}^{\alpha} \dot{u}^{\beta} - \hat{g}^{\alpha\beta} - h^{\alpha\beta} \right) \quad , \qquad (12.45)$$

where it m ust be remembered that the $\dot{u}^{\alpha}\dot{u}^{\beta}$ are also dependent upon this same sum. In the non-relativistic limit the effects of this sum of metric tensors appear as a sum in the stress tensor

$$\tau^{\alpha\beta} = -P\hat{g}^{\alpha\beta} - Ph^{\alpha\beta} \ , \ \alpha,\beta = 1,2,3 \ . \tag{12.46}$$

Recall that the $\hat{g}^{\alpha\beta}$ refer to the three dimensional physical space viewed from the five dimensional manifold. The $h^{\alpha\beta}$, however, contain the information about the surface embedded in the five dimensional space. If we then associate the tensor

$$t^{\alpha\beta} \equiv -Ph^{\alpha\beta} \tag{12.47}$$

with the viscous stresses, we are saying that the viscous stresses depend upon the geometric character of the hyper-surface.

In the limit of small displacements we write the strain velocity tensor as

$$\dot{e}_{\alpha\beta} = \frac{1}{2}\left(v_{\alpha,\beta} + v_{\beta,\alpha}\right) \ .$$

Then the first order coefficients of viscosity are related to the strain velocity tensor and viscous stresses according to

$$t^{\alpha\beta} = \frac{c^{\alpha\beta\delta n}}{2}\left(v_{\delta,n} + v_{n,\delta}\right) \ . \tag{12.48}$$

If we then use Equation (12.47) in Equation (12.48), we find that the relationship between the geometric character of the hyper-surface and the viscous coefficients is given by

$$-Ph^{\alpha\beta} = \frac{c^{\alpha\beta\delta n}}{2}\left(v_{\delta,n} + v_{n,\delta}\right) \ . \tag{12.49}$$

Equation (12.49) then expresses the functional dependence of the viscous coefficients upon the strain velocities, pressure, mass density, and their derivatives.

Relativistic Hydrodynamics.

By viewing classical hydrodynamics to be given by the conservation of mass embedding a four dimensional hyper-surface within a five dimensional manifold, the association Equation (12.49) between the geometric properties of the hyper-surface and the viscous coefficients could be tentatively made. We may now go back and develop this relationship more completely.

The hyper-surface, which becomes embedded in the five dimensional manifold by the conservation of mass restriction that $x^4 = x^4(x^0, x^1, x^2, x^3)$ is the four dimensional relativistic, entropy manifold. Thus, for the surface we may use the relativistic energy-momentum tensor, which is

$$T^{\mu\nu} = \gamma\, u^{\mu\nu} u + \frac{P}{c^2}\left(u^\mu u^\nu - g^{\mu\nu}\right) , \qquad (12.50)$$

where $u^\mu \equiv dx^\mu / dx^0$ and $\mu, \nu = 0,1,2,3$. The divergence of Equation (12.50) yields the flow equations for a fluid under the effects of its own internal pressure.

However, from the viewpoint of the Dynamic Theory, the surface metric coefficients may be written in terms of the metric coefficients of the first four space coordinates as given by Equations (12.42) and (12.43), or

$$A_{\alpha\beta} = \hat{g}_{\alpha\beta} \quad , \quad \alpha, \beta = 0, 1, 2, 3 \ .$$

Thus, the square of the arc length for the entropy manifold may be written as

$$\left(dq^0\right)^2 = A_{\alpha\beta} dx^\alpha dx^\beta = \hat{g}_{\alpha\beta} dx^\alpha dx^\beta + h_{\alpha\beta} dx^\alpha dx^\beta$$

or, if $u^\alpha \equiv dx^\alpha / dq^0$, then

$$A_{\alpha\beta} u^\alpha u^\beta = \hat{g}_{\alpha\beta} u^\alpha u^\beta + h_{\alpha\beta} u^\alpha u^\beta \ .$$

Then on the hyper-surface the energy-momentum tensor would become

$$T^{\alpha\beta} = \gamma\, u^\alpha u^\beta + \frac{P}{c^2}\left(u^\alpha u^\beta - A^{\alpha\beta}\right)$$

or

$$T^{\alpha\beta} = \gamma\, u^\alpha u^\beta + \frac{P}{c^2}\left(u^\alpha u^\beta - \hat{g}_{\alpha\beta} - h^{\alpha\beta}\right) . \qquad (12.51)$$

Since the surface coordinates, x^α, are the same as the first four coordinates of the surrounding space, the velocities u^α are the same whether considered as surface or space vectors. The difference between the surface view and a four dimensional space view appears in the metric coefficients. Thus, while the square of the arc element on the surface is unity, the square of the arc element in the surrounding space is not, or

$$1 = A_{\alpha\beta} u^\alpha u^\beta$$

but

$$\hat{g}_{\alpha\beta} u^\alpha u^\beta = 1 - h_{\alpha\beta} u^\alpha u^\beta \quad .$$

Classical Hydrodynamics.

Suppose we consider the metric given by $\hat{g}_{\alpha\beta}$ to be a flat space then because of Equation (12.51) we may write

$$T^{\alpha\beta} = \gamma\, u^\alpha u^\beta + \frac{P}{c^2}\left(u^\alpha u^\beta - \hat{g}^{\alpha\beta}\right) - \frac{P}{c^2} h^{\alpha\beta} \quad .$$

If we then form the space divergence

$$T^{0\nu}{}_{,\nu} = \frac{1}{c}\frac{\partial}{\partial t}\left\{\gamma - \frac{P}{c^2} h^{00}\right\} + \frac{\partial}{\partial x^\alpha}\left\{\gamma\frac{u^\alpha}{c} + \frac{P}{c^2}\left(\frac{u^\alpha}{c} - H^{0\alpha}\right)\right\} = 0 \quad ,$$

this may be written as

$$\frac{1}{c}\left[\frac{\partial\gamma}{\partial t} + \overline{\nabla}\bullet(\gamma\overline{v})\right] - \frac{1}{c^2}\frac{\partial(Ph^{00})}{\partial t} + \frac{1}{c^2}\overline{\nabla}\bullet(P\overline{v}) - \frac{1}{c^2}\overline{\nabla}\bullet(Ph^{-0}) = 0 \quad ,$$

where h^0 has components $h^{0\alpha}$, $\alpha = 1,2,3$. Therefore,

$$\frac{\partial\gamma}{\partial t} + \overline{\nabla}\bullet(\gamma\overline{v}) = \frac{1}{c}\overline{\nabla}\bullet(P\overline{v}) + \frac{1}{c^2}\left[\frac{\partial(Ph^{00})}{\partial t} + \overline{\nabla}\bullet(\overline{Ph_0})\right] \quad ,$$

so that if h^ν is a four-vector with components $h^{0\nu} \equiv h^\nu$, then

$$\frac{\partial\gamma}{\partial t} + \overline{\nabla}\bullet(\gamma\overline{v}) = -\frac{1}{c}\overline{\nabla}P\bullet\overline{v} - \frac{P}{c}v^\alpha{}_{,\alpha} + \frac{1}{c^2}(Ph^\nu)_{,\nu} \quad .$$

The remaining components of the divergence are given by

$$T^{\alpha\nu}{}_{,\nu} = \frac{1}{c}\frac{\partial}{\partial t}\left\{\gamma\frac{u^\alpha}{c} + \frac{Pv^\alpha}{c^3} - \frac{P}{c^2}h^{\alpha 0}\right\}$$

$$+\frac{\partial}{\partial x^\beta}\left\{\gamma\frac{v^\alpha v^\beta}{c^2} + \frac{P}{c^2}\left(\frac{v^\alpha v^\beta}{c^2} + \delta^{\alpha\beta}\right) - \frac{Ph^{\alpha\beta}}{c^2}\right\} = 0 \quad,$$

which may be rearranged to read

$$\gamma\left[\frac{\partial v^\alpha}{\partial t} + \bar{v}\bullet\Delta\, v^\alpha\right] = -\frac{\partial P}{\partial x^\alpha} - v^\alpha\left[\frac{\partial\gamma}{\partial t} + \bar{\Delta}\bullet(\gamma\,\bar{v})\right]$$

$$+\frac{1}{c}\frac{\partial(Ph^{\alpha 0})}{\partial t} + \frac{\partial(Ph^{\alpha\beta})}{\partial x^\beta}$$

$$-\frac{1}{c^2}\left[\frac{\partial(Pv^\alpha)}{\partial t} + \bar{\Delta}\bullet(P\,v^\alpha\,\bar{v})\right]\quad.$$

If we look at the non-relativistic limit, then, by neglecting the terms $P(v/c)$, we get

$$\frac{\partial\gamma}{\partial t} + \bar{\nabla}\bullet(\gamma\,\bar{v}) = \frac{1}{c^2}\left[\frac{1}{c}\frac{\partial(Ph^{00})}{\partial t} + \bar{\nabla}\bullet(Ph^{-0})\right]\quad.$$

The multiplicative factor $1/c^2$ on the right-hand side suggests that

$$\frac{\partial\gamma}{\partial t} + \bar{\nabla}\bullet(\gamma\,\bar{v}) \cong 0 \quad,$$

which is the statement of the conservation of mass and the assumption we chose to place our system on a particular surface. This corresponds to a cl assical system where conservation of mass is assumed. Therefore, on the surface of a curve specified by

$$T^{\mu\nu}{}_{,\nu} = 0 \quad,$$

we must then have

$$\gamma\left[\frac{\partial v^\alpha}{\partial t} + \bar{v}\bullet\bar{\nabla}\,v^\alpha\right] = -\frac{\partial P}{\partial x^\alpha} + \frac{1}{c}\frac{\partial(Ph^{\alpha)}}{\partial t} + \frac{\partial(Ph^{\alpha\beta})}{\partial x^\beta}$$

or

The Dynamic Theory

$$\gamma \, a^{\alpha} = -\frac{\partial P}{\partial x^{\alpha}} + \frac{1}{c}\frac{\partial(\,Ph^{\alpha 0}\,)}{\partial t} + \frac{\partial(\,Ph^{\alpha\beta}\,)}{\partial x^{\beta}}$$

$$= -\frac{\partial P}{\partial x^{\alpha}} + \frac{1}{c}\frac{\partial(\,Ph^{\alpha 0}\,)}{\partial t} + (\,Ph^{\alpha\beta}\,)_{,\beta} \quad , \quad \beta = 1,2,3 \ .$$

Thus, we may write

$$\gamma \, a^{\alpha} = \tau^{\alpha\beta}{}_{,\beta}$$

where

$$\tau^{\alpha\beta} = -P\hat{g}^{\alpha\beta} + Ph^{\alpha\beta} = -P(\,\hat{g}^{\alpha\beta} - h^{\alpha\beta}\,) \ . \qquad (12.52)$$

The term $\dfrac{1}{c}\dfrac{\partial(\,Ph^{\alpha 0}\,)}{\partial t}$ has been neglected in Equation (12.52).

Thus we see that the geometric character of the hyper-surface, contained in the term $Ph^{\alpha\beta}$, behaves as if it were a v iscous effect to be added to the normal viscous effects. Recalling Equation (12.43), it may be seen that the viscous like effects of the geometry of the hyper-surface depend upon the density gradient. If these terms exist, they must be very small in everyday phenomena. Yet if we consider phenomenon which involve very large density gradients, these terms could become large enough to see.

Shock Waves.

One field of physical phenomena that displays large density gradients is shock waves. Therefore let us take a quick look at the effect of these additional terms on t he description of a shock front for a steady, one dimensional shock.

The total stress in a steady, one dimensional shock would be given by

$$\sigma = P\left\{1 - \left(\frac{1}{a_0^{2}}\right)\left(\frac{\partial\gamma}{\partial x}\right)^{2}\right\} + \eta\left(\frac{du}{dx}\right) \ ,$$

when $g_{11}=1$ and h_{11} is evaluated using Equation (12.43). However, for a s teady shock we also have the jump conditions

$$\gamma u = k_1 \ ,$$

$$k_1 u + \sigma = k_2 \ ,$$

and

$$k_1^2 E - \frac{\sigma^2}{2} = k_3 \ .$$

These equations represent the conservation of mass, momentum, and energy in a shock wave. By using the conservation of mass relation we may write the total stress as

$$\sigma = P + \eta_{eff} \left(\frac{du}{dx} \right) \ ,$$

where

$$\eta_{eff} \equiv \eta - \frac{P k_1^2}{a_0^2 u^4} \left(\frac{du}{dx} \right) \qquad (12.53)$$

may be called the effective viscous coefficient. Since within the shock front the velocity gradient du/dx is negative, we see that the effective viscous coefficient acts so as to thicken the shock front when compared to the classical viscous coefficient.

Using the second jump condition, an expression for the velocity gradient is

$$\frac{du}{dx} = \frac{a_0^2 u^4 \eta}{2 P k_1^2} \left\{ 1 - \sqrt{1 + \frac{4 P k_1^2}{a_0^2 u^4 \eta^2} [P - k_2 + k_1 u]} \right\}, \quad (12.54)$$

which may be approximated by

$$\frac{du}{dx} = -\left(\frac{1}{\eta} \right) [P - k_2 + k_1 u] \left\{ 1 - \frac{P k_1^2}{a_0^2 u^4 \eta^2} [P - k_2 + k_1 u] \right\}. \quad (12.55)$$

The effect of the correction term on the velocity gradient is seen in Equation (12.55), because the multiplicative factor outside the brackets is the classical expression for the negative velocity gradient. The effect of the correction term lessens the negative velocity gradient and extends the shock front.

The effect of the correction term in Equation (12.53) is estimated by considering the strong shock dependence of

The Dynamic Theory

pressure upon shock velocities. For instance, the shock pressure, from the jump conditions, is

$$P = \gamma_o U u_p \ . \tag{12.56}$$

If the shock velocity is related linearly to the particle velocity as the assumed solid equation of state, $U = c_o + s u_p$, then Equation (12.56) becomes

$$P = \frac{\gamma_o U}{s}(U - c_o).$$

Thus, for strong shocks, P varies approximately as the square of the shock velocity.

Consider Equation (12.54) or (12.55). From either of these equations, the velocity gradient varies as the square of the shock velocity. Using these two conclusions in Equation (12.53)for the total viscosity η_{eff} and remembering that the integration constant k_1 is given by $-\gamma_o U$, the effective viscosity varies approximately as the square of the shock velocity or, essentially , as the pressure.

The conclusion is that if the effective viscosity varies with the pressure, an increase by the same factor of 10^3 must be accompanied by a viscosity increase by the same factor of 10^3. This explains the apparent discrepancy between the reported low and high pressure aluminum viscous effects. For instance, the Asay-Bertholf limits are:

$$P = 25 \text{ GPa} \qquad \eta > 40 \text{ poise}$$
$$P = 36 \text{ GPa} \qquad \eta < 2,500 \text{ poise} \ .$$

Another experiment places an upper limit of 10^3 poise for a shock pressure of 40 GPa. If 10^2 poise is considered representative of the viscosity when P 10 GPa, then from Equation (12.53), a pressure of 10^3-10^4 GPa must be accompanied by a viscous effect of 10^4-10^5 poise.

This total viscosity estimate is supported by numerical integration across the shock front using the Tillotson equation of state for aluminum. The classically predicted rise times for shocks of *40* GPa with $\eta=575$ p and $5x10^3$ GPa for $\eta=5x10^4$ p are duplicated by using the effective viscosity expression in Equation (12.53) with $\eta=1.0$ p and $a_0{\approx}365$ g/cm^4.

Thus, the Dynamic Theory correlates these data points that appear contradictory by classical theory. Further, these data points provide an estimate of the new universal constant appearing in the Dynamic Theory. This value of a_0 provides an estimate of other predictions of the theory in fields other than shock waves.

Mass Conservative Electrodynamics

One of the incentives for seeking to determine whether the five equations of motion were generalizations of the classical hydrodynamic equations was the possibility of shedding new light upon electromagnetically contained fusion plasmas. Now before mass conversion is accomplished the plasma must reach certain conditions. The attainment of these conditions involves electromagnetic fields not encountered in usual circumstances on earth. If the Dynamic Theory is to be believed, then perhaps it may provide new insight into the attainment of the appropriate conditions before mass conversion begins.

The following development still assumes conservation of mass in order to see the geometry of the hyper-surface for a s ystem under the influence of electromagnetic fields.

Suppose we now describe the behavior of charged matter under the influence of an electromagnetic field from the viewpoint of the Dynamic Theory. From this viewpoint the conservation of mass has the effect of restricting our system to a four dimensional hyper-surface which is embedded in the five dimensional manifold of space, time, and mass density.

The Dynamic Theory

Since we desire to consider the effects of an electromagnetic field we must consider a gauge function. When a gauge function exists, the square of the arc length in the entropy space is related to the square of the arc length in the sigma space by

$$\left(dq^{0}\right)^{2} = \hat{g}_{ij}\, dx^{i}\, dx^{j} = \left(\frac{1}{h_{00}}\right)\hat{g}_{ij}\, dx^{i}\, dx^{j} = \left(\frac{1}{h_{00}}\right)(d\sigma)^{2} \quad .$$

When the system is restricted to a hyper-surface by the relation $x^{4}=x^{4}(x^{0},x^{1},x^{2},x^{3})$, then the entropy surface may be written as

$$\left(dq^{0}\right)^{2} = \hat{a}_{\alpha\beta}\, du^{\alpha}\, du^{\beta} \quad ,$$

where

$$\hat{a}_{\alpha\beta} = \hat{g}_{ij}\, \frac{\partial x^{i}}{\partial u^{\alpha}}\, \frac{\partial x_{i}}{\partial u^{\beta}} = g_{ij}\, x^{i}_{\alpha}\, x^{j}_{\beta} \quad .$$

Likewise for the sigma surface

$$(d\sigma)^{2} = \hat{a}_{\alpha\beta}\, du^{\alpha}\, du^{\beta}$$

where

$$\hat{a}_{\alpha\beta} = \hat{g}_{ij}\, x^{i}_{\alpha}\, x^{j}_{\beta} \quad .$$

Thus, we have

$$\hat{a}_{\alpha\beta} = \left(\frac{1}{h_{00}}\right)\hat{a}_{\alpha\beta} \quad .$$

The principle of increasing entropy requires that the equations of motion be geodesics in the entropy space but they will appear as equations involving forces in the sigma space. We desire to expose these forces and, therefore, should work in the sigma space. Our objective then is to determine the effect of embedding a four dimensional surface given by $x^{4}=x^{4}(x^{0},\ x^{1},\ x^{2},\ x^{3})$ in the sigma space and thus obtain a sigma surface describing a system subjected to the classical conservation of mass restriction.

Hydrodynamic Systems

Having previously determined the metric coefficients for the entropy space by Equations (12.41) and (12.42) we may write the coefficients for the sigma surface as

$$\hat{a}_{\alpha\beta} = h_{00}\,\hat{a}_{\alpha\beta} = h_{00}\left[\hat{g}_{\alpha\beta} + h_{\alpha\beta}\right].$$

However by considering the effects of the electromagnetic field as a force we must first consider the space field tensor:

$$F_{ij} = \begin{vmatrix} 0 & E_1 & E_2 & E_3 & V_4 \\ -E_1 & 0 & B_3 & -B_2 & V_1 \\ -E_2 & -B_3 & 0 & B_1 & V_2 \\ -E_3 & B_2 & -B_1 & 0 & V_3 \\ -V_4 & -V_1 & -V_2 & -V_3 & 0 \end{vmatrix}$$

If we restrict ourselves to the classical field quantities E and B and for the moment assume that the field quantities V_4 and V are zero or negligible, then we obtain only the effects of the hyper-surface viewpoint. This assumption seems reasonable considering the interpretation of the new field quantities are gravitational effects. Under this assumption our field tensor becomes

$$F_{ij} = \begin{vmatrix} 0 & E_1 & E_2 & E_3 & 0 \\ -E_1 & 0 & B_3 & -B_2 & 0 \\ -E_2 & -B_3 & 0 & B_1 & 0 \\ -E_3 & B_2 & -B_1 & 0 & 0 \\ 0 & 0 & 0 & 0 & 0 \end{vmatrix}$$

We can now use this space field tensor to determine the appearance of the fields when viewed from the surface. The surface field tensor will be given by

$$F_{\alpha\beta} = F_{ij}\, x_\alpha^i\, x_\beta^j.$$

But since $x_\alpha^i = \delta_{\alpha i}$ for $i,\alpha = 0,1,2,3$ and $x_\alpha^4 = f_\alpha$, the surface field tensor of a purely electromagnetic space field tensor is only the four dimensional portion of the space field tensor since $F_{i4} = 0$ for $i=0,1,2,3,4$.

422

Thus when we use the relativistic energy-momentum tensor for the surface, we have

$$T^{\mu\nu} = \gamma u^{\mu} u^{\nu} + \frac{1}{c^2}\left[F^{\mu}_{\alpha} F^{\alpha\nu} + \frac{1}{4}\hat{a}^{\mu\nu} F^{\alpha\beta} F_{\alpha\beta} \right] , \qquad (12.57)$$

which is the relativistic energy-momentum tensor for matter under the influence of electromagnetic fields. But since $\hat{a}_{\alpha\beta} = \hat{g}_{\alpha\beta} + h_{\alpha\beta}$, then Equation (12.57) becomes

$$T^{\mu\nu} = \gamma u^{\mu} u^{\nu} + \frac{1}{c^2}\left[F^{\mu}\alpha F^{\alpha\nu} + \frac{1}{4}\left(\hat{g}^{\mu\nu} + h^{\mu\nu} \right) F^{\alpha\beta} F_{\alpha\beta} \right]$$

or

$$T^{\mu\nu} = T^{\mu\nu}_{rel} + T^{\mu\nu}_{geo} \qquad (12.58)$$

where

$$T^{\mu\nu}_{rel} \equiv \gamma u^{\mu} u^{\nu} + \frac{1}{c^2}\left[F^{\mu}\alpha F^{\alpha\nu} + \frac{1}{4}\hat{g}^{\mu\nu} F^{\alpha\beta} F_{\alpha\beta} \right]$$

is the four dimensional space relativistic energy momentum tensor and

$$T^{\mu\nu}_{geo} \equiv \left(\frac{1}{4c^2} \right) h^{\mu\nu} F^{\alpha\beta} F_{\alpha\beta}$$

is the portion of the energy-momentum tensor which contains the geometrical properties of the hyper-surface.

From Equation (12.58) we can say that the Dynamic Theory has the appearance of adding a term to the relativistic energy-momentum tensor. This term contains the geometrical character of the surface and represents the difference between the appearance of the energy-momentum tensor when viewed from the surrounding space as compared to the view from the hyper-surface.

If we take the divergence of the energy-momentum tensor Equation (12.58), we have

$$T^{\mu\nu}{}_{,\nu} = T^{\mu\nu}_{rel,\nu} + T^{\mu\nu}_{geo,\nu} .$$

The additional force terms from the surface geometry are given by

Hydrodynamic Systems

$$\left(\frac{1}{4c^2}\right)\left(h^{\mu\nu}\,F^{\alpha\beta}\,F_{\alpha\beta}\right)_{,\nu} = F^{\mu} \quad .$$

But if we define

$$F^{\alpha\beta}\,F_{\alpha\beta} \equiv -16\tau\xi \qquad (12.59)$$

as the electromagnetic energy density, where

$$\xi = \frac{1}{8\pi}\left(E^2 + B^2\right)$$

then the geometric energy-momentum tensor becomes

$$T^{\mu\nu}_{geo} = \frac{-4\pi\,h^{\mu\nu}\xi}{c^2}$$

and the additional forces are given by

$$F^{\mu} = -\frac{4\pi}{c^2}\left(h^{\mu\nu}\xi\right)_{,\nu}.$$

We may also look at the radiation pressure predicted by the Dynamic Theory to see how the surface restriction affects the relativistic prediction of radiation pressure.

The relativistic radiation pressure is taken as one third of the three dimensional Maxwell stress tensor which is the space portion of the energy-momentum tensor, or

$$T^{M}_{\alpha\beta} = \frac{1}{4\pi}\left(E_\alpha E_\beta + B_\alpha B_\beta\right) - \delta_{\alpha\beta}\xi$$

where $\alpha,\ \beta = 1,\ 2,\ 3$.

To get the equivalent stress tensor for the Dynamic radiation pressure we must add the space portion of Equation (12.59) so that the total stress tensor becomes

$$T_{\alpha\beta} = \frac{1}{4\pi}\left(E_\alpha E_\beta + B_\alpha B_\beta\right) - \delta_{\alpha\beta}\xi - 4\,h_{\alpha\beta}\xi$$

$$= \frac{1}{4\pi}\left(E_\alpha E_\beta + B_\alpha B_\beta\right) - \xi\left(\delta_{\alpha\beta} + h_{\alpha\beta}\right) \quad .$$

We can then obtain the negative of the trace by

$$-\{T\} = -\left[\frac{1}{4\pi}\left(E^2 + B^2\right) - 3\xi - \left(h_{11} + h_{22} + h_{33}\right)\xi\right]$$

$$= -\left[-\xi - \left(h_{11} + h_{22} + h_{33}\right)\xi\right] .$$

The radiation pressure is then given by

$$P = \frac{\xi}{3}[1 + h_{11} + h_{22} + h_{33}] . \tag{12.60}$$

The first term in Equation (12.60) is the classical radiation pressure in electrodynamics. The remaining three terms give the difference between the pressure predicted by the Dynamic Theory and the classical prediction. To determine what this difference is let us restrict our system to again be very near equilibrium so that the $g_{\alpha4} = 0$ for $\alpha = 0$, 1, 2, 3 and $g_{44} =$ constant. Thus, we have a flat space. For this space we have

$$h_{\alpha\alpha} = \frac{\hat{g}_{44}}{a_0^2}\left(\frac{\partial\gamma}{\partial x^\alpha}\right)^2$$

from Equation (12.60) and $g_{44} = -1$. Thus

$$h_{11} + h_{22} + h_{33} = -\frac{1}{a_0^2}\left[\left(\frac{\partial\gamma}{\partial x^1}\right)^2 + \left(\frac{\partial\gamma}{\partial x^3}\right)^2 + \left(\frac{\partial\gamma}{\partial x^3}\right)^2\right] . \tag{12.61}$$

By substituting Equation (12.61) into Equation (12.60) the pressure becomes

$$P = \frac{\xi}{3}\left\{1 - \frac{1}{a_0^2}\left[\left(\frac{\partial\gamma}{\partial x^1}\right)^2 + \left(\frac{\partial\gamma}{\partial x^2}\right) + \left(\frac{\partial\gamma}{\partial x^3}\right)^2\right]\right\} .$$

However, since the classical pressure is given by $P_c = \frac{\xi}{3}$, then the pressure predicted by the Dynamic Theory becomes

$$P_D = P_c\left\{1 - \left(\frac{1}{a_0^2}\right)\left[\left(\frac{\partial\gamma}{\partial x^1}\right)^2 + \left(\frac{\partial\gamma}{\partial x^2}\right)^2 + \left(\frac{\partial\gamma}{\partial x^3}\right)^2\right]\right\} .$$

We see then that the Dynamic Theory predicts a decrease in the radiation pressure as a result of viewing the system to be restricted to a four dimensional hyper-surface embedded in a five dimensional space by the conservation of mass. The amount of this decrease in pressure depends upon the gradient of the mass density and the constant a_0. Once the constant a_0 is determined, then the deviation in predicted pressures can be specified.

This prediction should appear in attempts to use electrodynamics to control ionized plasmas and perhaps there are large enough density gradients for these predictions to show up in cosmological events.

Part 3 Summary of What is New

In the foregoing there is a lot of detailed and complicated mathematical logic. This proliferation of equations may make it difficult to keep track of which concepts are old and which are new. This part of the book is added to summarize some of the new concepts that arise in the new theoretical development. Each of the topics listed in Chapter 13 m ay deserve a full chapter or book itself. However, there is no r oom for that detailed or comprehensive of coverage for each topic. Therefore, I have chosen to list several topics with a brief discussion and leave more detailed presentation for some later publication.

Before listing the topics of new concepts this research introduces I would like to offer some words with regard to whether this research suggests previous theories are wrong or are to be modified. When Einstein introduced his assumption of the constancy of the speed of light he introduced a modification to Newton's equations of motion to make them more consistent with the concept that all observers see the same physics. This brings up the question of whether Newton's equations of motion were wrong or were they to be considered right to the first order approximation? There are many who express the notion that if a theory is wrong in one detail then it is totally wrong regardless of any fact that it may be very nearly correct for all everyday uses. This notion of black or white, totally correct or wrong seems a little too stringent of a requirement on any attempt to formulate an understanding of natural phenomena that may not strictly follow our logic though we develop it to the best of our abilities.

Indeed the question as to whether man can formulate a theory that exactly explains natural phenomena is a question that philosophers might debate for years without reaching any conclusion. It is not my position here to enter into such a debate. Rather I would note that each of the theories such as the special and general theories of relativity and quantum mechanics are found in this research to have some aspect that is supported and some aspects or interpretations that are not supported.

I would not, then, say that this research proves any of these theories to be wrong. Nor would I say that this research prove any to be correct. Rather I would state that the research here supports some aspects of each while it does not support other aspects.

Chapter 13 What's New?

In this chapter a few of the foundations, interpretations, and conclusions that the Dynamic Theory introduces or arrives at that are new and different from those in the standard model of physics will be mentioned and/or briefly discussed. Each of these might be worthy of an article of their own, but a lack of space argues that a more lengthy discussion be done elsewhere. The order of these items will roughly be that of their appearance in the preceding chapters. This should represent the most logical sequence for them to appear.

Fundamental Laws

The first thing that is new in the Dynamic theory is the choice of the fundamental laws. During the development of the study of physics Newton's three laws were the first offered to describe the dynamics of the physical world. On top of his famous three laws Newton adopted the gravitational force law to use in his equations of motion.

The study of electric and magnetic phenomena developed laws which Maxwell incorporated into his four equations of electromagnetism. These equations were basically used as laws independent from Newton's three laws plus his gravitational force law.

The next in the line of introduction of laws of nature was the three laws of classical thermodynamics. The first law of thermodynamics was called the statement of conservation of energy and contained within it the forces that were considered to be used in the problem under consideration. While this law, under proper restrictions, produced Newton's first and second laws, it never seemed to have superseded Newton's laws in determining the motion of something. The second law of thermodynamics, while being stated in terms of the direction of heat flow, was generally accepted as the law that denied perpetual motion. The third law of thermodynamics established a cal ibration of the entropy by setting its value at the absolute temperature.

Einstein then stated his postulates for the foundation of his special theory of relativity. These postulates consisted of postulating that everyone should have the same physics and that the speed of light did not depend upon the motion of the source or the receiver of the light. By using his second postulate Einstein developed the famous transformations that were used to interpret the time dilation or material contraction and the mass increase for motion nearing the speed of light. He used these transformations and his postulate that all should see the same physics to modify Newton's second law for high velocity motion.

Einstein later introduced his general theory of relativity wherein he introduced a set of field equations that were interpreted to describe the motion of bodies under the influence of gravity. The general theory basically was thought to be a more general version of the combination of Newton's equation of motion and his gravitational force law.

Next in the historical sequence of the presentation of laws describing physical dynamics are the foundations of quantum mechanics. The foundations of quantum mechanics were never really called laws even though they have acquired the role of something that cannot be violated and are thought to be the foundations of all dynamics. These foundations include the assumption of all three of Newton's laws, but with the use of de Broglie's matter wave postulate to arrive at the Schrödinger wave equation. Later Dirac extended quantum into the realm of relativistic motions. Another aspect of the quantum mechanical foundation is Heisenberg's uncertainty principle that is interpreted to be universally applicable with a unique uncertainty.

Many writers following Einstein's lead have discussed the need and methods to seek a s imple set of

430

fundamental laws from which all the various aspects of the totality of physics might be determined. Perhaps the most famous of these attempts at unification is the quantization of gravitational phenomena. These attempts still leave a lot to be desired. Nowhere in the whole body of physics literature did writers suggest starting with the laws of classical thermodynamics. Indeed there existed a lot of discussion concerning the assumption that classical thermodynamics could be obtained using statistical methods. This is basically assuming that quantum mechanics with its probabilistic interpretations was more fundamental than classical thermodynamics.

So the adoption of the laws of classical thermodynamics as the foundational laws of nature from which all dynamics might be derived is indeed new. A few have even referred to classical thermodynamics as thermostatics because of its use of equilibrium states. However, the adoption of the First Law of thermodynamics leads quickly to Newtonian mechanics when the thermodynamics force, the pressure, vanishes and one assumes that the First Law is not path dependent. The path dependence of the First Law is a vital part of the foundational aspect of the First Law and to assume it away is to miss the richness of the this law when it is used in conjunction with the second law.

When the Second Law of thermodynamics is taken in the manner established by Caratheódory and applied to mechanical forces, its value is immediately seen when it may be used to show that there must exist an absolute velocity independent of the type of force considered. This leads to a determination that all forces must vanish at this absolute velocity. The vanishing of all forces at the absolute velocity is the new look to relativity and says that it is not the mass going to infinity that prevents a body from being accelerated beyond the limiting velocity, but that the force vanishes and it cannot accelerate anything to a h igher velocity. This change in apparent reason for a limit to the velocity to which a body can be accelerated leads to a m ajor change in interpretations.

The easiest method of arriving at the absolute velocity was the one used in Chapter 5 with constant velocity processes. In this derivation it was readily seen that since the absolute velocity must hold for everyone it must hold for all who are moving with constant velocity with respect to each other. This is Einstein's postulate concerning the constancy of the speed of light, but this is more general in that the constancy applies to all observers regardless of their relative motion.

The choice of the laws of classical thermodynamics as the foundational laws of physics is new and is totally different from all prior attempts to seek a simple set of laws for physics. The remainder of things presented as new in the Dynamic Theory are a direct result of this choice of laws.

Mechanical Entropy

The definition of mechanical entropy is new. It is defined just as the entropy of classical thermodynamics except that it arises when the thermodynamic force vanishes and only a mechanical force appears in the First Law. Also, the definition of mechanical entropy does not depend in any way upon a statistical basis or probabilities so an interpretation of the entropy as energy that becomes unavailable is virtually the only interpretation that can be applied. All the interpretations including chaos, probabilities, information, or statistics come only after certain restrictions are used.

Just as in classical thermodynamics where the temperature acts as an integrating denominator and is the function that provides the definition of the classical entropy, the relativistic square root is the integrating denominator for purely mechanical forces. It is the function that converts the

432

path dependent First Law to the path independent change in entropy.

Entropy and Geometry

Entropy has a new role in the Dynamic Theory for isolated systems. This new role stems from the fact that for isolated systems the entropy must not decrease. However, in order to see this role we must recognize that equations of motion determine how things proceed. In thermodynamics we may see how stable systems return to equilibrium by using the stability conditions. The stability conditions are second order differentials describing how the system will go. These second order differentials define a natural metric for determining change in entropy. Applying the stability conditions and the principle of increasing entropy leads to a relativistic looking metric where the entropy plays the role of Einstein's proper time. We will return to this later.

In determining how the stability conditions show the way the system will move we found that two metrics became involved. Both metrics were relativistic metrics in that they both had different signs on the space and time differentials. But they were not identical. One metric, which may be called the energy metric, was found to have a path dependent arc length. The other metric, the entropy metric, had a path independent arc length. This later was true because the arc length in the entropy metric is the entropy. The difference between the energy and the entropy metrics is a multiplicative function. If the path dependent energy metric is divided by this function, which can be called the gauge function, it produces the path independent entropy metric. This means the gauge function acts as a geometric integrating denominator.

Notice what is new here. Instead of having only one type of geometry to worry about we have three. Newton's use of geometry in his equations of motion initiated many discussions concerning whether or not an absolute coordinate system might exist. Einstein introduced the Riemannian geometry with his relativistic theories and this lead to discussions as to whether space was curved or not. What is really new here is that the existence of the three geometries changes the discussion drastically.

The coordinate system and geometry used for writing the First Law may be freely chosen. This means one can choose a coordinate system that makes the problem as simple as it can be made. The stability conditions and the entropy principle for an isolated system use this coordinate system from the First Law and generate the two relativistic metrics. This means that once the coordinate system is chosen for the First Law the relativistic metrics are set by the stability conditions and the entropy principle. If we called the coordinate system used in the First Law the physical coordinate system, the two relativistic metrics generated from that law through the stability conditions and the entropy principle cannot constrain the chosen coordinate system at all. This means that the physical coordinate system does not take on any characteristics of either of the relativistic metrics. If one, or both, of the relativistic metrics is/or curved that curvature does not apply to the physical system.

What is new here is that the coordinate system used in the First Law may be freely chosen. Both laws work together to produce two new metrics. These new metrics must be used to obtain a solution that maximizes the entropy during any dynamics that occurs. Therefore, the coordinate system in the First Law is the physical coordinate system with its chosen geometry and the two relativistic metrics are the solution metrics.

The Arrow of Time

Why nature causes things to run in one direction has been a topic for discussion for many centuries. What is new here is that we see that the two laws of thermodynamics specify which way the isolated system must go. The entropy principle requires the entropy to never decrease. Therefore, the entropy must increase or remain fixed. This means that in the entropy metric the arc length which is the entropy, or proper time, must never decrease. Proper time must go forward. This is the source of the arrow of time in nature.

Mechanical Enthalpy and other Energy Concepts

Entropy is a thermodynamic concept that also is useful in mechanical systems. This argues that other energy concepts currently confined to thermodynamics have mechanical counterparts. These concepts include mechanical enthalpy, mechanical Helmholtz free energy and Gibbs free energy. One new aspect that may be seen involving these energy concepts is that for non-isolated systems that exchange energy with their surroundings it is the free energy that must be minimized to determine how the dynamics must go. In all of the work presented within this book only isolated systems have been used. This leaves the non-isolated systems for a later presentation.

Space-Time-Matter

Einstein is said to have put time on an equal footing with space in that the relativistic metric developed to use on relativistic systems uses both time and space differentials in the metric. To put the three concepts of space, time and matter together implies that perhaps matter is being put on the same footing with space and time. How does this occur? When one teaches classical thermodynamics and they write the First Law on the board, they place the differential change in energy exchanged between the system and its surroundings, heat, equal to the sum of five terms, each with a differential. The first term is the system energy change. The next term is usually the thermodynamic work term, or the Pdv term. The last three terms consist of the three mechanical work terms and the differential change in position for each. This means that we have five differentials on the right hand side to set equal to the single differential on the left hand side. This means that fundamentally we have five independent variables and must produce five independent equations in order to find a solution. Now the instructor follows the time honored tradition and states that mass density, the reciprocal of the specific volume, may always be written as a function of space and time. Next the three equations of motion from Newton are introduced along with an equation of state. This is offered as the set of five independent equations that allow for a solution.

What happens then if someone should ask for a proof that mass density can always be written as a function of space and time? Can a proof be found? Of course the common usage of mass density leads to a feeling that a proof should be possible. However, if such a proof cannot be found then nature must fundamentally be five dimensional in space, time and mass density. The opposite approach may be taken. That is the mass density may be assumed to be independent of space and time and the results of this assumption may be compared with experiment. If the predictions made using the independent assumption do not compare with experiment then the assumption should be taken as incorrect. On the other hand, if all the predictions made using the independence assumption compare favorably with experiment then it may be taken that mass density is fundamentally independent of space and time.

In the preceding chapters we find numerous predictions made using the assumption that mass density is independent of space and time and all of these predictions are supported by experience more closely than predictions of the standard model. These results support the assumption that the universe is fundamentally five dimensional with the fifth dimension being the physically real mass density. This is new.

Weyl's Quantum Principle

The preceding chapters presented the historical development of Weyl's suggestion of using the gauge function to unify electromagnetism with gravity, Schrödinger's publication of his wave equations and London's derivation of the Schrödinger equations from the requirement of a unity scale factor. This effectively turns the approach to quantization around completely. In the standard model quantization is assumed as a fundamental assumption. Here quantization is the result of a restriction of some aspect of nature. That is quantization is not the fundamental property of nature that the standard model suggests it should be. Indeed the appearance of quantization as the result of a restriction places quantization in the same category as the quantization of the modes of vibration allowed for guitar strings when both ends are firmly fixed and only certain standing wave modes are allowed.

This leads naturally to the statement of a principle of quantization that was called the Weyl Quantum Principle. This was shown to result from the restriction to isentropic states. It was shown that by using the First and Second Laws plus the stability conditions the principle of increasing entropy for an isolated system produced two relativistic metrics. One of these metrics was an energy metric while the other was the entropy metric where the entropy was that arc length in that metric. This means that the very stable isentropic system must have a null trajectory in this metric. It was shown that isentropic systems satisfy the unity scale factor London used to derive the Schrödinger wave equations. This effectively displays the quantization through a restriction to isentropic states and led to the adoption of this restriction as the Weyl Quantum Principle.

The display that quantization comes as the result of a restriction of a continuous function is something new.

The standard model adopts quantization as a fundamental assumption. Here quantization is a subset of the total allowed possibilities for the dynamics of an isolated system.

5D QM, Magnetic Moments and Octets

Perhaps others have presented a five dimensional version of Dirac's relativistic quantum mechanics, though I have not seen it. However, in all the previous attempts of five dimensional fields the authors of those attempts did not consider the fifth dimension to have any physical reality and, by so doing, limited what they might see in the five dimensional quantization using these fields. When the fifth dimension is allowed a physical reality there are at least two things that come to light. First, physical reality of the field for the fifth dimension leads to the prediction of magnetic moments for a b ody without electric charge. Second, the spin states of the five dimensions include two more spin vectors in addition to the currently used spin states.

The prediction of a magnetic moment for electrically neutral, spinning bodies is new in the sense of the source of its prediction. This means that it is a new prediction that the fifth dimensional field, being the gravitational field, leads to a magnetic moment for a gravitating body that is electrically neutral. Spinning gravitational bodies have previously been thought to have magnetic moments. The source of the magnetic moment in the five dimensional fields is what is new.

Murray Gellman received a N obel Prize for his suggestion that particles come in octets. His suggestion came from the astute recognition of octets in the experimental data. It did not come from a t heoretical prediction such as is the case here. In the five dimensional Dynamic Theory the addition of the extra dimension leads to a new three component spin vector and a four component spin vector that are added to the usual three component spin vector appearing in Dirac's four dimensional equations. A well known theorem concerning the eigenstates of these spin vectors show that they may only appear in octets. The theoretical prediction of octets within particle physics is new.

WUP Fields

Newton assumed that the gravitational field diminishes with the square of the distance from a gravitational body. The same dependence upon separation was assumed for the electric field. Some may argue that this radial variation of the electric field may be derived using Maxwell's equations. However, in this procedure, shown in Chapter 7, one must use the non-physical Dirac function which diminishes the logical necessity of the conclusion. On the other hand, when the five dimensional fields are used we found not only did the long range dependence upon separation turn out to be the usual inverse square dependence, but we also learned that the field for fundamental particles had other, previously unsuspected dependences. This means there are several new things that may be seen in the fundamental particle fields.

Perhaps the first new aspect of the particle fields is the requirement that they were found for particles that satisfy the restriction to the Weyl unitary scale factor condition. This argues that these fields must apply to fundamental particles rather than composite particles. These fundamental particles were referred to as Weyl Unitary Particles, or WUPs. This is a new particle class distinction.

The Gauge Function is Fundamental

The historical development of electromagnetic fields has led to the approach that the field, being the measurable quantity is the fundamental property of electromagnetism. That is since the fields are measurable and the potentials are not the field is taken as the more fundamental property. Further, since the potentials have always been obtained by integrating the field the potential could not be uniquely specified because of the existence of the integrating constant. This means that a measurement of the field cannot specify a unique gauge potential and, since the gauge function is obtained by integration of the potential, an almost complete dismissal of the gauge function.

What the Dynamic Theory offers is a complete reversal of this approach to the gauge properties. The isolated system requires the principle of increasing entropy. This leads to the stability conditions producing a relativistic metric where the entropy is the arc length and, more importantly for this discussion, the entropy principle leads to the existence of a g auge function. From this gauge function the unitary scale factor condition, demanded by an isentropic state, produces the gauge potentials as its first order derivatives and the gauge fields as its second order derivatives. Since the gauge potentials are derivatives of the gauge function they are fully determined by the gauge function. Then the gauge fields being derivatives of the potentials are also fully determined.

The theoretical importance of fields, potentials and gauge function has been turned completely around. For the WUPs it is the gauge function that is fundamental. From this gauge function the potentials and fields are determined. This is new.

Quantized Electric Charge

Experimentally it has been seen that the electric charge on particles always comes as multiples of a fundamental unit of electric charge. However, there has been no t heoretical requirement for this quantization of electric charge until now. The theoretical requirement of the quantization of the electric charge is new. This requirement stems from the requirement that a fundamental particle be in an isentropic state which, in turn, requires that the Weyl scale factor be unitary and this requires the quantization. If the electrostatic field is given the unitary scale factor requires quantization of the allowed paths through the Schrödinger wave equation. The requirement that the fundamental particles, WUPs, retain their identifying characteristics as they move around in space and through time establishes the quantization.

Non-Singular Fields

The feature that the gauge potentials must be the logarithmic derivatives of the gauge function leads to a gauge function that depends exponentially upon space, time and mass. The gauge function's dependence upon space turns out to vanish both as the separation extends to infinity and when the separation goes to zero. It is the vanishing of the gauge potentials and fields as the separation reduces to zero, or that it is non-singular, that is new.

Previous assumptions that the electric and gravitational fields falls off as the inverse square of the separation requires that the field become infinite as the separation tends to zero. This feature of the electrostatic field to become infinite for vanishing separations necessitated the renormalization procedures developed to remove the infinities. The non-singular fields do not tend to infinity and, therefore, have no need of renormalization.

Time Dependence of Gravity

In a five dimensional manifold of space, time and mass density a gauge function must then depend upon each of these dimensions or one or more of the gauge potentials and fields will vanish. In Chapter 7 the gauge function was found to be dependent upon time through an exponential function that causes the gravitational field gets slowly weaker with time. This is new.

A time dependence of the Earth's gravitational field was reported in the 1970's by Van Flandern of the US Naval Observatory. At that time I found it interesting that the time dependence he obtained using the measured distance between the Earth and the moon over decades matched Hubble's constant very well. A recent discussion with Van Flandern before his death revealed that he had been told, and I believe convinced, that the apparent time dependence determined from the moon's orbit was due to a systematic error.

Mass Dependence of the Electric Field of WUPs

There is no surprise in finding that the gravitational field depends upon m ass. How it depends upon mass is new. The exponential dependence of the gauge function found for the mass dependence means that the electric field of the WUP depends upon the mass. This is new; however, I have not yet found a prediction that might be measured to verify this feature of the gauge function.

Electrostatic to Gravitational Force Ratio

The extraordinary difference between the strength of the electric force and gravitational force of particles has been the subject of many articles. Past scientists have speculated on the reason for this value of the ratio of these forces. Within the Dynamic Theory the fields of WUPs show that this ratio of forces is determined by the product of the time and mass dependences of the fields. The time dependence is roughly 10^{-18} and mass dependence is of the order of 10^{-22} for a ratio of electric to gravitational force of some 10^{-40}, as was shown in Chapter 7.

Though the time dependence is small it is sufficient to have measurable results over cosmological distances as will be seen later. The somewhat smaller dependence upon mass density makes it d ifficult to imagine thus far what conditions might be found in which this small effect might make itself know except in something such as the force ratio expression.

Proton-Proton Scattering

It was early proton-proton scattering that gave the first experimental indication that the electrostatic forces did not rule the interactions at nuclear separations. When the higher energies forced the protons closer together the scattering data begin to show a deviation from the scattering predictions of Coulomb forces. The scientific community then faced a decision point. On the one hand, the data might be an indication that the Maxwell equations did not tell the whole story concerning the interaction. This seemed to be a very undesirable conclusion as the Maxwell equations had been verified in many experiments. On the other hand, if the Maxwell equations were not to be messed with then there must be another force that would explain the data.

The standard model is the result of the communities' choice of a force independent of the electromagnetic force with which to explain the data. Because this data showed a rapid decrease in the repulsive forces between the protons at nuclear separations the force became referred to as the strong nuclear force. Though this notion has been researched for over three quarters of a century there is no s atisfactory display of the radial dependence of the strong nuclear force.

It was not long after this tacit assumption that the strong nuclear force was independent of the Maxwell's equations that scientists began seeking ways of unifying the strong force with the electromagnetic force. They sought unification without seeming to address the tacit assumption of independence.

The non-singular electrostatic force drastically changes the prediction of the data to be expected in proton-proton scattering. Because the non-singular nature of the electrostatic potential and force, the repulsive force between the protons begin to diminish exponentially as their separation drops to nuclear distances. This rapid reduction in the repulsive forces between the protons predicts data similar to the data obtained in the proton-proton scattering experiments. It remains to analyze the data using the non-singular force. However, the community has far too much invested in the notion of a strong nuclear force for funds to be made available for such an analysis in the near future.

The Dependence of Uncertainty upon the Gauge Function

We saw that the WUP fields come from a gauge function. We also have seen that quantum mechanics is required of an isentropic state. Along with the development of quantum mechanics there came the development of the Heisenberg uncertainty principle which has been taken to be a universal and constant value of uncertainty. In Chapter 8 we see that the quantum Poisson brackets depend upon geometry. Indeed when we use these brackets in the study of atomic phenomena we must include the effects of geometry in the covariant differentiation associated with these brackets. It then seems reasonable to assume that if the geometry is influenced such as the curving of space by mass as is done by general relativity to argue that within the space of an atom this curvature is too small to be seen.

But for the WUPs we find that the gauge function is the dominant feature of the geometry and the forces between the WUPs are due to this gauge function. Therefore, the gauge function controls the scale factor in the relativistic metric. When covariant differentiation is carried out in this Weyl space the gauge function enters into the differentiation. We then saw that value of the uncertainty, or the unit of action, varies with the gauge function. We saw that within nuclear phenomena that the units of action vary from that used in atomic phenomena.

This change in the unit of action then changes interpretations within nuclear phenomena. It is the dependence of the unit of action upon the gauge function that is new. This dependence results in different interpretations and predictions.

Phat Photons

When the question was asked as to what gauge potentials were allowed if the characteristics of the electrostatic potential was to be independent of time and position, we saw that these gauge potentials must be quantized. Further, we followed up on t he quantization of the electrostatic potential and found that the five dimensions led to a non-singular potential that depended upon space, time and mass. Now we may ask what the influence of the five dimensions and quantization of isentropic states has upon t he gauge vector potentials. In Chapter 8 we saw that the Weyl Gauge Principle required the vector potentials to lead to wave equations that predicted the propagation of light. We then investigated the results of the isentropic state upon t hese same vector potentials. The result was that there was a quantization of the energy of the light in these electromagnetic waves and that the energy of the photons was quantized by the square of the quantum number.

For decades the wave-particle duality has led scientists into discussions, debates and experimentation trying to understand why in some circumstances light behaves as a wave would and in other cases it acts as if it were quantized. There never seemed to a discussion of the potential that light might be a quantized wave that would exhibit these properties.

There are, then, two new things here. First, the wave-particle duality has an explanation in the quantized wave of the vector potentials. Next the wave energy is quantized as $\varepsilon = N^2 h\nu$. This means that the photons predicted by Einstein are those with *N=1*. Photons with quantum numbers of 2 or greater has more energy for a given frequency than does the Einstein photon and this is the reason they were given the name of phat photons to distinguish them from the Einstein photons.

Forces Violating Newton's Third Law

Virtually all physics taught and used since Newton's time has been done with Newton's third law enforced. But must it be enforced for all forces? Using the First and Second Law of thermodynamics there is no requirement that Newton's third law be enforced. These laws hold whether Newton's third holds or not. Indeed within these laws the form of the forces themselves determines whether Newton's third law holds.

We saw that the non-singular forces for different particles do not satisfy Newton's third law when their separations approached those of nuclear phenomena. Therefore, Newtonian, non-relativistic, and relativistic quantum mechanics were developed without requiring Newton's third law. Though the development of the equations of motion requires that the problem be solved as a two-body problem rather than allowing a reduction to a one-body problem, solutions were still available.

When we investigated the relativistic quantum mechanics for unlike particles when Newton's third law does not hold we found the equations produced the Yang-Mills equations. This means that the weak nuclear forces are the non-singular forces between unlike particles at nuclear separations. This is indeed new.

Using the same relativistic quantum mechanical equations for forces that do not necessarily satisfy

Newton's third law for three particles we found that equations typically displayed for the SU3 group for the strong nuclear forces. This, too, is new.

The Neutron

When we applied the non-singular forces between unlike particles we found that a proton in orbit around an electron becomes a neutron. Of course one needs to use the fact that the unit of action is control by the gauge function and we found that for the protons orbit around the electron within the neutron to be roughly 2/3 of Planck's constant. On the other hand we found that all of the spin and magnetic moment relations were satisfied if the unit of action for the electron within the neutron was approximately that of the proton reduced by the ratio of the electron mass to the proton mass.

It is not new to suggest that the neutron consist of a proton and an electron because that has been done before. Indeed every disintegration of a neutron results in the appearance of a proton and an electron and motion of the center of mass. This would suggest strongly that a neutron consist of a proton and an electron. However, arguments such as using the Heisenberg's uncertainty principle to forbid the possibility of an electron being confined to such a small space eventually won since there was no theoretical means of countering this argument. Also, with only the tacit assumption that the force holding the neutron together obeyed Newton's third law there was no basis for understanding why the center of mass should move upon disintegration. What we learned in Chapter 9 is that the forces holding the proton and electron together within the neutron do not obey Newton's third law and the center of mass is itself in orbit around a space point. A disintegration of the neutron with the center of mass in motion will, therefore, be accompanied by motion of the center of mass.

There is another reason to commend the non-singular forces that do not obey Newton's third law in the neutron. The standard model does not provide a functional form for the weak force that may be used to predict the life time of the neutron. In the Dynamic Theory the fact that the center of mass in trapped in a positive energy well means that the concept of quantum tunneling may be used to calculate the lifetime with the result comparing favorably with the experimentally determined value. This ability to predict the lifetime of the neutron is new.

Nuclear Model

A follow-up concept to the proton orbiting around an electron within the neutron is two protons orbiting around the electron. This gives a nucleus with a mass number of 2 and a charge of + one. That means it is the deuterium nucleus. This led to the notion of a nuclear model of protons orbiting around a sub-nuclear core with an overall negative charge. This is new.

We saw how such a model may be used to calculate the masses of the low mass number nuclei with a smaller RMS error than the semi-empirical mass formula now used in the standard model. (A private communication with Pat McDaniel reveals that he has found that the model can be made more accurate than I indicated.) we also saw that when two deuterium nuclei fused together to form a helium nucleus the notion of protons in orbit around a negatively charged core was displayed. However, since the helium nucleus consists of four protons and two electrons making up a six-body problem, determining the orbital solutions to larger nuclei will become even more difficult.

Fusion

In Chapter 9 we saw that when the spin axes of two deuterium nuclei are aligned we may calculate the fusion barrier by solving a six-body problem that involves the four protons and two electrons of the two deuterium nuclei. The non-singular, repulsive forces between the protons are reduced significantly when the separation of the protons approach nuclear separations. Also the repulsive force between the two electrons is reduced when their separations reach the sub-nuclear separations achieved within the helium nucleus. This reduction in the repulsive forces between protons and between the electrons allows the very strong remaining attractive force the protons have for the electrons to reduce the fusion barrier tremendously when compared with the standard nuclear model. The standard nuclear model requires the energy required for fusion to overcome the repulsive force of the Coulombic force between the protons and, therefore, the fusion barrier is expected to be much, much higher than the non-singular forces require.

Further, it is the trick of aligning the spin axes of the two deuterium nuclei that establishes this greatly reduced fusion barrier. If the two deuterium nuclei approach each other with their spin axes in any other orientation with respect to each other the fusion barrier increases to the fusion barrier of the standard model or greater. It is this alignment of the spin axes of the deuterium together with the non-singular forces which cause the significant reduction in the fusion barrier that is new.

Protons

The experimental data from high energy electrons scattered from protons indicate that the proton consists of three smaller particles. This data gives support to the standard model suggestion that the proton is made up of three quarks. However, the data does not indicate what the constituent particles are. They might be positrons and electrons for all the data cares.

What's New

The equations that provide the solution to the two protons in orbit around an electron within the deuterium nucleus are the same form of the equations needed to calculate the orbits of two positrons orbiting around an electron. The only difference between these sets of equations would be the lambdas of the protons and positrons. In Chapter 10 we found that the lambdas are proportional to the mass. Then by reducing the lambda of the proton by using the positron to proton mass ratio one obtains the lambda of the positron which is also the lambda of the electron. With this value of the lambda for the positron we may calculate orbits of two positrons in orbit around an electron which satisfy all the experimental data.

The solution showing two positrons orbiting around an electron within a proton is new.

The solution showing the constituents of the proton also allow us to calculate mass of the virtual particles thought to carry the forces within the proton. This calculation shows that the mass energy of the virtual particles within the proton is 7,254 MeV, or approximately the 7,000 MeV output of the new Large Hadron Collider. The mass energy of the virtual particles of the proton being only slightly greater than the output energy of the Large Hadron Collider argues that some of the protons should be busted up to produce positrons and electrons.

Derivation of Einstein's Field Equations

Einstein developed his field equations from the requirement that inertial mass equal gravitational mass and his concept of relativity. In Chapter 10 the requirement of conservation of mass imposed upon the five dimensional manifold of space, time and mass density embeds a four dimensional hyper surface of space and time in the five dimensional manifold. The equations specifying the curvature of the four dimensional hyper surface are Einstein's field equations.

450

Of course the only case that Einstein considered was the case wherein mass was conserved. Also, within the Dynamic Theory gravitational and inertial mass are equal. The ability to derive Einstein's field equations for the curvature of the space-time hyper surface is new. The concept of Einstein's general theory being a subset of a more fundamental theory is also new.

Magnetic Moments

Within the development of the five dimensional quantum mechanics we saw the necessity for an electrically neutral, spinning body to have a magnetic moment. When the charge to mass ratio, found in Chapters 7 and 10, is used with the parameters for the Earth's angular momentum the predicted magnetic moment of the Earth is verified by the experimental value. As far as I know the accuracy of this prediction is new.

Cosmological Red Shift

The fact that light from distant stars is red shifted is old news. The fact that the red shift is due to the distance to the star and its gravitational field has also been known for some time. What is new here is that the red shift depends exponentially upon both the star's distance and the gravitational field. The exponential dependence provides predictions of high red shifts for objects that are closer than would be the case for a linear dependence. This means the quasars may not be as queer as previously thought. At least the light output can now be attributed to natural nuclear energy rather than needing to conjure up a new energy source to explain what had appeared as prodigious energy output for their apparent distance using the linear distance red shift prediction.

Another aspect of the cosmological red shift is the prediction that by comparing the red shift obtained on the Earth's surface and that obtained at heights above the Earth one may separate the red shift due to distance from the red shift due to the star's gravitational field. This means that these data can provide distance and gravitational field strength of the cosmological objects.

Big Bang

Many authors have written about their disbelief in the big bang beginning of the universe. However, few have offered a t heoretical alternative to the singularity in Newton's gravitational force law that predicts the big bang origin. Here we find the non-singular gravitational potential and field describes all the measureable cosmological data without the need of a singularity in the beginning. Indeed there is no singularity anywhere. The existence of a theoretical alternative to the big bang is new.

Of course the reader may be thinking, "What about the cosmological background radiation? Is that explained by the non-singular gravitational field?" The answer is no, but we will see later that the background radiation has an explanation in the zero pressure radiation required by the five dimensional gauge fields.

Black Holes

The non-singular gravitational field does not allow for a prediction of a black hole due to a singularity in the gravitational field. Just as in the argument for a big bang beginning to the universe, the existence of a theoretical alternative is new.

Quantum Gravity

Scientists have been searching for a long time for something called quantum gravity. All of their searches appear to be limited to searching on a microscopic scale. Here we find that since quantum mechanics is required for isentropic systems it must apply to all isentropic systems regardless of size. Then we used this quantum mechanical approach with the gravitational potential and solved them for a solution for the planetary orbits. This solution shows that not all quantum numbers lead to a solution to the gravitational Schrödinger wave equation. This appears to be the case for the sun's planetary orbits as past researchers have reported gaps in the apparent quantum numbers.

There is one big difference in this approach to quantum gravity that may not be seen immediately. You may recall that the unit of action depends upon the gauge function. We used this feature of the Dynamic Theory in the solution for the orbits of the proton around the electron within the neutron. It also came into play when looking for solutions for the proton orbits within the deuterium nucleus. The gauge function for the gravitational field is such that the unit of action is very large compared with the unit of actions for the atomic and nuclear orbits. I don't think this was expected to be the case for quantum gravity. Therefore, while it is an easy task to put the gravitational potential into the Schrödinger wave equation, what is new is the unit of action provided by the gravitational gauge function.

Dark Matter

Dark matter was hypothesized to account for the data which showed that the tangential velocity of stars in the arms of spiral galaxies do not fall off as would be predicted by Newton's gravitational force law. Newton's gravitational force is time independent. The gravitational field in the Dynamic Theory is time dependent. It shows that the gravitational field strength diminishes in time. This means that for gravitational effects that travel at the speed of light any object at distance away from the gravitating body would be responding to a gravitational field at was stronger when it left the gravitating body. Therefore, the stars in the arms of the spiral galaxies are responding to a gravitational field strength that was greater in the past when its effects would have left the galaxy center.

Therefore, in the Dynamic Theory the time dependent gravitational field shows that the tangential velocity of the stars is the results of a t ime dependent gravitational field. The first order approximation to this time dependent effect is the same as the Modified Newtonian Dynamics (MOND) that has been hypothesized as an explanation for the effect.

An explanation of the dark matter data using a time dependent gravitational field is new.

Dark Energy

Dark energy has been hypothesized to help explain data that, when analyzed using general relativistic procedures and a time independent gravitational field, argues that the expansion of the universe is increasing rather than decreasing. The time dependence of the gravitational field enters into the analysis in three ways. First, the weakening gravitational field changes the analytical procedure. Second, the different exponential form for the red shifts depends upon the time dependence of the field. However, it is the third influence that seems to have the greatest effect on the analysis.

In the study of the expansion of the universe the use of a standard candle becomes important. This is the reason the particular supernovas used were selected. The type aI supernovas are thought to produce the same luminosity regardless of where that might be found or their distance from Earth. This may be true for a time independent gravitational field where Chandrasekhar calculated the mass required to overcome the electron gas pressure and cause the collapse that becomes the supernova. However, for a time dependent gravitational field this limiting mass becomes time dependent and the mass of the type aI supernovas are not constant in time. This means that the luminosity of new supernovas is not the same as the luminosity of older supernovas. In other words the luminosity of the standard candle is not constant as it was expected and needed to be. However, this time variation of the limiting mass may be used in the analysis of the data.

The time dependence of the Chandrasekhar limiting mass is new.

Five Dimensional Waves

Perhaps one of the previous researchers who looked into five dimensions as a means for unification of forces may have looked into the five dimensional wave equations. However, I have not read anything of such research. For me then the four wave equations that may be derived from the five dimensional gauge fields is new. Also new is the one of these wave equations is for a scalar wave while the other three are vector waves.

Though there is a curious interdependence between the gravitational field and the electric field, solutions to these waves equations come in two virtually independent sets of waves. We are accustomed to the transverse waves used in all radio, TV, and other communication systems. However, the transverse wave solution to the five dimensional wave equations has three components. Of course two of these components are the familiar electric and magnetic components. The new component is a gravitational field component. However, the energy in the gravitational component is a small percentage of the total wave energy while the electric and magnetic components share the majority of the energy.

The second set of solutions also has three components. Two of the components are the electric and the gravitational field component. The third component to this new wave is a scalar gravitational component that is somewhat like a gravitational potential, though its precise nature has not been determined. The primary difference between these waves and the first set of transverse wave solutions are is that the vector components in the new wave are in the direction of the wave propagation, or longitudinal. I am reluctant to call these wave longitudinal waves because of the scalar component that accompanies the wave. If the waves were called longitudinal then the implication would be that all components are in the direction of propagation and they are not. Therefore, I referred to these waves as non-transverse waves because none of the three components are transverse to the direction of propagation.

This non-transverse wave is new. Perhaps it should be called an electrogravitic wave since it is made up on electric and gravitational components. But what are they?

Neutrinos

Previously we showed that the quantization required by the Weyl Quantum Principle led to the non-singular potentials and fields when properties of fundamental particles were sought. Also, when we looked into the vector potentials we found that the energy there turned out to be quantitized and a generalization of Einstein's photon energy. But there is one more gauge potential in the five dimensional gauge fields; the gravitational like potential associated with the fifth dimension. When this potential is considered in the Weyl Quantum Principle we find that it too is quantized very much like the photon energy is quantized. That is, the energy is proportional to the frequency and the square of the quantum number just as the photon. However, the constant of proportionality is different from Planck's constant and this particle is associated only with the scalar aspect of the non-transverse wave. So this means the non-transverse wave's energy is also quantized.

This alone doesn't identify the non-transverse wave. However, when I looked into the skin depth of this new wave I found the skin depth in copper was some 10^7 m. That is a remarkably long distance. This is more like the neutrino as it apparently can travel long distances through almost any material. The association of the neutrino with non-transverse wave is indeed something new.

Zero Pressure Radiation

In classical electromagnetism the radiation energy density and the radiation pressure are both the sum of the squares of the electric and magnetic wave components. This means that if one quantity is non-zero so is the other. However, in the five dimensional fields the radiation energy density id the sum of the squares of all wave components, electric, magnetic and gravitational. The radiation pressure is not the same sum of the squares of all components. Here the radiation pressure is the sum of the squares of the electric, magnetic and gravitational scalar components minus the square of the gravitational component.

This means that when the frequency is such that the radiation pressure is zero there must be non-zero radiation energy density. In cosmological situations where there is virtually no m aterial present which can act to support a pressure this says there must be a non-zero radiation energy density. Such a prediction is new. Even though cosmology presents us with just such a condition and this has been known for some time. The zero pressure of cosmological space must have a non-zero radiation energy density. This is a prediction that a cosmological background radiation must exist. While others have used the cooling universe to predict the existence of the background radiation, this represents a new prediction of a background radiation. With such a prediction a big bang beginning of the universe is not needed to have a background radiation.

Converting Wave Energy

While the solutions sets of the five dimensional wave equations show that the transverse and the non-transverse waves may propagate independently, at boundaries between two different materials the boundary conditions may cause a transfer of energy from one type of wave to the other. We showed that the boundary may be set up to transfer energy from one wave type to the other and then back again. The ability to transfer energy between transverse and non-transverse wave is new.

Self Energy of Charged Particles

In the standard model the fact that the electrostatic potential tends to infinity as the radius tends to zero prohibits integrating over the volume of a charged particle to calculate the self energy of the particle. Instead within the standard model the mass energy of the particle is used to calculate the size of the particle such as an electron. This means the classical electron radius has been obtained in this manner.

The thermodynamic basis for the Dynamic Theory means that one may use the concept of minimizing the free energy to calculate a self energy of a charged particle. The use of the non-singular potential allows for integrating over the volume of the body to obtain the self energy. This is new.

Shock Waves

Shock waves are used extensively in the study and use of explosives. In the computer codes that predict the propagation and effects of shock waves an artificial viscosity is used to stabilize the computer code that would otherwise become unstable if only the symmetric Newtonian viscosity is used. In the Dynamic Theory this instability of the computer code may be eliminated because the five dimensional aspect of nature imposes a limitation upon the rate at which the mass density may change just as it requires that there be a limit to the rate of change of position. The limiting rate of change in the mass density in turns creates a n on-symmetric viscosity to exist that removes the need for an artificial viscosity in the computer codes. More importantly, the appearance of a viscous effect due to a limiting rate of change in mass density is new.

What's New

460

Bibliography

Aharnonov, Y. B. (1959). *Phys. Rev.* , 485.

Caratheodory, C. (1909). *Math. Ann.* , 355.

Collins, I. (2003). *Virial Theorem in Stellar Astrophysics.* Harvard.

Compton, A. (1923). A Quantum Theory of the Scattering of X-Rays of Light Elements. *Physical Review* , 483-502.

Dirac, P. (1928). *Proc. Roy. soc.*, (p. 610).

Dirac, P. (1958). *Quantum Mechanics.* Oxford: Oxford U. P.

Einstein, A. (1905). *Annalen der Physik* , 132.

Einstein, A. (1917). *Phys. Zeitschr.* , 121.

Einstein, A. (1924). *Berliner Tageblatt* .

Einstein, A. B. (1938). *Ann. of Math* , 683.

Einstein, A. M. (1931). *Berl. Ber.* , 541.

Einstein, A. (1905). On a Heuristic Viewpoint Concerning the Production and ransformation of Light. *Annalen der Physik* .

Eisberg, R. R. (1985). *Quantum Pysics of Atoms, Molecules, Solids, Nuclei, and Particles.* John Wiley & Sons.

Kaluza, T. (1921). *Sitzungber. d. Preuss. Akad. d. Wiss.* , 966.

Lewis, G. (1926). The Conservation of Photons. *Nature* , 874.

London, F. (1927). Quantum Mechanical Interpretation of Weyl's Theory. *Zeit F. Phys.*

Luchak, G. (1951). *Canadian J. of Phys.*

McConnell, A. J. (1931). *Absolute Differential Calculus.* London.

Milgrom, M. (1983). *ApJ* , 365.

Milgrom, M. (1983b). *ApJ* , 371.

Milgrom, M. (1983c). *ApJ* , 384.

Newburgh, R. G. (1970). A Relativistic Time and the Principle of Carathéodory. *Il Nuovo Cimento* , 84-102.

O'Raifeartaigh, L. (1997). *The Dawning of Gauge Theory.* Princton, NJ: Princeton Press.

Pauli, W. (1933). *Ann. D. Physik* , 305-337.

Planck, M. (1900). *Verh. Deutsch. Phy.* , 237.

Riess, e. A. (1998). Observation Evidence from Supernovae for an Accelerating Universe and a Cosmological Constant. *Astron, J.* , 1009-1038.

Schrödinger, E. (1926). *Annalen der Physik* , 162.

Schrödinger, E. (1922). On a Remarkable Property of the Quantum Orbits of a Single Electron . *Zeit. F. Phys.*

Schrödinger, E. (1927). Quantizations an Eigenvalue Problem. *Annalen der Physik* , 162.

Utiyama, R. (1956). *Phys. Rev.* , 1957.

Veblen, O. (1933). *Projecktive Rellativitatsthorie.* Berlin: Springer.

Walker, E. H. (2000). *The Physics of Consciousness. The Quantum Mind and the Meaning of Life.* Perseus.

Weyl, H. (1929). Electron and Gravitation. *Zeit. f. Phys.* , 56.

Weyl, H. (1918). Gravitation and Electricity. *Sitzungsber. Preuss. Akad. berlin* , 465.

Weyl, H. (1918). *Space Time Matter.*

Williams, P. E. (2007). Alternate Communications for Space Travel. *Space Technology and Applications International Forum (STAIF-2007).* Albuquerque, NM.

Williams, P. E. (2007). Compact Reactor. *Space Technology and Applications International Forum.* Albuquerque, NM.

Williams, P. E. (2002). Energy and Entropy as the Fundaments of Theoretical Physics. *Entropy* , 128-141.

Williams, P. E. (2009). Fusion for Earth and Space. *Space Propulsion and Energy Sciences International Forum (SPESIF 2009).* Huntsville, AL.

Williams, P. E. (2001). Mechanical Entropy and Its Implications. *Entropy* , 76-115.

Williams, P. E. (2008). New Time Dependent Gravity Displays Dark Matter and Dark Energy Effects. *Apeiron* .

Williams, P. E. (1976). *On a Possible Formulation of Particle Dynamics in Terms of Thermodynamic Conceptualizations and the Role of Entropy in it.* Monetrey, CA: U.S. Naval Postgraduate School.

Williams, P. E. (1998). Quantum Measurement, Gravitiaton, and Locality in the Dynamic Theory. *Causality and Locality in Modern Physics*, (pp. 261-268).

Williams, P. E. (2009). Superluminal Space Craft. *Space Propulsion and Energy Sciences International Forum (SPESIF 2009).* Huntsville, AL.

Williams, P. E. (1981). *The Arrow of Time in the Dynamic Theory.* Los Alamos, NM: Los Alamos National Labortory.

Williams, P. E. (1981). *The Dynamic Theory - Some Shockwave and Energy Implications.* Los Alamos, NM: Los Alamos National Laboratory.

Williams, P. E. (1980). *The Dynamic Theory: A New View of Space, Time, and Matter.* Los Alamos, NM: Los Alamos Scientific Laboratory.

Williams, P. E. (1983). The Nuclear Model from the Dynamic Theory. *Third Annual Southwest Theoretical Physics Conference.* Boulder, CO.

Williams, P. E. (1983). *The Possible Unifying Effect of the Dynamic Theory.* Los Alamos, NM: Los Alamos Scientific Laboratory.

Williams, P. E. (1977). *The Principles of the Dynamic Theory.* Annapolis, MD: U.S. Naval Academy.

Williams, P. E. (1997). Thermodynamic Basis for the Constancy of the Speed of Light. *Modern Physics Letters A* , 2725-2738.

Williams, P. E. (2010). Thermodynamics as the Foundation of Physics. *VII Vigier Symposium.* London, UK.

Williams, P. E. (2001). Using the Hubble Telescope to Determine the Split of a Cosmological Object's Redshift into its Graviational and Distance Parts. *Apeiron* .

Yang, C. N. (1954). *Phys. Rev.* , 63.
Zwicky, F. (1937). On the Masses of Nebulae and of
Clusters of Nebulae. *Astrophysical Journal* .

Index

Made in United States
North Haven, CT
14 January 2023

31085441R00271